Universitext

For other titles in this series, go to
http://www.springer.com/series/223

Arnold L. Rosenberg

The Pillars of Computation Theory

State, Encoding, Nondeterminism

 Springer

Arnold L. Rosenberg
Research Professor
Colorado State University
 and Distinguished University Professor Emeritus
University of Massachusetts, Amherst
rsnbrg@colostate.edu

ISBN 978-0-387-09638-4   e-ISBN 978-0-387-09639-1
DOI 10.1007/978-0-387-09639-1
Springer New York Dordrecht Heidelberg London

Library of Congress Control Number: 2009937878

Mathematics Subject Classification (2000): 03D05, 03D10, 03D15, 03D25, 03D45, 68Q01, 68Q05, 68Q10, 68Q15, 68Q17, 68Q45

Printed on acid-free paper

Springer is part of Springer Science+Business Media (www.springer.com)

To my wife, Susan, for her infinite patience during my extended episodes of "cerebral absence."

Contents

Preface xi

List of Acronyms xv

I PROLEGOMENA 1

1   Introduction 3
    1.1   The Three Pillars of Computation Theory 3
          1.1.1   State 3
          1.1.2   Encoding 4
          1.1.3   Nondeterminism 7
    1.2   The Nature of Computation Theory 10

2   Mathematical Preliminaries 13
    2.1   Sets and Their Operations 13
    2.2   Binary Relations 17
          2.2.1   The Formal Notion of Binary Relation 17
          2.2.2   Equivalence Relations 18
    2.3   Functions 20
    2.4   Formal Languages 22
          2.4.1   The Notion of Language in Computation Theory 22
          2.4.2   Languages as Metaphors for Computational Problems .. 24
    2.5   Graphs and Trees 25
    2.6   Useful Quantitative Notions 26

II STATE 31

3   Online Automata: Exemplars of "State" 33
    3.1   Online Automata and Their "Languages" 33
    3.2   A Myhill–Nerode-like Theorem for OAs 41
    3.3   A Concrete OA: The Online Turing Machine 43

4 Finite Automata and Regular Languages 51
   4.1 Introduction 51
   4.2 Preliminaries 53
   4.3 The Myhill–Nerode Theorem for FAs.......................... 56
      4.3.1 The Theorem: States Are Equivalence Classes 57
      4.3.2 What Do Equivalence Classes Look Like? 59

5 Applications of the Myhill–Nerode Theorem 63
   5.1 Proving that Languages Are *Not* Regular 64
   5.2 On Minimizing Finite Automata 66
   5.3 Finite Automata with Probabilistic Transitions 70
      5.3.1 PFAs and Their Languages 70
      5.3.2 PFA Languages and Regular Languages 72
   5.4 State as a Memory-Constraining Resource...................... 79
      5.4.1 $\mathscr{A}_M(n)$ for Two Specific Infinite OAs 80
      5.4.2 A Bound on $\mathscr{A}_M(n)$ for Any OA M with Nonregular $L(M)$ .. 81
   5.5 State as a Time-Constraining Resource.......................... 83
      5.5.1 Online TMs with Multiple Complex Tapes 84
      5.5.2 An Information-Retrieval Problem as a Language 86
      5.5.3 The Impact of Tape Structure on Memory Locality 87
      5.5.4 Tape Dimensionality and the Time-Complexity of L_{DB} 88

6 Enrichment Topics 91
   6.1 Pumping in Formal Languages 91
      6.1.1 The Phenomenon of Pumping in Finite, Closed Systems..... 91
      6.1.2 Pumping in Regular Languages 93
      6.1.3 Pumping in Nonregular Languages 96
   6.2 Closure Properties of the Regular Languages 101
   6.3 Systems of Linear Equations with Languages as Coefficients 105

III ENCODING 111

7 Countability and Uncountability: The Precursors of "Encoding" 113
   7.1 Encoding Functions and Proofs of Countability 116
   7.2 Diagonalization: Proofs of Uncountability 121
   7.3 Where Has (Un)countability Led Us?........................... 123

8 Enrichment Topic: "Efficient" Pairing Functions, with Applications .. 125
   8.1 Background 126
   8.2 The Prettiest Pairing Function(s) 127
      8.2.1 The Diagonal PF $\mathscr{D}(x,y)$ 127
      8.2.2 Is $\mathscr{D}(x,y)$ the *Only* Polynomial PF? 129
   8.3 Pairing Functions and the Storage of Extendible Arrays/Tables 130
      8.3.1 Array-Storage Mappings via Pairing Functions 131
      8.3.2 Pursuing *Compact* Pairing Functions 133
   8.4 Pairing Functions and Volunteer Computing 139

8.4.1   A Methodology for Designing *Additive* Pairing Functions ... 141
8.4.2   A Sampler of Explicit APFs 142

9   Computability Theory 147
9.1   Introduction and History................................. 147
9.2   Preliminaries 151
9.2.1   Representing Computational Problems as Formal Languages . 151
9.2.2   Functions and Partial Functions 153
9.2.3   Self-Referential Programs: Interpreters and Compilers 155
9.3   The Halting Problem: The "Oldest" Unsolvable Problem 155
9.3.1   The Halting Problem Is Semisolvable but Not Solvable 156
9.3.2   Why We Care about the Halting Problem—An Example 158
9.4   Mapping Reducibility 160
9.4.1   Basic Properties of m-Reducibility...................... 162
9.4.2   The *s-m-n* Theorem: Where Does One Find Encodings?..... 163
9.5   The Rice–Myhill–Shapiro Theorem 165
9.6   Complete, or "Hardest," Semidecidable Problems............... 169
9.7   Some Important Limitations of Computability 172
9.8   (Online) Turing Machines and the Church–Turing Thesis 174
9.8.1   Simplifying an OTM without Diminishing Its Power........ 176
9.8.2   Augmented TMs That Are No More Powerful Than OTMs .. 195

IV   NONDETERMINISM 209

10   Nondeterministic Online Automata 211
10.1   Nondeterministic OAs 211
10.2   Nondeterminism as Unbounded Search, 1 213
10.3   An Overview of Nondeterminism in Computation Theory 215

11   Nondeterministic FAs... 217
11.1   Nondeterministic FAs vs. Deterministic FAs 217
11.1.1   NFAs Are No More Powerful Than DFAs................. 217
11.1.2   Does the Subset Construction Waste DFA States? 219
11.2   An Application: The Kleene–Myhill Theorem 221
11.2.1   A Convenient Enhancement of NFAs..................... 221
11.2.2   The Kleene–Myhill Theorem 223

12   Nondeterminism in Computability Theory 233
12.1   Introduction 233
12.2   Nondeterministic Turing Machines 234
12.2.1   The NTM Model 234
12.2.2   Deterministic Simulation of Nondeterminism:
NTMs and OTMs 236
12.3   Nondeterminism as Unbounded Search, 2 241

13 **Complexity Theory** 245
    13.1 Introduction 245
    13.2 Time and Space Complexity 252
        13.2.1 On Measuring Time Complexity 253
        13.2.2 On Measuring Space Complexity 256
    13.3 Reducibility, Hardness, and Completeness in Complexity Theory . . . 263
        13.3.1 A General Look at Resource-Bounded Computation 263
        13.3.2 *Efficient* Mapping Reducibility 264
        13.3.3 Hard Problems and Complete Problems 268
        13.3.4 An **NP**-Complete Version of the Halting Problem 269
        13.3.5 The Cook-Levin Theorem: The **NP**-Completeness of SAT . . . 276
    13.4 Nondeterminism and Space Complexity 285
        13.4.1 Simulating Nondeterminism Space-Efficiently: Savitch's
               Theorem 287
        13.4.2 Beyond Savitch's Theorem 297

V Sample Exercises 299

References 311

Index 315

Preface

The abstract branch of theoretical computer science that we shall call "computation theory" typically appears in undergraduate academic curricula in a form that obscures both the mathematical concepts that are central to the various components of the theory and the relevance of the theory to the typical student. This regrettable situation is due largely to the thematic tension among three main competing principles for organizing the material in the course.

1. *One can organize material to emphasize underlying mathematical concepts.*
   The challenge with this approach is that it often violates boundaries mandated by computation-theoretic themes. A good example of this dilemma is seen wth our Section 5.5, which studies a computation that can be viewed as an abstraction of the following. Say that you have a database D, viewed abstractly as a sequence of binary strings. Say that you also have a (possibly very long) sequence of membership queries about the contents of D. What are the consequences, in terms of overall processing time, of demanding that a program respond to each of your queries as it arrives (the "online" scenario), in contrast to reading in all of your queries, preprocessing the sequence, and responding to all queries at once (the "offline" scenario)?
   The pedagogical dilemma here is that the computational model of Section 5.5 is a variant of the traditional Turing machine—material that traditionally comes quite a way into a course on computation theory—but the technical argumentation uses techniques that were developed originally for studying finite automata—material that traditionally comes at the very beginning of a course on computation theory.

2. *One can organize material to emphasize underlying computation-theoretic themes.*
   The challenge with this approach is almost the mirror image of the preceding one. Referring to Section 5.5, if one decides to cover this (quite illuminating!) material, but to place it in a chapter devoted to "powerful" computational models such as Turing machines, it will be quite challenging to expose the student to the fact that the mathematical underpinnings of this study actually hark back to material on finite automata that she has not seen since the earliest part of the course.

3. *One can organize material to emphasize the relevance of the Theory's concepts to real computational (hardware and software) artifacts.*

   Since theoretical computer science is, ostensibly, a branch of the more general field of computer science, arguing against this approach is almost like denying one's roots. That said, this approach would force one to cover material in a way that largely obscures the "pure" concepts that underlie the various artifacts, both the mathematical concepts and the computation-theoretic ones.

So, what is one to do? Almost all undergraduate texts on computation theory opt for the second of these alternatives; a very few opt for the third. We opt here for the first alternative! We are motivated by the belief that a deep understanding of—*and operational control over*—the few "big" mathematical ideas that underlie the theory is the best way to enable the typical student to assimilate the "big" ideas of the theory into her daily computational life.

Why do we need a new computation theory text? In order to answer this question, we must agree on what we want an upper-level undergraduate, lower-level graduate computation theory course to accomplish. In my opinion, the course should impart to the as yet uninitiated student of computer science:

1. the need for theoretical/mathematical underpinnings for what is predominantly an engineering discipline. This should include an appreciation of the need to think (and argue) rigorously about the artifacts and processes of "practical" computer science.
2. the rudiments of the "theoretical method," as it applies to computer science. This should include an *operational* command of the basic mathematical concepts and tools needed for the rigorous thinking of item 1.
3. a firm foundation in the most important concepts of theoretical computer science. This foundation should be adequate for subsequent navigation of (large portions of) advanced theoretical computer science.
4. topics from computation theory that have a clear path to major topics in general computer science. It is crucial that the student recognize the relevance of these topics to her professional development and, ultimately, her professional life.

(To me, a corollary of this presumed agenda is that an appropriately designed course in computation theory should be mandatory for all aspiring computer scientists.) With some regret, I would argue that most current curricula for computation theory courses—as inferred from the contents of the standard texts—neither focus on nor satisfy these objectives. Standard texts typically prescribe a two-module approach to the subject.

Module 1 comprises a smattering of topics that provide a formal-language-theory approach to the mathematical theories of automata and grammars. The main justification for much of the material in this module seems to be the long histories of these theories. Within the context of this module, I part ways with the major texts along two axes: (1) the inclusion of several topics of largely historical interest and the omission of several topics of central conceptual importance; (2) the way that they present certain topics. Most of the material in this module and the approaches to that material seem to be passed from one generation of texts to the next, without a critical analysis of what is relevant to the general student of computer science.

Module 2 (which is usually the larger one) provides an intense study of one specific topic, complexity theory, preceded by some background on its (historical and intellectual) precursor, computability theory. This is indisputably important material, which does expose aspects of the intrinsic nature of computation by (digital) computers and does establish the theoretical underpinnings of important topics relating to the theory of algorithm design and analysis. That said, I feel that much of what is typically included in this module goes beyond what is essential for, or even relevant to, the general computer science student (as opposed to the aspiring theoretical computer scientist); moreover, these topics preclude (because of time demands) the inclusion of several topics that are more relevant to the development of embryonic computer scientists. Additionally, I am troubled by the typical presentation of much of the material via artificial, automata-theoretic models that arose during the heyday of automata theory in the 1970s.[1]

My proposed alternative to the preceding material is a "big-ideas" approach to computation theory that is based on the three computation-theoretic "pillars" that name this book. The mathematical correspondents of these concepts underlie much of the basic development of theoretical computer science; and the concepts themselves underlie many of the intellectual artifacts of practical computer science. Such an approach to the theory allows one to expose students to all of the major introductory-level ideas covered by present texts and courses, while augmenting these topics with others that are (in my opinion) at least as relevant to an aspiring computer scientist. I contend that, additionally, this approach gives one a chance to expose the student to important mathematical ideas that do not arise within the context of the topics covered in most current texts. I thus view the proposed "big-ideas" approach as strictly improving our progress toward all four educational goals enumerated previously. We thereby (again, in my opinion) enhance students' preparations for their futures, in terms of both the material covered and the intellectual tools for thinking about that material.

While my commitment to the proposed "big-ideas" approach has philosophical origins, it has been evolving over several decades, as I have taught versions of the material in this book to both graduate and undergraduate students at (in chronological order) Polytechnic Univ. (formerly, Brooklyn Poly.), NYU, Duke, and UMass Amherst. Each time I have offered the course, I have made further progress toward my goal of a "big-ideas" presentation of the material. My (obviously biased) perception is that my students (who have been statistically *very* unlikely to become computer theorists) have been leaving the course with better perspectives and improved technical abilities as the transition to this approach has progressed.

My dream is that this book, which has been developed around the just-stated philosophy, will make the goals and tools of computation theory as accessible to the "computer science student on the street" as David Harel's well-received book [35] has achieved with the algorithmic component of theoretical computer science.

---

[1] This position echoes that espoused in [32] and in the classical computability theory text [80].

I end this preface with expressions of gratitude to the many colleagues who have debated this educational approach with me and the even greater number of students who have suffered with me through the growing pains of the "big-ideas" approach. Both groups are too numerous to list, and I shall not attempt to do so, for fear of missing important names. I also wish to thank the UMass Center for Teaching for a grant that contributed to the costs of preparing this text.

Falmouth, MA, and Denver, CO   *Arnold L. Rosenberg*
                                                                                               October 1, 2009

List of Acronyms

$\Longleftrightarrow$	"if and only if"
$\overset{\text{def}}{=}$	"equals, by definition": used to define notions
$u \to v$	arc in a digraph; parenthood in a rooted tree
$u \Rightarrow v$	ancestry in a rooted tree
$u \overset{d}{\Rightarrow} v$	depth-d ancestry in a rooted tree
$\mathbb{N}$	the set of nonnegative integers
$\mathbb{N}^+$	the set of positive integers
$\mathbb{Z}$	the set of all integers
$\mathbb{O}$	the set of positive odd integers
$|S|$	the cardinality of set S: its number of elements when S is finite
$\emptyset$	the empty set, which has no elements: $|\emptyset| = 0$
$\mathscr{P}(S)$	the power set of set S: the set of all of S's subsets
κ_S	the characteristic function of the set S; $\kappa_S(w) = $ **if** $w \in S$ **then** 1 **else** 0.
κ'_S	the semicharacteristic function of the set S; $\kappa'_S(w) = $ **if** $w \in S$ **then** 1 (undefined otherwise).
Σ	a finite set of "letters:" an *alphabet*; usually the *input* set of an automaton
Σ^k	the set of length-k strings of letters from Σ, where $k \in \mathbb{N}$, i.e., is a nonnegative integer
$\Sigma^\star$	the set of all finite-length strings of letters from Σ
$\ell(w)$	the length of word $w \in \Sigma^\star$; for all $w \in \Sigma^k$, $\ell(w) = k$
w^R	the reversal of word $w \in \Sigma^\star$
ε	the (unique) null string, of length 0; $\ell(\varepsilon) = 0$
$L_1 \cdot L_2$	the concatenation of languages L_1 and L_2
L^k	the kth power of language L: $L \cdots L$ (k occurrences of "L")
$L^\star$	the star-closure of language L
$\mathscr{R}$	a regular expression
$\mathscr{L}(\mathscr{R})$	the language denoted by regular expression $\mathscr{R}$
Γ	a finite set of "letters": usually the *working alphabet* of a Turing machine (TM)

$\boxed{B}$	the *blank* symbol, which resides in the working alphabet of every Turing machine (TM)
$o(1)$	any function $f(n)$ that tends to the limit 0 as n grows without bound
$O(f(n))$	the class of functions $g(n)$ whose graphs eventually (for large n) stay below that of $c_g \cdot f(n)$ for some constant $c_g > 0$
$\Omega(f(n))$	the class of functions $g(n)$ whose graphs eventually (for large n) stay above that of $c_g \cdot f(n)$ for some constant $c_g > 0$
$\Theta(f(n))$	the intersection of the classes $O(f(n))$ and $\Omega(f(n))$
$|x|$	the absolute value, or magnitude, of the number x
$\log_b x$	the base-b logarithm of positive number x
$\log x$	$\log_2 x$: the base-2 logarithm of positive number x
$\exp2(n)$	an alternative notation for 2^n
R_k	the (perforce, integer) contents of register R_k of a Register Machine
CFG	context-free grammar
CFL	context-free language
FA	finite automaton
IRTM	input-recording Turing machine
NFA	nondeterministic finite automaton
NTM	nondeterministic Turing machine
OA	online automaton
OTM	online Turing machine
PFA	probabilistic finite automaton
TM	Turing machine
G	a context-free grammar (CFG)
Q	the set of states of an automaton
q_0	the initial state of an automaton
F	the set of final, or accepting, states of an automaton
$E(q)$	the ε-reachability set of an NFA's state q
A-POP	the register machine analogue of the POP operation on a stack
A-PUSH	the register machine analogue of the PUSH operation on a stack
POP	the operation that removes a symbol from a stack
PUSH	the operation that adds a symbol to a stack
δ	the state-transition function of an automaton. For an OA, δ maps $Q \times \Sigma$ into Q.
$\equiv_M$	the (right-invariant) equivalence relation on $\Sigma^\star$ "defined" by an OA: $x \equiv_M y$ just when $\delta(q_0, x) = \delta(q_0, y)$.
$\equiv_M^{(t)}$	a time-parameterized version of $\equiv_M$ that is studied in Section 5.5.
$\equiv_\delta$	the equivalence relation on Q, the set of states of an OA: $p \equiv_\delta q$ just when for all $z \in \Sigma^\star$, either both $\delta(p, z)$ and $\delta(q, z)$ are accepting states, or neither is.
$\equiv_L$	the (right-invariant) equivalence relation on $\Sigma^\star$ "defined" by a language $L \subseteq \Sigma^{l}star$: $x \equiv_L y$ just when, for all $z \in \Sigma^\star$, either both xz and yz belong to L, or neither does.
$\widehat{\delta}$	the function δ extended to strings rather than single letters.
$L(M)$	the language accepted by automaton M

$L(G)$ the language generated by CFG G

$\mathcal{T}_M(x)$ the computation tree generated by the automaton M when processing input word x

$\widehat{\mathcal{T}_M}(x)$ the analogue of $\mathcal{T}_M(x)$ generated by a *space-bounded* automaton M, truncated to eliminate nonhalting branches

$L(M,\theta)$ the language accepted by the PFA M with acceptance threshold θ

L_{DB} the database language of Section 5.5

$\mathcal{A}_M(n)$ the number of states in the smallest FA that is an order-n approximation of the OA M.

DHP   the "diagonal" halting problem: the set of strings x such that program x halts on input x

EMPTY   the set of programs that do not halt on any input

HP   the halting problem: the set of program-input pairs $\langle x,y \rangle$ such that program x halts on input y

$HP^{(\mathrm{poly})}$ the poly-time version of the halting problem

SAT   the satisfiability problem: the set of CNF formulas that can be made TRUE by some assignment of truth values to logical variables

TOT   the set of programs that halt on all inputs

NP   the family of languages accepted by nondeterministic Turing machines that operate in (nondeterministic) polynomial time

P   the family of languages accepted by deterministic Turing machines that operate in polynomial time

$\ominus$ positive subtraction

$\leq_m$ "is m-reducible to"; "is mapping reducible to"

$\leq_{\mathrm{poly}}$ "is polynomial-time-reducible to"

$\leq_R$ "is mapping-reducible to" within resource bound R

PART I
PROLEGOMENA

"The longest journey begins with a single step."

(folk saying)

This portion of the book is dedicated to laying out the road we shall follow in developing the rudiments of computation theory and to developing the tools we shall need during our journey. There are myriad more-or-less equivalent ways to approach the study of Computation Theory. That said, I believe that certain ways—not surprisingly, including the approach we take in this text—make it easier for the student to gain operational command over the fundamentals of the Theory and to assimilate all that the Theory has to say about "real" computation into her professional kitbag. I attempt to describe—and motivate, and justify—my approoach in Chapter 1. We begin our journey for real in Chapter 2, which is devoted to developing the mathematical tools that will uncover for the student the secrets—and, hopefully, the beauty—of Computation Theory. One feature of our approach will be to emphasize the many ways that one can look at various phenomena relating to computation. Technical issues that are rather complicated when looked at from one perspective become almost transparent from another perspective. This point is illustrated as early as Chapter 2, but it will really gather steam in the later chapters. It must be admitted that some of the diversity of terminology and perspective can be frustrating at times, but we attempt to focus mainly on the benefits we can harvest from the diversity. Even when no such benefits are obvious, it is often interesting to note how the Theory has absorbed terminolgy and viewpoints from so many historically diverse sources.

Chapter 1
Introduction

This book is intended as an introduction to computation theory for upper-level undergraduate students and lower-level graduate students. We develop the underpinnings of the theory by studying three mathematical concepts that underlie much of the subject, followed by a number of fundamental applications of each concept. This chapter attempts to render concrete the rather abstract philosophical "manifesto" of the preface, focusing specifically on the "three pillars" that give the book its title and that give rise to the "big ideas" that anchor the book.

1.1 The Three Pillars of Computation Theory

In a famous Talmudic story, Rabbi Hillel is challenged to encapsulate all of the voluminous laws of Judaism while standing on one leg. (His response was, "What you find hateful, do not unto others.") What would a computation theorist respond when similarly challenged? This book attempts, in some way, to answer this question. It turns out that virtually every major result in elementary computation theory—the portion of the theory that (in my opinion) every computer scientist should have in her conceptual kit bag—refers in some fundamental way to one or more of the following three notions: state, nondeterminism, and encoding. This book uses these three notions as the pillars upon which we develop an introductory course in elementary computation theory.

1.1.1 State

Myriad computational systems, both hardware and software, are organized as state-transition systems. Such a system evolves over time (or, as we shall say, computes) by continually changing state in response to one or more discrete stimuli (typically

A.L. Rosenberg, *The Pillars of Computation Theory*, Universitext,
DOI 10.1007/978-0-387-09639-1_1, © Springer Science+Business Media, LLC 2010

termed "inputs"). When in a "stable" configuration/situation, the system is in one of its well-defined *states*. (This condition really defines "state.")

> In a hardware system, for instance, the "stable" configurations are typically those in which the bistable devices (say, flip-flops or transistors) used to build the system have attained a stable logic level. In a software system, "stability" may reside in a given thread's having completed execution and returned a result.

At any such moment, in response to any valid stimulus, the system goes through some process, ending up in another "stable" configuration/situation, i.e., in another well-defined state. John Myhill and Anil Nerode jointly produced one of the conceptual gems of finite-automata theory, the Myhill–Nerode theorem (Section 4.3, Theorem 4.1), which offers a complete mathematical characterization (via basic algebraic notions) of the concept "state" within the genre of state-transition system that we are discussing. Although the theorem focuses solely on *finite* state-transition systems, one can fruitfully formulate a version of the theorem that applies to arbitrary such systems, not just finite ones; we do so in Chapter 3. The theorem's characterization of "state" allows one to analyze many diverse aspects of state-transition systems, with an eye toward improving their designs and/or exposing and quantifying their limitations. Indeed, the applications of the mathematical characterization that we offer in Chapters 3 and 4 involve several diverse aspects of systems, ranging from their sizes, to their computational memory requirements, to their computing times for special computations.

> To whet the reader's appetite for the advertised "range" of applications of Theorem 4.1: In Section 5.2, we demonstrate how the Myhill–Nerode theorem supplies the mathematical underpinnings of the state-minimization algorithm for finite-state machines that every student of logic design learns about. In Section 5.3, we use the theorem to establish a surprising computational limitation relating to certain current approaches to machine learning. In Section 5.4, we exploit the theorem to obtain the strongest possible general lower bound on the memory requirements of a surprisingly broad class of computations. In Section 5.5, we use the theorem to bound from below the computation time of an abstract class of programs on an abstract database computation—a bound that has important lessons for the theories of both data structures and algorithms.

1.1.2 Encoding

Arguably, the most fundamental results in computability theory and complexity theory depend on the ability to encode one computational problem, call it A, as another, often quite different, computational problem, call it B, via a mapping (called a *reduction*) that translates any solution for (an instance of) B to a solution for (the corresponding instance of) A.

> One simple example will illustrate how dramatically problems A and B can differ from one another, yet still be encodings of one another.
>
> Let problem A be presented via a finite sequence of positive integers, $m_1, m_2, \ldots, m_n$, together with an $(n+1)$th positive integer N. The "problem" is to determine whether there is a subset of the integers m_i that sum to N. (A is often called the *subset-sum problem*.)

Let problem B be presented by: (a) a set of locations; (b) a matrix of nonnegative numbers whose (i, j)th entry is the "cost" of traveling from location i to location j—it makes no difference how this "cost" is measured: it is just a number; (c) a target cost C. The "problem" is to determine whether there is a tour that visits each location precisely once and that incurs aggregate cost $\leq C$. (B is often called the *traveling salesman problem*.)

While it is not intuitively obvious, there is in fact a way of encoding instances of problem A as instances of problems B that maps "yes" instances of A to "yes" instances of B and "no" instances of A as "no" instances of B. Just as surprising: There is also an encoding that goes in the other direction, from instances of problem B to instances of problem A.

Within computability theory (which we introduce in Chapter 9), one demands that the encoding that maps instances of problem A to instances of problem B be supplied/specified via a *program* that produces an instance of problem B from each instance of problem A. In other words, one insists that the encoding *can actually be computed*. Within complexity theory (which we introduce in Chapter 9), one demands additionally that the program that performs the encoding be *efficient*, in some sense. The specific notion of efficiency that one insists on depends on the notion of computational complexity being studied. For instance, if one is studying the time efficiency of a given class of computing devices on a given class of computations, then one would insist on encodings that could be computed quickly on these computing devices; if one is studying the memory efficiency of a given class of computing devices on a given class of computations, then one would insist on encodings that could be computed succinctly on these devices. You can easily extrapolate from these two sample complexity measures. An even more basic use of encodings is found in Alan Turing's original study of the inherent limitations of any "reasonable"[1] digital computing system [104]. Turing's work builds on the encodings used in Gödel's seminal work on "incompleteness" in logical systems [30]; this work shows that no "reasonable" logical system can capture through the notion of proof all true arithmetic facts. (This is, essentially, the meaning of the term "incompleteness.") Both of these intellectual *tours de force* use encodings to demonstrate rigorously the stark distinctness of two notions that are easily—and fallaciously—identified in common discourse; the notions were truth and theoremhood for Gödel, and functions and programs for Turing. Importantly for the viewpoint espoused here, the encodings in both Gödel's and Turing's work are based on the relatively simple mathematics underlying the following results of Georg Cantor [9]. (1) There exist one-to-one associations (based on computationally simple *pairing functions*) between the positive integers and the positive rational numbers. (2) There can be no one-to-one association between the rational numbers and the real numbers—even the real numbers between 0 and 1. The second of these results employs Cantor's well-known *diagonal argument*, which itself lives on as a basic tool in studies of reductions. By going rather far from Cantor's set-theoretic theme, one can use pairing functions to show that simple integer arithmetic (addition and multiplication) suffices to encode elaborate finite structure—e.g., finite data structures, arithmetic expressions, or strings of integers—as single integers. Also relevant to the viewpoint underlying this book,

---

[1] Reasonableness here and in Kurt Gödel's work precludes systems that have answers "wired in."

even the original, unembellished notion of pairing function has meat to chew on that retains juice to this day, as we show in Section 7.1.

> As just one example, we employ pairing functions in Section 7.1 to develop storage mappings for multidimensional arrays or tables that allow one to expand and shrink the array/table at will without remapping any already-stored array entries; you can see easily that the dimension-order array-storage mappings and their ilk that are used by compilers for standard programming languages do not allow such stability of array entries.

The (un)encodability demonstrated by Cantor's work was crucial to Gödel and Turing, for it showed that, quite remarkably, even primitive formal systems can contain encodings of sentences that are self-referential—in much the way that the paradoxical sentence, "This sentence is false," is. Within Gödel's world, therefore, an integer that occurs, apparently "harmlessly," in a logical sentence S could in fact be an encoding of a sentence—perhaps even of S itself! Thus, sentences that appear to be making statements about integers can be construed as making statements about sentences. And within Turing's world, an integer input to a program P could actually be an encoding of a program—perhaps even of P itself! Thus—and here is the rub!—programs that appear to perform even simple computations on integers can be encodings of programs that effect complex transformations of program. We study the relevant mathematical underpinnings of the notion of encoding in Chapter 7, and we observe their ripples throughout computability theory and complexity Theory throughout the remainder of Part III (Encoding). Not surprisingly, the relevant notions of encoding gather complexity as we make our way through Part III, from the pairing functions of Chapter 7, through the mapping-reductions of Chapter 9 and the polynomial-time mapping-reductions of Chapter 13. One amazing outgrowth of the reductions in Parts III and IV (Nondeterminism) is that there often exist within a class of computational problems individual problems that are the "hardest" ones in the class, in the sense that every problem in the class reduces to each of them. (Such problems are typically said to be *complete* for the class.)

> We mention two examples here.
>
> The *halting problem* of Section 9.3, which is generally considered the prototypical computationally unsolvable problem, is to decide, given a program P and an input x for P, whether program P ever halts if it is started on input x. The fact that the halting problem is complete for the class of "semidecidable" or "partially solvable," problems (as defined in Section 9.2) means, for instance, that any program that could solve the halting problem can be transformed into a program that decides the truth or falsity of statements about a broad range of relations among positive integers.
>
> **Historical aside**. The halting problem is often known as the "halting problem for Turing machines" and is almost universally attributed to Alan M. Turing's seminal paper [104]. Citing a published letter of Thorkil Naur [71], Andrew Pitts of Cambridge University has pointed out (in private communication) that the property of Turing machines studied in [104], which Turing called "circularity," is actually rather different from halting. That said, Turing's proof of the undecidability of circularity puts one well on the road to a proof of the undecidability of the halting problem. But the two problems are distinct!
>
> The subset-sum and traveling salesman problems (problems A and B) that we referred to when beginning our discussion of encodings are both complete for the class **NP** of problems that can be solved in "nondeterministic polynomial time" (see the next subsection, on nondeterminism, and see Part IV). A polynomial-time (deterministic) solution for either

problem would, therefore, settle the well-known **P**-vs.-**NP** problem by showing that every problem that could be solved in "nondeterministic polynomial time" actually admits a computationally tractable deterministic solution. Widely acknowledged as the most significant computation-theoretic open problem, the **P**-vs.-**NP** problem forms the centerpiece of the portion of complexity theory that we develop in Chapter 13.

The encoding-related notions of reduction and completeness underlie most of the material in Parts III and IV.

1.1.3 Nondeterminism

The state-transition systems—both hardware and software—that one encounters in "real life" are typically, but not universally, *deterministic*, in the sense that the current state of the system, coupled with the specific relevant discrete stimuli that the system can sense, uniquely determines the next state of the system. We really rely in "real life" on the fact that if one runs a program repeatedly (on the same system, with the same stimuli), then one will always get the same answer. The absence of such determinism—which is called *nondeterminism*—would make it all but impossible to design verifiably correct and efficient digital logic and would make it even harder than it already is to craft verifiably correct and efficient software. Since deterministic behavior seems, thus, to be fundamental to the desired functioning of the systems that form the stock in trade of the computer professional, why should one even contemplate systems that are nondeterministic? The surprising answer comes in two installments, one based in the late 1950s, the other in the early 1970s. It came as quite a surprise in the late 1950s when the physically unrealizable mathematical idealization of a "nondeterministic machine"—it is not *really* a machine because of its nondeterministic behavior—was shown, in [75, 79], to yield dramatically simplified algorithms for representing the behavior of finite state-transition systems, as exposed by the Kleene–Myhill theorem (Theorem 11.3, the "regular-expression theorem") of finite-automata theory. This surprise became an intellectual supernova in the early 1970s with the discovery of the **P**-vs.-**NP** problem and its attendant theory of **NP**-completeness [18, 56]. Nondeterminism was exposed in those sources as much more than just a mathematical/algorithmic convenience; indeed, it was shown to be a fundamental computational notion that explains the apparent computational difficulty of myriad important computational problems. Subsequent studies (cf. the early encyclopedic review of [28]) have exposed seemingly countless computational problems, in areas ranging from constraint satisfaction,[2] to structure mapping,[3] to scheduling,[4] to logic

---

[2] Sample instance: Can a given set of disjunctive logical contraints be satisfied simultaneously?

[3] Sample instance: What is the most efficient way to simulate a (logical) communication network of structure *A* on a (physical) network of structure *B*?

[4] Sample instance: What is the most efficient way to schedule final exams in a fixed set of classrooms whose seating capacities are fixed?

minimization,[5] and beyond, that would admit simple, computationally efficient solutions in a computing platform that implemented true nondeterministic behavior
but that, to this day, defy efficient solution on any known deterministic computing
platform.

> To lend some quantitative texture to the preceding few sentences, many of the problems re-
> ferred to admit solutions that take (nondeterministic) "time" that is *linear* in the size of the
> problem description (the set of constraints, or the networks A and B, or the class and room
> sizes, or the circuit description) upon a nondeterministic computing platform, whereas their
> only known solutions on a deterministic computing platform take time that is *exponential* in
> the input size.

Interestingly, the benefits of nondeterminism can be explained in a nutshell, and have
been well known for decades. Nondeterminism in an "algorithm"—as with "machines" in our earlier discussion, we put the word in quotes because a nondeterministic "algorithm" is not implementable as a "real" algorithm on a real (hence, deterministic!) computing platform—can be viewed as abbreviating a possibly lengthy, ardu-
ous search that is part of an algorithm, by means of a conceptual mechanism that is
embodied in a one-step superalgorithmic primitive of the form "Search for x." The important *conceptual* role of nondeterminism in specifying an "algorithm" is to expose
the existence of the search explicitly, which many (deterministic) algorithms' specifications do not do. This exposure has successfully explained—but has not explained
away!—the observed computational intransigence of the **NP**-complete computational
problems one finds in compendia such as [28]. The exposure—when coupled with
other important computational concepts, notably *completeness*—has additionally allowed us prove that finding a speedy algorithm for any of myriad important intransi-
gent problems will automatically provide speedy algorithms for all of the problems.
The seminal work underlying this branch of complexity theory was done independently by Stephen A. Cook [18] and Leonid Levin [56], with extremely important
followup work by Richard M. Karp [48] and many others. Within our coverage of
computation theory, nondeterminism is an indispensable technical/algorithmic tool
in Section 11.2, but it is even more important in Chapter 13, where it provides the
intellectual raw material for much of the theory of computational complexity.

   This book builds the elements of computation theory upon the preceding three
pillars (State, Encoding, Nondeterminism) via a part devoted to each. Within this
nonstandard organization, we develop the rudiments of three classical topics, the
theories of finite automata (Chapter 4), computability (Chapter 9), and complexity
(Chapter 13). Our organization allows us to explore interrelationships among these
three branches of computation theory that are not typically exposed in introductory
texts; it also allows us to expose certain common mathematical roots of the branches.
Some of the interrelationships we expose are thematic, such as the role of "state"
in complexity theory (Sections 5.4 and 5.5), while others are more technical, such
as the development of computational reductions, in Part III (Encoding), beginning

---

[5] Sample instance: What is the smallest logic circuit built using nand gates that is functionally
equivalent to a given circuit?

with Cantor's countability arguments (Chapter 7). Our approach singles out certain concepts and results of the three classical branches of computation theory as "big ideas"—and we view this as a strength of the approach—but we *do* go beyond these "big ideas" as we develop the branches, so that the "big ideas" emerge as major signposts in a rich landscape rather than as a collection of isolated topics. We develop finite automata theory up to, and including, the two basic theorems that characterize finite automata by, respectively, definitively elucidating the pillar notion, "state," (the Myhill–Nerode theorem, Theorem 4.1) and establishing the "equivalence" of finite automata and regular expressions (the Kleene–Myhill theorem, of Stephen C. Kleene and John Myhill, Theorem 11.3).

> Only the former of these two results is designated a "big idea" here. Our choice is dictated by the widespread applications of Theorem 4.1 within computation theory (as exemplified throughout Part II (State)), in contrast to the relatively narrower shadow that Theorem 11.3—basically an artifact of finite-automata theory—casts on computation theory as a whole. Because Theorem 11.3 becomes algorithmically accessible only in the presence of the pillar "nondeterminism," we develop the theorem in Part IV, which is dedicated to that pillar.

We develop computability theory from the underlying notions of (non)encodability [of one computational system as another] and reducibility [of one computational problem to another]. The culmination of this development is the notion of the completeness of certain individual computational problems within certain classes of such problems. Complete problems are singled out for their ability to encode, in a quite precise sense, any other problem in the class. One of the most exciting applications of the notion of completeness (at least within computability theory) is the Rice–Myhill–Shapiro theorem of Henry G. Rice, John Myhill, and Norman Shapiro (Theorem 9.5). Informally, the theorem asserts the impossibility of effectively [i.e., algorithmically] determining any properties of the dynamic behavior of a program from its static description.[6]

> The "big ideas" in computability theory all arise within the context of encoding, so we develop the theory in Part III, which is devoted entirely to this "pillar."

We develop complexity theory as an outgrowth of computability theory, built upon (computational) resource-bounded versions of the encoding-based notions of Part III, coupled with the very important added "pillar" ingredient of nondeterminism.

> The importance of nondeterminism within complexity theory has led us to develop the rudiments of the subject in Part IV of the book.
>
> **A philosophical aside.** In this book, we view complexity theory as an outgrowth of computability theory. Many complexity theorists would reject this view, arguing (with more than a little truth) the central role of complexity theory within the field of algorithmics—a role which is not shared by computability theory. Even taking that observation into account, I still feel that this book's approach to the material is the appropriate one pedagogically, for it presents the essential "pillar" ingredients of encodings and nondeterminism in versions of increasing complication, thereby easing the student's assimilation of the ideas.

---

[6] Sample questions: Does program P halt on all inputs? Does it halt on any inputs? Does it ever produce a designated output (such as an "error" message)?

Our development of complexity theory culminates, in Section 13.3.5, with the underpinnings of the **P**-vs.-**NP** problem in the Cook–Levin theorem (Theorem 13.2). The proof of this seminal result identifies, by example, an algorithmic structure in computational problems that explains the problems' efficient solvability on nondeterministic computing platforms in the face of their (apparent) inherent inefficiency on any deterministic computing platform. Aside from its intrinsic usefulness—which is amply manifest in the world of modern algorithmics—this identification exposes the centrality of nondeterminism as a putative explanation of the inherent time requirements of myriad diverse "real" computational problems. We finish our coverage of complexity-via-nondeterminism with the classic theorem of Walter J. Savitch (Theorem 13.3) about *space complexity*, which broadens our perspective on the computational implications of nondeterminism. As a counterpoint to the apparent need for deterministic exponential time to simulate nondeterministic linear time, at least in the worst case, Savitch's theorem demonstrates that simulating a nondeterministic computing platform on a deterministic one can at worst *square* the amount of memory that one needs for a computation. Nondeterminism thus affects the space requirements of computations much more modestly than it (apparently) affects their time requirements. The qualifier "apparently" here exposes the exciting fact that complexity theory is very much a living discipline: Most of its most sought-after secrets remain to be uncovered.

In an effort to get the reader to view what we are terming the "big ideas" of computation theory within a broader context, we include throughout the book digressions and "enrichment" topics that branch out from our central development of the theory. Some of this extra material applies concepts in quite different settings from those in which they were developed; others supplement the "big ideas" with material that may be less "big" in their impact on computation but that round out the reader's perspective on the material.

1.2 The Nature of Computation Theory

Computation theory is a mathematical subject that arose from a variety of disparate sources, mathematics (including mathematical logic), engineering, and linguistics being the main three. Studying different problems and facing different conceptual and algorithmic challenges, all of these sources attempt to understand the power and limitations of various formal systems: mathematical logic and proof systems, digital computing systems, and formal mechanisms for specifying and analyzing the syntaxes of natural and artificial languages. The most exciting aspects of the theory are as follows:

The dynamic nature of computation theory. *Computation theory models objects that "do things."*

The systems (e.g., machines and circuits) studied within the theory are dynamic, in the sense that they evolve over time and "do things," such as compute. This contrasts sharply with the static systems encountered in most mathematical theories.

Any student of mathematics has encountered multicomponent mathematical systems such as graphs, groups, rings, fields. As complex as (some of) these systems are, their specifications are "one dimensional": Each system comprises a fixed set S of objects that can be acted on by a fixed set of operations to yield other objects from the set S; this process is governed by a fixed set of rules. Thus, if one understands the objects, the operations, and the rules, then one can—in principle, at least!—understand everything about the system. The systems in computation theory differ from the preceding picture, in that they add to the "syntactic" triumvirate of objects, operations, and rules a "semantic" component that describes how the system *evolves over time*. The notion of time thereby sneaks into computation theory as a guest that is unexpected because it does not appear explicitly when one specifies the formal systems that the theory uses as models for computational devices and processes. It may take some time getting used to systems that have both semantic and syntactic components, but the reader will view the time as well spent as this dynamic theory begins to unfold.

Robustness. *Several of the fundamental models studied by computation theory retain their basic (computational) properties even when their features are changed significantly.*

Many of the computational models studied in the theory have been developed by practitioners in quite distinct fields, using quite different intuitions and formalisms.

Numerous variants of the finite automaton model, for instance, were invented independently by electrical engineers studying synchronous sequential circuits, by linguists studying language acquisition in children, by neural scientists seeking models for the behavior of the brain, by programming theorists studying various constructs in programming languages, and by computation theorists seeking a "high-level" model for digital computers. While the resulting models were quite distinct in form, they all turned out to be behaviorally equivalent. Thus one can perturb many features of the finite automaton model without changing the underlying theory.

Computability theory evinces an even greater degree of robustness. For many decades of the twentieth century, people tried to devise mathematical models that captured the notion "computable by a digital computer." Models too numerous to list here, often differing dramatically in form, were proposed. Yet no proposed model ever exceeded the computing power of the rather primitive model invented by Turing in his original paper [104]. Indeed, this fact led people to formulate the (extra-mathematical) Church-Turing thesis, which asserts that a vast array of such models (including Turing's original) actually do capture the target notion. In other words, the thesis posits the *coincidence* of the preceding concept—which is unformalizable because we have not yet seen all possible digital computers—and the quite formal concept "computable by a Turing machine." (We discuss the thesis at some length in Section 9.1.)

Robustness of the sort just exemplified has (at least) two impacts. First, it enhances our ability to navigate the relevant portion of the theory: As we strive to understand various phenomena, we can switch from one formulation to another in a search for perspicuity. Second, it enhances our faith in having discovered something "fundamental": A superficiality is unlikely to retain its inherent nature in so many dramatically different guises.

Applicability. *Many aspects of computation theory have important computational applications in quite distinct computing-related fields.*

The computational applications to which the preceding assertion refers are those that transcend the fundamental conceptual applications of the theory. We note two examples: We find far-reaching implications in the theory of finite automata for activities as disparate as the design of compilers, of digital circuits (especially *sequential* circuits, which have state), and of programming languages. Computability theory has far-reaching consequences in mathematical logic, as well as in every aspect of the design and use of digital computers and their programs. Complexity theory has incisive messages for any field that is concerned with optimizing complex processes, as well as for many approaches to designing and analyzing complex systems.

As we introduce each topic throughout the book, we put the development of the topic's individual corner of computation theory into its historical context. Most obviously, such context helps one appreciate the nature of the theory and its underlying culture. Less obviously, such context helps one understand (and tolerate) some of the often obscure and sometimes conflicting terminology and notation in which much of the theory is couched.

Enough introduction! Let us begin our journey.

Chapter 2
Mathematical Preliminaries

"If your only tool is a hammer ... "

This chapter is devoted to reviewing a broad range of mathematical concepts that are central to our approach to developing computation theory. As we develop these concepts, we shall repeatedly observe instances of the following "self-evident truth" (which is what "axiom" means).

> **The conceptual axiom**. *One's ability to think deeply about a complicated concept is always enhanced by having more than one way to think about the concept.*

We shall harvest only small benefits from this axiom within this chapter, but we shall gather an abundant harvest in the remaining chapters.

2.1 Sets and Their Operations

Sets are probably the most basic object of mathematical discourse. We assume, therefore, that the reader knows what a set is and recognizes that some sets are finite, while others are infinite. Sample finite sets are, for example, the set of words in this book or the set of characters in any JAVA program. Some familiar infinite sets that will appear somewhere in our discussions in this book are:[1]

- the set of *nonnegative integers*, which we denote by $\mathbb{N}$,
- the set of *positive integers*, which we denote by $\mathbb{N}^+$,
- the set of *all integers*, which we denote by $\mathbb{Z}$,
- the set of nonnegative *rational numbers*—which are quotients of integers,
- the set of nonnegative *real numbers*—which can be viewed computationally as the set of numbers that admit infinite decimal expansions,

---

[1] We assume prior familiarity with all of these sets. We include them here just to establish notation and terminology.

A.L. Rosenberg, *The Pillars of Computation Theory*, Universitext, DOI 10.1007/978-0-387-09639-1_2, © Springer Science+Business Media, LLC 2010

- the set of nonnegative *complex numbers*—which can be viewed as ordered pairs of real numbers,
- the set of *all* finite-length binary strings—i.e., strings of 0's and 1's—which we denote $\{0,1\}^*$.

When discussing computer-related matters, one often calls each 0 and 1 that occurs in a binary string a *bit*, (for *binary digit*), which leads to the term *"bit string"* as a synonym of "binary string." With respect to general set-related notions, a source such as [34] will supply more than enough background for the topics we discuss in this book. Despite this assumption, we devote this short section to reviewing some basic concepts concerning sets and operations thereon. (Others will appear as needed throughout the book.)

For any finite set S, we denote by $|S|$ the *cardinality* of S, which is the number of elements in S. Finite sets having three special cardinalities are singled out with special names. If $|S| = 0$—i.e., if S has no elements—then we call S the *empty set* and denote it $\emptyset$. The empty set will reappear myriad times throughout the book, as a limiting case of set-defined entities. If $|S| = 1$—i.e., if S has just one element— then we call S a *singleton*; and if $|S| = 2$—i.e., if S has precisely two elements—then we call S a *doubleton*. In many of our discussions throughout the book, the sets of interest will be subsets of some fixed "universal" set U.

> We use the term "universal" as in "universe of discourse," not in the self-referencing sense of a set that contains all other sets. (Bertrand Russell has shown us in [91] [Chapter X, section 100] that the latter notion leads to mind-bending paradoxes.)

Two universal sets that will appear often are the two sample infinite sets mentioned earlier, $\mathbb{N}$ and $\{0,1\}^*$. Given a universal set U and a *subset* $S \subseteq U$ (the notation meaning that every element of S—if there are any—is also an element of U), we note that the set inequalities

$$\emptyset \subseteq S \subseteq U$$

always hold.

It is often useful to have a convenient term and notation for *the set of all subsets of a set S*. This bigger set—it contains $2^{|S|}$ elements when S is finite—is denoted by $\mathscr{P}(S)$ and is called the *power set* of S.[2] You should satisfy yourself that the biggest and smallest elements of $\mathscr{P}(S)$ are, respectively, the set S itself and the empty set $\emptyset$.

> Let's pause for a moment. *Why does the power set $\mathscr{P}(S)$ of a finite set S contain $2^{|S|}$ elements?*
>
> The **conceptual axiom** will help us answer this question. We begin by taking an arbitrary finite set S—say of n elements—and laying its elements out in a line. We thereby establish a correspondence between S's elements and positive integers: there is the first element, which we associate with the integer 1, the second element, which we associate with the integer 2, and so on, until the last element along the line gets associated with the integer n.
>
> Next, let's note that we can specify any subset S' of S by specifying a length-n *binary string*, i.e., a string of 0's and 1's. The translation is as follows. If an element s of S appears in the subset S', then we look at the integer we have associated with s (via our linearization of S),

---

[2] The name "power set" arises from the relative cardinalities of S and $\mathscr{P}(S)$ for finite S.

and we set the corresponding bit-position of our binary string to 1; otherwise, we set this bit-position to 0. In this way, we get a distinct subset of S for each distinct binary string, and a distinct binary string for each distinct subset of S. In particular: *the number of length-n binary strings is the same as the number of elements in the power set of S!*

The binary string that we have constructed to represent each set of integers $N \subseteq \{0, 1, \ldots, n-1\}$ is called the *(length-n) characteristic vector of the set N*. Of course, the finite set N has characteristic vectors of all finite lengths. Generalizing this idea, *every* set of integers $N \subseteq \mathbb{N}$, whether finite or infinite, has an *infinite* characteristic vector, which is formed in precisely the same way as are finite characteristic vectors, but now using the set $\mathbb{N}$ as the base set.

We are making progress, but let's look at an example before pressing onward. Let us focus on the set $S = \{a, b, c\}$. Just to make life more interesting, let us lay S's elements out in the order b, a, c, so that b has associated integer 1, a has associated integer 2, and c has associated integer 3. We depict the elements of $\mathscr{P}(S)$ and the corresponding binary strings in the following table.

Binary string	Set of integers	Subset of S
000	$\emptyset$	$\emptyset$
001	$\{3\}$	$\{c\}$
010	$\{2\}$	$\{a\}$
011	$\{2,3\}$	$\{a,c\}$
100	$\{1\}$	$\{b\}$
101	$\{1,3\}$	$\{b,c\}$
110	$\{1,2\}$	$\{a,b\}$
111	$\{1,2,3\}$	$\{a,b,c\} = S$

So, we need now only establish that there are 2^n binary strings of length n. This is accomplished most simply by noting that there are always twice as many binary strings of length n as there are of length $n-1$. This is because we can form the set of binary strings of length n by taking the set A of binary strings of length $n-1$, duplicating A to obtain two equinumerous sets A_1 and A_2, and appending 0 to every string in A_1 and appending 1 to every string in A_2. The thus-amended sets A_1 and A_2 collectively contain all binary strings of length n.

We now have the desired result.

Given two sets S and T, we denote by:

- $S \times T$ the *direct product* of S and T, which is the set of all ordered pairs whose first coordinate contains an element of S and whose second coordinate contains an element of T.
- $S \cap T$ the *intersection* of S and T, which is the set of elements that occur in *both* S and T.
- $S \cup T$ the *union* of S and T, which is the set of elements that occur in S, or in T, *or in both*. (Because of the "or both" qualifier, this operation is sometimes called *inclusive union*.)
- $S \setminus T$ the *difference* of S and T, which is the set of elements that occur in S but not in T. (Particularly in the United States, one often finds "$S - T$" instead of "$S \setminus T$.")

We exemplify the preceding operations with the sets $S = \{a,b,c\}$ and $T = \{c,d\}$. For these sets:

$$S \times T = \{\langle a,c \rangle, \langle b,c \rangle, \langle c,c \rangle, \langle a,d \rangle, \langle b,d \rangle, \langle c,d \rangle\},$$
$$S \cap T = \{c\},$$
$$S \cup T = \{a,b,c,d\},$$
$$S \setminus T = \{a,b\}.$$

When studying the several contexts that involve a universal set U that all other sets are subsets of, we include also the operation

- $\overline{T} = U \setminus T$, the *complement* of T (relative to the universal set U).

  For instance, the set of odd positive integers is the complement of the set of even positive integers, relative to the set of all positive integers.

We note a number of basic identities involving sets and operations on them. Working on verifying them will cement your understanding:

- $S \setminus T = S \cap \overline{T}$,
- If $S \subseteq T$, then

  1. $S \setminus T = \emptyset$,
  2. $S \cap T = S$,
  3. $S \cup T = T$.

Note, in particular, that[3]

$$[S = T] \text{ iff } \big[[S \subseteq T] \text{ and } [T \subseteq S]\big] \text{ iff } \big[(S \setminus T) \cup (T \setminus S) = \emptyset\big].$$

The operations union, intersection, and complementation—and operations formed from them, such as set difference—are usually called the *Boolean (set) operations*, acknowledging the seminal work of the nineteenth-century English mathematician George Boole.[4] There are several important identities involving the Boolean set operations. Among the most frequently invoked are the two "laws" attributed to the nineteenth-century French mathematician Auguste De Morgan:

$$\text{For all sets } S \text{ and } T: \quad \begin{cases} \overline{S \cup T} = \overline{S} \cap \overline{T}, \\ \overline{S \cap T} = \overline{S} \cup \overline{T}. \end{cases} \tag{2.1}$$

While we have focused here on Boolean operations on *sets*, there are "logical" analogues of these operations for logical sentences and their logical "truth values" 0 and 1:

---

[3] "iff" abbreviates the common mathematical phrase, "if and only if."

[4] One often encounters the lowercase adjective "boolean." Such is the price of fame.

- The logical analogue of complementation is (logical) not, which we shall denote by an overline;[5] e.g., $[\overline{0} = 1]$, and $[\overline{1} = 0]$.
- The logical analogue of union is (logical) or, which is also called *disjunction* or *logical sum*. Texts often denote "or" in expressions by "$\vee$"; e.g., $[X \vee Y = 1]$ iff $[X = 1]$ or $[Y = 1]$ or both.
- The logical analogue of intersection is (logical) and, which is also called *conjunction* or *logical product*. Texts often denote "and" in expressions by "$\wedge$"; e.g., $[X \wedge Y = 1]$ iff both $[X = 1]$ and $[Y = 1]$

We end this section with a set-theoretic definition that recurs often throughout our study. Let $\mathscr{C}$ be any (finite or infinite) collection of sets, and let S and T be two elements of $\mathscr{C}$. (Note that $\mathscr{C}$ is a set whose elements are sets.) Focus, just for example, on the set-theoretic operation of intersection; you should be able to extrapolate easily to other operations. We say that $\mathscr{C}$ is *closed* under intersection if whenever sets S and T (which could be the same set) both belong to $\mathscr{C}$, the set $S \cap T$ also belongs to $\mathscr{C}$. As one instance of the desired extrapolation, $\mathscr{C}$'s being closed under union would mean that the set $S \cup T$ belongs to $\mathscr{C}$.

2.2 Binary Relations

2.2.1 The Formal Notion of Binary Relation

Given sets S and T, a *relation on S and T* (in that order) is any subset

$$R \subseteq S \times T.$$

When $S = T$, we often call R a *binary relation on (the set) S* ("*binary*" because there are *two* sets being related). Relations are so common that we use them in every aspect of our lives without even noticing them. The relations "equal," "less than," and "greater than or equal to" are simple examples of binary relations on the integers. These same three relations apply also to other familiar number systems such as the rational and real numbers; only "equal," though, holds (in the natural way) for the complex numbers. Some subset of the three relations "is a parent of," "is a child of," and "is a sibling of" probably are binary relations on (the set of people constituting) your family. To mention just one relation with distinct sets S and T, the relation "A is taking course X" is a relation on (the set of all students) $\times$ (the set of all courses).

We shall see later (Section 7.1) that there is a formal sense in which binary relations are all we ever need consider: 3-set (*ternary*) relations—which are subsets of $S_1 \times S_2 \times S_3$—and 4-set (*quaternary*) relations—which are subsets of $S_1 \times S_2 \times S_3 \times S_4$—and so on (for any finite "arity"), can all be expressed as binary relations of binary relations ...of binary relations. As examples: For ternary relations, we can

---

[5] Context will always make it clear when we are talking about set complementation and when we are talking about logical not.

replace any subset R of $S_1 \times S_2 \times S_3$ by the obvious corresponding subset R' of $S_1 \times (S_2 \times S_3)$: for each element $\langle s_1, s_2, s_3 \rangle$ of R, the corresponding element of R' is $\langle s_1, \langle s_2, s_3 \rangle \rangle$. Similarly, for quaternary relations, we can replace any subset R'' of $S_1 \times S_2 \times S_3 \times S_4$ by the obvious corresponding subset R'''' of $S_1 \times (S_2 \times (S_3 \times S_4))$: for each element $\langle s_1, s_2, s_3, s_4 \rangle$ of R'', the corresponding element of R'''' is $\langle s_1, \langle s_2, \langle s_3, s_4 \rangle \rangle \rangle$.

> You should convince yourself that we could achieve the desired correspondence also by replacing $S_1 \times (S_2 \times S_3)$ with $(S_1 \times S_2) \times S_3$ and by replacing $S_1 \times S_2 \times S_3 \times S_4$ by either $((S_1 \times S_2) \times S_3) \times S_4$ or $(S_1 \times S_2) \times (S_3 \times S_4)$.

By convention, with a binary relation $R \subseteq S \times T$, we often write "sRt" in place of the more conservative "$\langle s, t \rangle \in R$." For instance, in "real life," we write "$5 < 7$" rather than the strange-looking (but formally correct) "$\langle 5, 7 \rangle \in <$."

The following operation on relations occurs in many guises, in almost all mathematical theories. Let P and P' be binary relations on a set S. The *composition* of P and P' (in that order) is the relation

$$P'' \stackrel{\text{def}}{=} \left\{ \langle s, u \rangle \in S \times S \mid (\exists t \in S) \Big[[sPt] \text{ and } [tP'u] \Big] \right\}.$$

(Note how we have used both of our notational conventions for relations here. Note also a new notational device that will recur frequently throughout the book: We use the compound symbol "$\stackrel{\text{def}}{=}$" as a shorthand for introducing notation. The sentence "$X \stackrel{\text{def}}{=} Y$" should be read "$X$ is, by definition, Y.")

There are two special classes of binary relations that play such a central role in computation theory—and elsewhere!—that we must single them out immediately, in the next two subsections.

2.2.2 Equivalence Relations

A binary relation R on a set S is an *equivalence relation* if it enjoys the following three properties:

1. R is *reflexive:* for all $s \in S$, we have sRs.
2. R is *symmetric:* for all $s, s' \in S$, we have sRs' whenever $s'Rs$.
3. R is *transitive:* for all $s, s', s'' \in S$, whenever we have sRs' and $s'Rs''$, we also have sRs''.

Sample familiar equivalence relations are:

- The equality relation, $=$, on a set S which relates each $s \in S$ with itself but with no other element of S.
- The relations $\equiv_{12}$ and $\equiv_{24}$ on integers, where[6]

---

[6] As usual, $|x|$ is the *absolute value*, or, *magnitude* of the number x. That is, if $x \geq 0$, then $|x| = x$; if $x < 0$, then $|x| = -x$.

1. $n_1 \equiv_{12} n_2$ if and only if $|n_1 - n_2|$ is divisible by 12.
2. $n_1 \equiv_{24} n_2$ if and only if $|n_1 - n_2|$ is divisible by 24.

We use relation $\equiv_{12}$ (without formally knowing it) whenever we tell time using a 12-hour clock and relation $\equiv_{24}$ whenever we tell time using a 24-hour clock.

Closely related to the notion of an equivalence relation on a set S is the notion of a *partition* of S. A partition of S is a nonempty collection of subsets $S_1, S_2, \ldots$ of S that are

1. *mutually exclusive:* for distinct indices i and j, $S_i \cap S_j = \emptyset$;
2. *collectively exhaustive:* $S_1 \cup S_2 \cup \cdots = S$.

We call each set S_i a *block* of the partition.

One verifies as follows that *a partition of a set S and an equivalence relation on S are just two ways of looking at the same concept.* To see this, we note the following.

Getting an equivalence relation from a partition. Given any partition $S_1, S_2, \ldots$ of a set S, define the following relation R on S:

$s R s'$ if and only if s and s' belong to the same block of the partition.

Relation R is an equivalence relation on S. To wit, R is reflexive, symmetric, and transitive because collective exhaustiveness ensures that each $s \in S$ belongs to some block of the partition, while mutual exclusivity ensures that it belongs to only one block.

Getting a partition from an equivalence relation. For the converse, focus on any equivalence relation R on a set S. For each $s \in S$, denote by $[s]_R$ the set

$$[s]_R \stackrel{\text{def}}{=} \{s' \in S \mid s R s'\};$$

we call $[s]_R$ *the equivalence class of s under relation R.*
The equivalence classes under R form a partition of S. To wit: R's reflexivity ensures that the equivalence classes collectively exhaust S; R's symmetry and transitivity ensure that equivalence classes are mutually disjoint.

The *index* of the equivalence relation R is its number of classes—which can be finite or infinite.

Let[7] $\equiv_1$ and $\equiv_2$ be two equivalence relations on a set S. We say that the relation $\equiv_1$ *is a refinement of* (or *refines*) the relation $\equiv_2$ just when each block of $\equiv_1$ is a subset of some block of $\equiv_2$. A couple of basic facts:

- The equality relation, $=$, on S refines every equivalence relation on S. (In this sense, it is the finest equivalence relation on S.)
- Say that the equivalence relation $\equiv_1$ refines the equivalence relation $\equiv_2$ and that $\equiv_2$ has finite index I_2. Then either $\equiv_1$ also has finite index $I_1 \geq I_2$, or $\equiv_1$ has infinite index.

---

[7] Conforming to common usage, we typically use the symbol $\equiv$, possibly with an embellishing subscript, to denote an equivalence relation.

2.3 Functions

One learns early in school that a function from a set A to a set B is a rule that assigns a unique value from B to every value from A. Yet, as one grows in (mathematical) sophistication, one finds that this notion of function is more restrictive than necessary. A simple example will illustrate our point. Our first example concerns division. We learn that division, like multiplication, is a function that assigns a number to a given pair of numbers. Yet we are warned almost immediately not to "divide by 0": The quotient upon division by 0 is "undefined." So, division is not quite a function as envisioned in the definition that begins this section. Indeed, in contrast to an expression such as "$4 \div 2$," which should lead to the result 2 in any programming environment,[8] expressions such as "$4 \div 0$" will lead to wildly different results in different programming environments. How can one deal with this situation? As presenters of computation theory, we are going to use an approach that is quite distinct from those of programming environments. We are going to broaden the definition of "function" in a way that behaves like the definition that begins this section in "well-behaved" situations and that extends the notion in an intellectually consistent way within "ill-behaved" situations. Let us begin to get formal.

A *(partial) function from set S to set T* is a relation $F \subseteq S \times T$ that is *single-valued:* for each $s \in S$, there is *at most* one $t \in T$ such that sFt. We traditionally write "$F : S \to T$" as shorthand for the assertion, "F is a function from the set S to the set T"; we also traditionally write "$F(s) = t$" for the more conservative "sFt." (The single-valuedness of F makes the nonconservative notation safe.) We often call the set S the *source (set)* and T the *target (set)* for function F. When there is always a (perforce, unique) $t \in T$ for each $s \in S$, then we call F a *total* function. Note that our terminology is a bit unexpected: *Every total function is a partial function;* that is, "partial" is the generic term, and "total" is a special case.

You may be surprised that we make partial functions our default domain of discourse. This is because most of the functions you deal with daily are *total* functions. Our mathematical ancestors had to do some fancy footwork in order to make your world so neat. Their choreography took two complementary forms.

1. They expanded the target set T on numerous occasions. As just two instances:

   - They appended both 0 and the negative integers to the preexisting positive integers[9] in order to make subtraction a total function.
   - They appended the rationals to the preexisting integers in order to make division (by nonzero numbers!) a total function.

   The irrational algebraic numbers, the nonalgebraic real numbers, and the nonreal complex numbers were similarly appended, in turn, to our number system in order to make certain (more complicated) functions total.

---

[8] We are, of course, ignoring demons such as round-off error.

[9] The great mathematician Leopold Kronecker said, "God made the integers, all else is the work of man"; cf. [3]. Kronecker was referring, of course, to the *positive* integers.

2. They adapted the function. In programming languages, in particular, undefinedness is anathema, so such languages typically have ways of making functions total, via devices such as "integer division" (so that odd integers can be "divided by 2") as well as various ploys for accommodating "division by 0."

We are going to be less pragmatic than our ancestors, because computation theory is traditionally a theory of functions on nonnegative integers (or, as we shall see, some transparent encoding thereof). The price for such "pureness" is that we must allow functions to be undefined on some arguments. Simple examples of such *nontotal* functions are "division by 2" and "taking square roots." Both of these functions are defined only on subsets of the positive integers (the even integers and the perfect squares, respectively).

Three special classes of functions merit explicit mention. For each, we give both a down-to-earth name and a more scholarly Latinate one.

A function $F : S \to T$ is:

1. *one-to-one* (or *injective*) if for each $t \in T$, there is at most one $s \in S$ such that $F(s) = t$;
   Example: "multiplication by 2" is injective; "integer division by 2" is not (because, e.g., 3 and 2 yield the same answer).
   An injective function F is called an *injection*.
2. *onto* (or *surjective*) if for each $t \in T$, there is at least one $s \in S$ such that $F(s) = t$;
   Example: "subtraction of 1" is surjective, as is "taking the square root"; "addition of 1" is not (because, e.g., 0 is never the sum), and "squaring" is not (because, e.g., 2 is not the square of any integer).
   A surjective function F is called a *surjection*.
3. *one-to-one, onto* (or *bijective*) if for each $t \in T$, there is precisely one $s \in S$ such that $F(s) = t$.
   Example: The (total) function $F : \{0,1\}^\star \to \{0,1\}^\star$ defined by:

$$(\forall w \in \{0,1\}^\star)\ F(w) = \text{(the reversal of } w\text{)}$$

is a bijection. The (total) function $F' : \{0,1\}^\star \to \mathbb{N}$ defined by

$$(\forall w \in \{0,1\}^\star)\ F(w) = \text{(the integer that is represented by } w \text{ viewed as a numeral)}$$

is *not* a bijection, due to the possibility of leading 0's.

> A *numeral* is a sequence of digits that is the "name" of a number. The numerical value of a numeral x depends on the *number base*, which is a positive integer $b > 1$ that is used to create x. Much of our focus will be on *binary* numerals—which are binary strings—for which the base is $b = 2$. For a general number base b, the integer denoted by the numeral $\beta_n \beta_{n-1} \ldots \beta_1 \beta_0$, where each $\beta_i \in \{0, 1, \ldots, b-1\}$, is
>
> $$\sum_{i=0}^{n} \beta_i b^i.$$
>
> We say that bit β_i has *lower order* in the numeral than does β_{i+1}, because β_i is multiplied by b^i in evaluating the numeral, whereas β_{i+1} is multiplied by b^{i+1}.

A bijective function F is called a *bijection*.

2.4 Formal Languages

2.4.1 The Notion of Language in Computation Theory

Let Σ be a finite set of (atomic) symbols. Reflecting the linguistic antecedents of computation theory (one of the theory's many ancestors), we often call the set Σ an **alphabet** and its constituent symbols **letters**. For each nonnegative integer k, we denote by Σ^k the set of all length-k strings—or sequences—of elements of Σ. For instance, if $\Sigma = \{a, b\}$, then:

$$\Sigma^0 = \{\varepsilon\} \quad (\varepsilon \text{ is the } null \ string\text{: the unique string of length 0}),$$
$$\Sigma^1 = \Sigma = \{a, b\},$$
$$\Sigma^2 = \{aa, ab, ba, bb\},$$
$$\Sigma^3 = \{aaa, aab, aba, abb, baa, bab, bba, bbb\}.$$

We denote by $\Sigma^\star$ *the set of all finite-length strings of elements of Σ*; symbolically,

$$\Sigma^\star = \bigcup_{k \in \mathbb{N}} \Sigma^k.$$

Again nodding to the theory's linguistic antecedents, we often call elements of $\Sigma^\star$ *words*, although we also often call them *strings*.

Notes. (*a*) If $\Sigma \neq \emptyset$, then $\Sigma^\star$ is infinite.

(*b*) Because $\Sigma^0 \subseteq \Sigma^\star$, $\Sigma^\star$ is never empty. Indeed, $\Sigma^\star$ is finite iff $\Sigma = \emptyset$, in which case $\Sigma^\star$ contains the single word ε.

(*c*) Be careful when reasoning about the null string ε (just as you should be careful when reasoning about the null list as a data structure). Specifically, despite ε's lack of letters, it *is* an object, so, for instance, the set $\{\varepsilon\}$ is *not* empty.

(*d*) The alphabet $\Sigma = \{\sigma_1, \sigma_2, \ldots, \sigma_n\}$ is a *set*; hence, it has no *intrinsic* order. However, in many situations, Σ is endowed with an *extrinsic* order. For instance, if Σ is the Latin alphabet, then we all "know" that "a" precedes "b," which precedes "c," and so on. Similarly, if $\Sigma = \{0, 1\}$, then we all "know" that "0" precedes "1." For such ordered alphabets, there is the important notion of *lexicographic order*, which is a total order on $\Sigma^\star$. Given any two words from $\Sigma^\star$,

$$x = \sigma'_1 \sigma'_2 \cdots \sigma'_k \quad \text{and} \quad y = \sigma''_1 \sigma''_2 \cdots \sigma''_\ell,$$

we say that x *precedes y in lexicographic order* precisely when one of the following holds:

- x is a *proper prefix* of y (meaning that $y = xz$ for some nonnull $z \in \Sigma^\star$);
- there exists an index $i \leq \min(k, \ell)$ such that
  - $\sigma'_j = \sigma''_j$ for all $j < i$
  - $\sigma'_i < \sigma''_i$ in the extrinsic order on Σ.

A **language over** the alphabet Σ is *any* subset $L \subseteq \Sigma^\star$.

A language L over Σ can be as "small" as $\emptyset$ or as "big" as $\Sigma^\star$, since, being a set, L satisfies

$$\emptyset \subseteq L \subseteq \Sigma^\star.$$

We denote by $\ell(w)$ the *length* of the word $w \in \Sigma^\star$. Hence, $\ell(\varepsilon) = 0$, and $\ell(w) = k$ for all $w \in \Sigma^k$.

The *concatenation* of words $x \in \Sigma^\star$ and $y \in \Sigma^\star$, which we denote by juxtaposition of x and y—namely, xy—is obtained by appending the string y after the string x. For instance, given two strings $x = 01001$ and $y = 110111$ over the alphabet $\{0, 1\}$, the concatenation of x and y is the string $xy = 01001110111$. Occasionally—but only occasionally—for emphasis, we actually insert an operation symbol to denote concatenation, by writing $x \cdot y$ in place of xy.

> The operation of concatenation is often called the *complex product* within an algebraic setting. In our context, the underlying algebra is the so-called *free semigroup* over the alphabet Σ, which is just an esoteric way of talking about the semigroup of words over Σ, viewing concatenation as a type of multiplication.

The operation of concatenation is *associative*, which means that for all strings x, y, and z from $\Sigma^\star$, we have

$$x \cdot (y \cdot z) = (x \cdot y) \cdot z.$$

We leave the inductive argument that establishes this fact to the reader.

> Associativity allows us to write long expressions without parentheses. We have been doing this "forever" with binary operations such as addition and multiplication. We are now just noting that we can do this also with this new, string-oriented, type of multiplication.

Equivalence relations on $\Sigma^\star$, specifically "right-invariant" ones, cast a broad shadow in the theory.

An equivalence relation $\equiv$ on $\Sigma^\star$ is **right-invariant** if for all $z \in \Sigma^\star$, $[xz \equiv yz]$ whenever $[x \equiv y]$.

Two simple examples illustrate right-invariance. (1) Consider first the finest equivalence relation $\equiv_1$, namely, equality:

$$[x \equiv_1 y] \text{ if and only if } [x = y].$$

This relation is right-invariant because if x and y are identical, then appending the same string z to both leaves you with identical strings, xz and yz. (2) Consider next the equivalence relation $\equiv_2$ that "identifies" binary strings that have the same number of 1's:

$$[x \equiv_2 y] \text{ if and only if the number of 1's in } x \text{ equals the the number of 1's in } y$$

(You should prove that $\equiv_2$ is indeed an equivalence relation.) This relation is right-invariant because if x and y share the same number of 1's, then so also do xz and yz, no matter what string z is.

A major focus in our development of the theory will be the following specific right-invariant equivalence relation on $\Sigma^\star$, which is defined in terms of a given language $L \subseteq \Sigma^\star$:

$$\text{For all } x, y \in \Sigma^\star : \quad [x \equiv_L y] \text{ iff } (\forall z \in \Sigma^\star)[[xz \in L] \Leftrightarrow [yz \in L]]. \tag{2.2}$$

The following important result is left as an exercise.

Lemma 2.1. *For any alphabet Σ and language $L \subseteq \Sigma^\star$, the equivalence relation $\equiv_L$ is right-invariant.*

2.4.2 Languages as Metaphors for Computational Problems

This section is devoted to an important example of how one can think about computations in nonobvious ways—a somewhat subtler instance of the **conceptual axiom** than we have observed to this point.

Every language $L \subseteq \Sigma^\star$ has an associated function that allows us to step back and forth between the world of functions and the world of languages.

We may initially be a bit uncomfortable hopping in this way between formal notions that are quite unrelated in day-to-day discourse. However, the historical antecedents of computation theory more or less force us to, especially if we want access to primary sources in the development of the theory.

The *characteristic function* of the set/language L is the function κ_L defined as follows:

$$(\forall x \in \Sigma^\star) : \quad \kappa_L(x) = \begin{cases} 1 \text{ if } x \in L, \\ 0 \text{ if } x \notin L. \end{cases}$$

Dually, every function $f : \Sigma^\star \to \{0, 1\}$ has an associated language L_f defined as follows:

$$L_f = \{x \in \Sigma^\star \mid f(x) = 1\}.$$

One can study a large range of computational issues involving two-valued functions by focusing on the languages associated with the functions; and one can study a large range of computational issues involving languages by focusing on the languages' characteristic functions. One thus finds three distinct notions talked about interchangeably within the theory:

1. a *language L*;
2. the *computational problem*: to compute L's characteristic function;
3. the *system property*: to decide, given $x \in \Sigma^\star$, whether $x \in L$.

Interestingly, we shall encounter situations in which we shall be able to compute only L's *semicharacteristic function* κ_L', which is a partial function that tells us when a given $x \in \Sigma^\star$ belongs to L but gives no response when $x \notin L$:

$$(\forall x \in \Sigma^\star): \quad \kappa'_L(x) = \begin{cases} 1 & \text{if } x \in L, \\ \text{undefined} & \text{if } x \notin L. \end{cases}$$

Alan M. Turing's world-changing demonstration of the existence of computational problems that cannot be solved algorithmically [104] in fact exhibited a language L whose *semicharacteristic function* is computable but whose *characteristic function* is not.

A more concrete example of the duality between functions and languages involves an arbitrary function

$$g : \{0,1\}^\star \times \{0,1\}^\star \to \{0,1\}^\star. \tag{2.3}$$

(Think of g as being addition or multiplication, for instance.) One often studies the problem of computing g via the following language-recognition problem. We define the language[10] $L(g)$ as follows. $L(g)$ is a language over the alphabet $\Sigma \overset{\text{def}}{=} \{0,1\} \times \{0,1\}$ whose letters are *ordered pairs* of bits. For each $n \in \mathbb{N}$, the n-letter word

$$\langle \alpha_0, \beta_0 \rangle \langle \alpha_1, \beta_1 \rangle \cdots \langle \alpha_{n-1}, \beta_{n-1} \rangle$$

in Σ^n belongs to the language $L(g)$ precisely when the nth bit of the bit-string

$$g(\alpha_{n-1} \cdots \alpha_0, \beta_{n-1} \cdots \beta_0)$$

is a 1.

Note that we reverse the orders of bit-strings so that the index of a bit-position equals the power of 2 that we use to convert the bit-string to an integer. Using this notational convention, the bit-string $\alpha_{n-1} \cdots \alpha_0$ is the numeral[11] for the integer $\sum_{i=0}^{n-1} \alpha_i 2^i$.

2.5 Graphs and Trees

A *directed graph* (*digraph*, for short) $\mathcal{G}$ is given by a set of *nodes* $\mathcal{N}_{\mathcal{G}}$ and a set of *arcs* (or *directed edges*) $\mathcal{A}_{\mathcal{G}}$. Each arc has the form $(u \to v)$, where $u, v \in \mathcal{N}_{\mathcal{G}}$; we say that this arc goes *from u to v*. A *path* in $\mathcal{G}$ is a sequence of arcs that share adjacent endpoints, as in the following path from node u_1 to node u_n:

$$(u_1 \to u_2), (u_2 \to u_3), \ldots, (u_{n-2} \to u_{n-1}), (u_{n-1} \to u_n). \tag{2.4}$$

It is sometimes useful to endow the arcs of a digraph with labels from an alphabet Σ. When so endowed, the path (2.4) would be written

$$(u_1 \overset{\lambda_1}{\to} u_2), (u_2 \overset{\lambda_2}{\to} u_3), \ldots, (u_{n-2} \overset{\lambda_{n-2}}{\to} u_{n-1}), (u_{n-1} \overset{\lambda_{n-1}}{\to} u_n),$$

---

[10] We avoid the notation "L_g" to avoid any confusion with languages and their characteristic functions.

[11] Recall that a *numeral* is the string-name for a number.

where the λ_i denote symbols from Σ. If $u_1 = u_n$, then we call the preceding path a *cycle*.

An *undirected graph* is obtained from a directed graph by removing the directionality of the arcs; the thus-beheaded arcs are called *edges*. Whereas we say:

the *arc* (u,v) goes *from* node u *to* node v

we say:

the undirected edge $\{u,v\}$ goes *between* nodes u and v

or, more simply:

the undirected edge $\{u,v\}$*connects* nodes u and v.

Undirected graphs are usually the default concept, in the following sense: *When $\mathscr{G}$ is described as a "graph," with no qualifier "directed" or "undirected," it is understood that $\mathscr{G}$ is an undirected graph.*

One specific genre of digraph merits separate mention: *rooted trees*, which are a class of *acyclic* digraphs. Paths in trees that start at the root are often called *branches*. The *acyclicity* of a tree $\mathscr{T}$ means that for any branch of $\mathscr{T}$ of the form (2.4), we cannot have $u_1 = u_n$, for this would create a cycle. Each rooted tree $\mathscr{T}$ has a designated *root node* $r_{\mathscr{T}} \in \mathscr{N}_{\mathscr{T}}$. A node $u_n \in \mathscr{N}_{\mathscr{T}}$ that resides at the end of a branch (2.4) that starts at $r_{\mathscr{T}}$ (so $u_1 = r_{\mathscr{T}}$) is said to reside at *depth* $n-1$ in $\mathscr{T}$; by convention, $r_{\mathscr{T}}$ is said to reside at depth 0. $\mathscr{T}$'s root $r_{\mathscr{T}}$ has some number (possibly 0) of arcs that go from $r_{\mathscr{T}}$ to its *children*, each of which thus resides at depth 1 in $\mathscr{T}$; in turn, each child has some number of arcs (possibly 0) to its children, and so on. (Think of a family tree.) For each arc $(u \to v) \in A_{\mathscr{T}}$, we call u a *parent* of v, and v a *child* of u, in $\mathscr{T}$; clearly, the depth of each child is one greater than the depth of its parent. Every node of $\mathscr{T}$ except for $r_{\mathscr{T}}$ has precisely one parent; $r_{\mathscr{T}}$ has no parents. A childless node of a tree is a *leaf*. The transitive extensions of the parent and child relations are, respectively, the *ancestor* and *descendant* relations. The *degree* of a node v in a tree is the number of children that the node has, call it c_v. If every nonleaf node in a tree has the same degree c, then we call c the *degree of the tree*.

It is sometimes useful to have a symbolic notation for the ancestor and descendant relations. To this end, we write $(u \Rightarrow v)$ to indicate that node u is an ancestor of node v, or equivalently, that node v is a descendant of node u. If we decide for some reason that we are not interested in really distant descendants of the root of tree $\mathscr{T}$, then we can *truncate* $\mathscr{T}$ at a desired depth d by removing all nodes whose depths exceed d. We thereby obtain the *depth-d prefix* of $\mathscr{T}$. (We encounter in Theorem 13.3 a situation in which we truncate a tree.)

Figure 2.1 depicts an arc-labeled rooted tree $\mathscr{T}$ whose arc labels come from the alphabet $\{a,b\}$. $\mathscr{T}$'s arc-induced relationships are listed in Table 2.1.

2.6 Useful Quantitative Notions

Although our main focus will be on logical relationships among computation-theoretic concepts, we shall now and then have occasion to discuss quantitive concepts. This section reviews a couple of basic definitions involving such concepts.

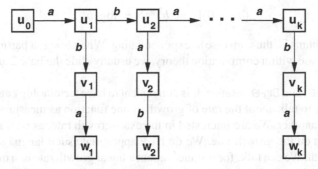

Fig. 2.1 An arc-labeled rooted tree $\mathscr{T}$ whose arc labels come from the alphabet $\{a,b\}$. (Arc labels have no meaning; they are just for illustration.)

The arc-labeled rooted tree $\mathscr{T}$ of Figure 2.1				
Node	Children	Parent	Descendants	Ancestors
$r_{\mathscr{T}} = u_0$	u_1	none	$u_1, u_2, \ldots, u_k, v_1, v_2, \ldots, v_k, w_1, w_2, \ldots, w_k$	none
u_1	u_2, v_1	u_0	$u_2, \ldots, u_k, v_1, v_2, \ldots, v_k, w_1, w_2, \ldots, w_k$	u_0
u_2	u_3, v_2	u_1	$u_3, \ldots, u_k, v_2, \ldots, v_k, w_2, \ldots, w_k$	u_0
$\vdots$	$\vdots$	$\vdots$	$\vdots$	$\vdots$
u_k	v_k	u_{k-1}	v_k, w_k	$u_0, u_1, \ldots, u_{k-1}$
v_1	w_1	u_1	w_1	u_0, u_1
v_2	w_2	u_2	w_2	u_0, u_1, u_2
$\vdots$	$\vdots$	$\vdots$	$\vdots$	$\vdots$
v_k	w_k	u_k	w_k	$u_0, u_1, \ldots, u_k$
w_1	none	v_1	none	u_0, u_1, v_1
w_2	none	v_2	none	u_0, u_1, u_2, v_2
w_k	none	v_k	none	$u_0, u_1, \ldots, u_k, v_k$

Table 2.1 A tabular description of the rooted tree $\mathscr{T}$ of Figure 2.1.

Floors and ceilings. Given any real number x, we denote by $\lfloor x \rfloor$ the *floor* (or *integer part*) of x, which is the largest integer that that does not exceed x. Symmetrically, we denote by $\lceil x \rceil$ the *ceiling* of x, which is the smallest integer that is at least as large as x. For any nonnegative integer n,

$$\lfloor n \rfloor = \lceil n \rceil = n;$$

for any positive rational number $n + p/q$, where n, p, and q are positive integers and $p < q$,

$$\lfloor n + p/q \rfloor = n, \text{ and } \lceil n + p/q \rceil = n + 1.$$

Logarithms and exponentials. Given any integer $b > 1$ (for "base"), the *base-b log-arithm* function $\log_b(\bullet)$ maps positive reals to reals and is defined by either of the following inverse relations:

$$(\forall x > 0)[x = b^{\log_b x} = \log_b b^x].$$

Taking logarithms is, thus, inverse to exponentiating. When $b = 2$, a particularly common special case within computation theory, we usually elide the base 2 and just write $\log x$.

Big-O, Big-Ω, and Big-Θ notation. It is convenient to have terminology and a notation that allows us to talk about the rate of growth of one function as measured by the rate of growth of another. We are interested in the exact growth rate, as well as upper and lower bounds on the growth rate. We do have appropriate such language for certain rates of growth. We can talk, for instance, about a linear growth rate or a quadratic rate or an exponential rate, to name just a few—and we get the desired bounds using the prefixes "sub" or "super," as in "subexponential" and "superlinear"—but our repertoire of such terms is quite limited. Mathematicians working in the theory of numbers in the late nineteenth century established a notation that gives us an unlimited repertoire of descriptors for growth rates, via what has come to be called the big-O, big-Ω, and big-Θ notations, which are collectively sometimes called *asymptotic notation*.

Let f and g be total functions from the nonnegative real numbers to the real numbers. We define the following notation:

$f(x) = O(g(x))$ means $(\exists c > 0)(\exists x^\#)(\forall x > x^\#)[f(x) \leq c \cdot g(x)]$
$f(x) = \Omega(g(x))$ means $f(x) = O(g(x))$,
$\quad$ i.e., $(\exists c > 0)(\exists x^\#)(\forall x > x^\#)[f(x) \geq c \cdot g(x)]$
$f(x) = \Theta(g(x))$ means $[f(x) = O(g(x))]$ and $[f(x) = \Omega(g(x))]$,
$\quad$ i.e., $(\exists c_1 > 0)(\exists c_2 > 0)(\exists x^\#)(\forall x > x^\#)[c_1 \cdot g(x) \leq f(x) \leq c_2 \cdot g(x)]$

Note that all three of the rate specifications are *eventual*—or *asymptotic*—because of the "$(\forall x > x^\#)$" quantifier. Thus, in contrast to the more familiar completely determined assertions such as "$f(x) \leq g(x)$," the assertion "$f(x) = O(g(x))$" has built-in uncertainty, regarding both the size of the scaling factor $c > 0$ and the threshold $x^\#$ at which the asserted relationship between $f(x)$ and $g(x)$ kicks in. In mathematical terms, it is best to think about these three asymptotic bounding assertions as establishing *envelopes* for $f(x)$:

- Say that $f(x) = O(g(x))$. If one draws the graphs of the functions $f(x)$ and $c \cdot g(x)$, then as one traces the graphs going rightward (letting x increase), one eventually reaches a point $x^\#$ beyond which the graph of $f(x)$ never enters the territory *above* the graph of $c \cdot g(x)$.
- Say that $f(x) = \Omega(g(x))$. This situation is the up-down mirror image of the preceding one: just replace the highlighted "*above*" with "*below*."
- Say that $f(x) = \Theta(g(x))$. We now have a two-sided envelope: beyond $x^\#$, the graph of $f(x)$ never enters the territory *above* the graph of $c_1 \cdot g(x)$ and never enters the territory *below* the graph of $c_2 \cdot g(x)$.

In addition to allowing one to make familiar growth-rate comparisons such as "$n^{14} = O(n^{15})$" and "$1.001^n = \Omega(n^{1000})$," we can now also make assertions such as "$\sin x = \Theta(1)$," which are much clumsier to explain in words.

There are "small"-letter analogues of the preceding "big"-letter asymptotic notations, but they are not encountered frequently in computation theory (although they do arise in the analysis of algorithms). In order to prepare the reader for Section 8.4.2, the one place in this book that employs the small-o notation, we note that the notation $o(1)$ refers to any function $f(n)$ that tends to the limit 0 as n grows without bound.

We refer the reader to a text such as [20] for the full repertoire of asymptotic notations that are useful when studying algorithms.

PART II
STATE

"This is a state of mind we live in..."
(Jerry Herman, in "Milk and Honey")

The notion "state" is fundamental to the design and analysis of myriad sophisticated systems. Consider, as one familiar example, an elevator. The elevator must "know" when it is in a *stable* situation—sitting at a floor—and when it is in an *unstable* situation—in transit between floors. It must respond to a fixed repertoire of external stimuli—signals from call buttons and floor identifiers—by moving from one stable situation to another. It is the stable situations that we call *states*.

Of particular interest to us, given the goal of this book is that the notion "state" is fundamental to the design and analysis of virtually all computational systems, from the sequential circuits that underlie sophisticated electronic hardware, to the semantic models that enable optimizing compilers, to leading-edge machine-learning concepts, to the models used in discrete-event simulations.

Decades of experience with state-based systems have taught that all but the simplest display a level of complexity that makes them hard—conceptually and/or computationally—to design and analyze. One brilliant candle in this gloomy scenario is the Myhill–Nerode theorem, which supplies a *rigorous mathematical* analogue of the following informal characterization of the notion "state":

> The state of a system comprises that fragment of its history that allows it to behave correctly in the future.

This part of the book is devoted to developing our first pillar, the notion "state." We begin with a formal model that we call an *online automaton*, a very abstract computational model that allows us to isolate the notion "state" and its role in computation. We then specialize online automata to their well-known finite submodel, *finite automata*. The conceptual pinnacle of this part is the Myhill–Nerode theorem, which we develop in two versions, a weak version for online automata and a strong one for finite automata. The remainder of our study of "state" focuses on several applications of the Myhill–Nerode theorem. We close this part with a brief development of so-called pumping lemmas for two classes of languages. Our inclusion of this topic is intended both for the enrichment of the reader and to contrast the very popular pumping lemma for regular languages, which seems to appear in all texts, with the much stronger and more perspicuous Myhill–Nerode theorem, which seems to have disappeared from almost all texts.

Since we are embarking on a rather mathematical development, this is a good time to review the material in Sections 2.1, 2.2, 2.3, and 2.4.1. *Throughout our discussion of online automata and their finite versions, finite automata, all functions will be total.*

Chapter 3
Online Automata: Exemplars of "State"

3.1 Online Automata and Their "Languages"

An *online automaton*[1] (*OA*, for short) is an abstract device that is a "pure" state-transition system. To hone your intuiton for the formal specification of OAs, consider Figure 3.1, which depicts a finite OA[2]—which we shall later (in the next chapter, in

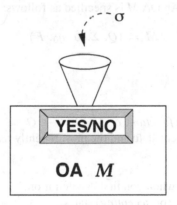

Fig. 3.1 A cartoonish depiction of a finite Online Automaton M.

fact) call a *finite automaton* (*FA*, for short and *finite automata* in the plural). One can think about an OA M as a simple machine that communicates with the world via:

---

[1] The word "automaton," being Greek in origin, forms its plural as "automata." It shares these singular/plural endings with more familiar English words of Greek ancestry, such as "phenomenon"/"phenomena" and "criterion"/"criteria."

[2] Throughout the book, we abbreviate the phrase "finite-state OA" (resp., "infinite-state OA") to "finite OA" (resp., "infinite OA).

A.L. Rosenberg, *The Pillars of Computation Theory*, Universitext,
DOI 10.1007/978-0-387-09639-1_3, © Springer Science+Business Media, LLC 2010

- an input port that can be thought of as a funnel, which admits symbols from some (finite) *input alphabet*, which we shall always denote by Σ_M (omitting the subscript M whenever the OA being discussed is clear from context)
- a bistable output mechanism that can be thought of as a light that flashes either "YES" or "NO."

When one activates an OA M, i.e., "turns it on," that action puts M into its designated *initial state*. Once M is "on," one can drop letters that are chosen from M's input alphabet Σ into M's funnel one at a time.

> While the repertoire of letters that you can drop into M's funnel is restricted to the finite set Σ, you do have access to as many instances of each letter as you want; i.e., for any $\sigma \in \Sigma$, you will always have access to another (instance of) σ whenever you want one. While the abstract development of this chapter views each letter as atomic—i.e., indivisible and unstructured— when we design actual OAs in Chapter 4, we actually employ letters with complex structure that elucidates their roles in the computation.

When M's internal logic "settles," so that it is ready to process a new input letter, M's output light flashes either "YES" or "NO," thereby announcing M's decision on the string of letters that it has seen to that point, since having been switched on. This process, or as we shall call it, *computation*, continues as long as you keep dropping letters into M's funnel. We now formalize this definition of OA in two ways, which will hopefully be synergistically helpful as you think about the model.

OAs as algebraic systems. An OA M is specified as follows:

$$M = (Q, \Sigma, \delta, q_0, F),$$

where

- Q is a set of *states*.
- Σ is a finite *alphabet*.
- δ is the *state-transition function:* $\delta : Q \times \Sigma \longrightarrow Q$.
  On the basis of the current state and the most recently read input symbol, δ specifies the next state of M.
- q_0 is M's *initial state*.
  q_0 is the state M enters when you first "switch it on."
- $F \subseteq Q$ is the set of *final* (or *accepting*) states.

OAs as labeled digraphs. One can view M as a labeled directed graph (*digraph*, for short), in a natural way.

- The nodes of the graph are the states of M.
  We shall represent each final state $q \in F$ of M by a double square, and each nonfinal state $q \in Q \setminus F$ by a single square.[3] A "tailless" arrow point to M's initial state q_0. These conventions are illustrated via some finite OAs in Figure 3.2.

---

[3] One usually finds circles instead of squares in these depictions. We use squares because they are easier to draw.

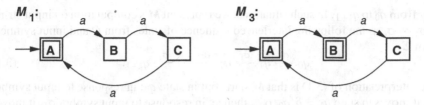

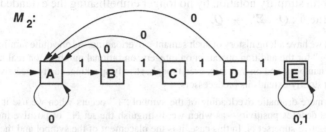

Fig. 3.2 Graph-theoretic representations of three simple OAs.

- The labeled arcs represent the state transitions. For each state $q \in Q$ and each alphabet symbol $\sigma \in \Sigma$, there is an arc labeled σ leading from state/node q to state/node $\delta(q,\sigma)$.

The "behavior" of an OA: the language it accepts. In order to make the OA model *dynamic*, we need to talk about how an OA M responds to strings of input symbols, not just to single symbols. We must therefore *extend* the state-transition function δ to operate on $Q \times \Sigma^*$, rather than just on $Q \times \Sigma$. It is of the utmost importance that our extension truly *extend* δ, i.e., that it agree with δ when applied to a string of length 1, so pay attention to make sure that it does. We call our extended function $\widehat{\delta}$ and define it via the following induction. For all $q \in Q$:

- $\widehat{\delta}(q,\varepsilon) = q$.
  If you give M no stimulus/input, then it gives you no response.
- $(\forall \sigma \in \Sigma, \forall x \in \Sigma^*) \; [\widehat{\delta}(q,\sigma x) = \widehat{\delta}(\delta(q,\sigma),x)]$.
  If you give M a multiletter stimulus/input—say the string σx that consists of the letter σ followed by the string x—then it begins by responding to the first letter (σ in this example)—as indicated by the "δ" on the right-hand side of the equation—and then it responds to the suffix x of the input.

Many times—as in most of this chapter—our interest in the behavior of M on input string $\sigma_0\sigma_1 \cdots \sigma_k \in \Sigma^k$ is satisfied by a "summary" behavioral equation of the form

$$\widehat{\delta}(q_0, \sigma_0\sigma_1 \cdots \sigma_k) = q_{k+1}.$$

There are times, however—notably in the chapters on computing devices that are more structured than OAs—when we want a more detailed description of M's trajectory from q_0 to q_{k+1}. In such situations, we represent M's computation on input string $\sigma_0\sigma_1\cdots\sigma_k$ as the following interleaved sequence of states from Q and input symbols from Σ:

$$q_0 \xrightarrow{\sigma_0} q_1 \xrightarrow{\sigma_1} q_2 \xrightarrow{\sigma_2} \cdots \xrightarrow{\sigma_k} q_{k+1}. \tag{3.1}$$

The interpretation of (3.1) is that M starts out in state q_0; in response to input symbol σ_0, it moves to state $q_1 = \delta(q_0, \sigma_0)$; thence, in response to input symbol σ_1, it moves to state $q_2 = \delta(q_1, \sigma_1)$; and so on.

Because it can cause no confusion to "overload" the semantics of the symbol "δ," we henceforth simplify notation by no longer embellishing the extended δ with a hat and just write $\delta : Q \times \Sigma^* \longrightarrow Q$.

Note that we have a long history of such semantic overloading. As one quite familiar example, we use "+" for the addition operation on integers, on rational numbers, on real numbers, on complex numbers, and on matrices, even though, strictly speaking, each successive operation in ths list strictly extends its predecessors.

An even more dramatic overloading of the symbol "+" occurs when we use it as a superscript that denotes positivity—as when we distinguish the set $\mathbb{N}^+$ of positive integers from its nonnegative superset $\mathbb{N}$. In this case, it is the placement of the symbol and the absence of immediate right and left neighbors that precludes ambiguity.

Although we are not yet ready to explain the significance of this natural extension of δ to $Q \times \Sigma^*$, we take a first step in this direction with a simple, yet basic, result that we call the continuation lemma.

Lemma 3.1. (The continuation lemma) *If $\delta(q_0, x) = \delta(q_0, y)$ for strings $x, y \in \Sigma^*$, then for all $z \in \Sigma^*$, $\delta(q_0, xz) = \delta(q_0, yz)$.*

In words: If strings $x, y \in \Sigma^*$ lead M from its initial state $\mathcal{Q}_o$ to the same state ($\delta(q_0, x) = \delta(q_0, y)$), then no "continuation" string $z \in \Sigma^*$ can help M distinguish x from y.

Proof. The lemma is immediate from the following chain of equalities:

$$\delta(q_0, xz) = \delta(\delta(q_0, x), z) = \delta(\delta(q_0, y), z) = \delta(q_0, yz).$$

The chain follows by the way we have extended δ to strings: M reads xz (resp., yz) by first reading x (resp., y) and then reading z. $\square$

Finally, we are ready to define the *language $L(M)$ that is accepted* (or *recognized*) by the OA M (sometimes called the "behavior" of M). The language $L(M)$ is the following subset of Σ^*:

$$L(M) \overset{\text{def}}{=} \{x \in \Sigma^* \mid \delta(q_0, x) \in F\}.$$

In analogy with the equivalence relation $\equiv_L$ of (2.2), which is associated with a language L, we associate with each OA M the following equivalence relation on Σ^*:

For all $x, y \in \Sigma^*$: $[x \equiv_M y]$ if and only if $[\delta(q_0, x) = \delta(q_0, y)]$.   (3.2)

(You should verify that $\equiv_M$ is always an *equivalence relation* on the set $\Sigma^\star$, i.e., that it is reflexive, symmetric, and transitive.)

Recall (from Sections 2.2.2 and 2.4.1) that for any equivalence relation $\equiv$ on $\Sigma^\star$:

- For each $x \in \Sigma^\star$, the $\equiv$-*class* that x belongs to is $[x]_{\equiv} \overset{\text{def}}{=} \{y \in \Sigma^\star \mid x \equiv y\}$.
  (When the subject relation $\equiv$ is clear from context, we simplify notation by writing $[x]$ for $[x]_{\equiv}$.)
- The *classes* of $\equiv$ are the blocks of a partition of $\Sigma^\star$.

The following basic facts about the equivalence relation $\equiv_M$ play a significant role in exposing the nature of the concept "state."

Lemma 3.2. *For each OA* $M = (Q, \Sigma, \delta, q_0, F)$:
  (a) *the equivalence relation* $\equiv_M$ *is right-invariant;*
  (b) $L(M)$ *is the union of some of the equivalence classes of relation* $\equiv_M$.

Proof. (a) The assertion that relation $\equiv_M$ is right-invariant is just a rewording of the continuation lemma (Lemma 3.1).

(b) $L(M)$ is the union of the classes of relation $\equiv_M$ that correspond to strings that lead M from q_0 to a state in F. $\square$

Reinforcing the preliminaries. Before continuing with our development, we illustrate our definitions with three simple finite OAs and one infinite one. We hope that these examples will hone the reader's intuition and serve as concrete hooks to stabilize our rather quick journey into the land of abstraction.

Figure 3.2 presents digraph representations of three finite OAs whose structures are specified in Table 3.1. One verifies by inspection that[4]

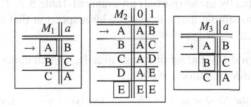

M_1	a
$\rightarrow$ A	B
B	C
C	A

M_2	0	1
$\rightarrow$ A	A	B
B	A	C
C	A	D
D	A	E
E	E	E

M_3	a
$\rightarrow$ A	B
B	C
C	A

Table 3.1 Tabular representations of the OAs of Figure 3.2.

1. $L(M_1) = \{a^k \mid k \equiv 0 \bmod 3\}$, the set of strings of a's (i.e., composed of instances of the letter a) whose lengths are divisible by 3.
2. $L(M_2) = \{x \in \{0,1\}^\star \mid (\exists y \in \{0,1\}^\star)(\exists z \in \{0,1\}^\star) [x = y1111z]\}$, the set of binary strings (i.e., strings of 0's and 1's) that contain four or more consecutive 1's in at least one place.

---
[4] We intentionally use diverse terminology in describing these languages, so that the reader will get familiar with alternative modes of describing the same concept.

3. $L(M_3) = \{a^k \mid k \not\equiv 2 \bmod 3\}$, the set of strings of a's whose length is congruent to either 0 or 1 modulo 3.

These three languages provide simple examples of a finite OA's ability to make finitely many discriminations regarding the structure of the string of inputs it has seen thus far.

> Note the roles that these three OAs' states play in making the discriminations that decide whether the OA should accept the string of inputs it has seen thus far.

OAs that have *infinitely* many states can be viewed as abstractions of programs that can make *infinitely* many *potential* discriminations regarding the structure of a set of potential input strings.

> The word "*potential*" in the preceding sentence is of critical importance. OAs with infinitely many states do not represent actual machines or programs. A human could never write an infinite program—she would die before completing it (although most of us have written programs that would run for an infinite number of steps if not stopped). One should, rather, think of infinite OAs as abstract representations of all potential finite behaviors of finite programs. Such an abstract representation is useful in many "real" situations. Consider, as a simple instance, a program—call it P—that executes a single loop some (finite) number of times that is specified by an integer input. Since there are infinitely many integers, P can, in principle, exhibit infinitely many distinct start-to-finish sequences of "states"—even though each such sequence is finite. It is often convenient to analyze such a program by conceptually (but not physically, of course) "unrolling" its loops, to create an associated quasiprogram that is infinite. (Sophisticated compilers do this as a matter of course.) If the program being "unrolled" is well-structured, then its infinite quasiprogram can be quite amenable to analysis—as our two simple sample infinite OAs will illustrate.

We present the following two simple infinite OAs, M_4 and M_5, via the (partial) digraph representations in Figure 3.3, augmented by the (partial) tabular specifications that appear schematically in the (partial) programs of Table 3.2. The OAs M_4 and M_5 are (intentionally) so simple that one can see by inspection that they recognize the following languages:

1. $L(M_4) = \{a^n b^n \mid n \in \mathbb{N}\}$, the set of strings over $\{a, b\}$ that consist of a block of a's followed by a *like-length* block of b's.

> Note the role of the "dead state" C of M_4. This state is entered just when the input string seen thus far has strayed irretrievably from the structure demanded by $L(M_4)$, either by not being a block of a's followed by a block of b's, or by having too long a block of b's. If either of these iniquitous conditions occurs in the prefix $x \in \{a, b\}^*$ read thus far, then no subsequent input string $y \in \{a, b\}^*$ can, when appended to x, lead to a string $xy \in L(M_4)$; i.e., the form of x cannot be completed to a string xy of the form demanded by $L(M_4)$.

The significance of this example resides in our imminent demonstration that the highlighted condition, "like-length," forces M_4 to have infinitely many states.

2. $L(M_5) = \{a^{n^2} \mid n \in \mathbb{N}\}$, the set of strings consisting of a block of a's whose length is a perfect square, i.e., a nonnegative integer k of the form $k = n^2$.

> In contrast to M_4, M_5 possesses no irretrievably flawed inputs—so M_5 does not need a "dead state."

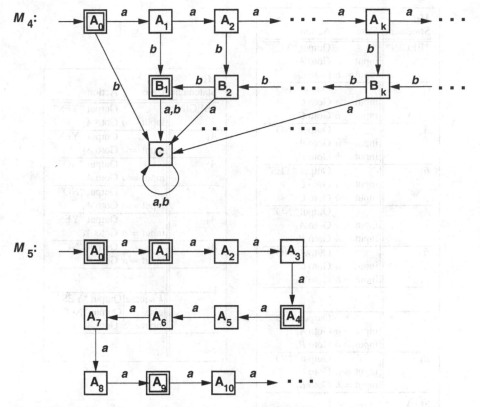

Fig. 3.3 Graph-theoretic representations of two simple infinite OAs.

The significance of this example resides in our imminent demonstration that the structure of the set of integers that are perfect squares forces M_5 to have infinitely many states.

We now argue intuitively and informally that the kind of discrimination that M_4 and M_5 make when deciding whether or not to accept an input string—namely, M_4's matching the lengths of the blocks of a's and b's by M_4, and M_5's detection of perfect squareness—*cannot* be made by any *finite* OA. Speaking *very* intuitively—but with an intuition that is quite useful—such discriminations require unbounded memory. We return to these examples with a formal treatment in Section 5.1.

We argue first that no finite OA recognizes the language $L_1 = L(M_4)$. The simplest way of seeing this begins by assuming, for contradiction, that there is a finite OA $M = (Q, \{a, b\}, \delta, q_0, F)$ such that $L_1 = L(M) = L(M_4)$.

This is our first example of a technique of argumentation called "proof by contradiction"—"*reductio ad absurdum*" in Latin. The technique consists in assuming the contrary of what you want to prove, and showing that this contrary assumption leads to "a contradiction," i.e., something that you know to be false. The only danger with this technique of argumentation is that when you recognize the contradiction, you had better check carefully that your contrary assumption is the *only* possible source of contradiction in your argument!

M_4:

Statement	Case	Action
BEGIN		Output "YES"
A_0	input $= a$ Goto A_1	
	input $= b$ Goto C	
C		Output "NO"
	input $= a$ Goto C	
	input $= b$ Goto C	
A_1		Output "NO"
	input $= a$ Goto A_2	
	input $= b$ Goto B_1	
B_1		Output "YES"
	input $= a$ Goto C	
	input $= b$ Goto C	
A_2		Output "NO"
	input $= a$ Goto A_3	
	input $= b$ Goto B_2	
B_2		Output "NO"
	input $= a$ Goto C	
	input $= b$ Goto B_1	
$\vdots$	$\vdots$	$\vdots$
A_k		Output "NO"
	input $= a$ Goto A_{k+1}	
	input $= b$ Goto B_k	
B_k		Output "NO"
	input $= a$ Goto C	
	input $= b$ Goto B_{k-1}	
$\vdots$	$\vdots$	$\vdots$

M_5:

Statement	Case	Action
BEGIN		Output "YES"
A_0	input $= a$ Goto A_1	
A_1		Output "YES"
	input $= a$ Goto A_2	
A_2		Output "NO"
	input $= a$ Goto A_3	
A_3		Output "NO"
	input $= a$ Goto A_4	
A_4		Output "YES"
	input $= a$ Goto A_5	
A_5		Output "NO"
	input $= a$ Goto A_6	
$\vdots$	$\vdots$	$\vdots$
A_k	k square	Output "YES"
	else	Output "NO"
	input $= a$ Goto A_{k+1}	
$\vdots$	$\vdots$	$\vdots$

Table 3.2 Representations of the OAs of Figure 3.3 as schematic programs.

One notes that since there are infinitely many finite-length strings of a's, some two distinct ones, say a^i and a^j, must be "confused" by M, in the sense that $\delta(q_0, a^i) = \delta(q_0, a^j) = q$; in other words, $a^i \equiv_M a^j$. Because δ is a function, we know that

$$\delta(q, b^i) \ = \ \delta(q_0, a^i b^i) \ = \ \delta(q_0, a^j b^i).$$

The preceding string of equations means that either both $a^i b^i$ and $a^j b^i$ are accepted by M or neither is. If both strings are accepted by M, then it accepts strings that do not belong to L_1; if neither string is accepted by M, then it fails to accept all strings that belong to L_1. In either case, M does not accept the language L_1.

We argue next that no finite OA recognizes the language $L_2 = L(M_5)$. The simplest way of seeing this begins by assuming, for contradiction, that there is a finite OA $M = (Q, \{a, b\}, \delta, q_0, F)$ such that $L_2 = L(M)$. One then notes that since there are infinitely many strings of a's of the form a^{k^2}—i.e., whose length is a perfect square—some two distinct ones, say a^{i^2} and a^{j^2}, where $j > i$, must be "confused" by M, in the sense that $\delta(q_0, a^{i^2}) = \delta(q_0, a^{j^2}) = q$; i.e., $a^{i^2} \equiv_M a^{j^2}$. On the one hand, we note that

$$\delta(q, a^{2i+1}) = \delta(q_0, a^{i^2} a^{2i+1}) = \delta(q_0, a^{i^2 + 2i + 1}) = \delta(q_0, a^{(i+1)^2}),$$

which M *should* accept, because $a^{(i+1)^2} \in L_2$. On the other hand, we note that

$$\delta(q, a^{2i+1}) = \delta(q_0, a^{j^2} a^{2i+1}) = \delta(q_0, a^{j^2 + 2i + 1}),$$

which M *should not* accept, because

$$j^2 < j^2 + 2i + 1 < j^2 + 2j + 1 = (j+1)^2,$$

so that $a^{j^2 + 2i + 1} \notin L_2$ (the exponent of a falls strictly between two adjacent perfect squares). We thus see that the state $\delta(q, a^{2i+1})$ (which is a unique state because δ is a function) should be an accepting state of M in order to accept $a^{(i+1)^2}$, but it should be a nonaccepting state in order not to accept $a^{j^2 + 2i + 1}$. We conclude that the OA M cannot exist.

> Note that we have intentionally worded the conclusions of the preceding two arguments a bit differently from one another. Our intention was to give the reader a sample of the variety of ways of saying that the posited finite OA does not behave correctly.

What is the common thread in the arguments about L_1 and L_2? In both cases, we argued that there must be two strings, x and y, that lead the putative OA M to the same state, even though M must distinguish x from y in order to accept all and only strings in the desired language. This inability to distinguish x from y is the (not yet quite formal) analogue of our saying that there are discriminations M must make if it is to function correctly. The argument is that simple, but its simplicity should not obscure the principle that it suggests: *The state of an OA M embodies what it "remembers" of its set of past histories.* In particular, past histories that M must discriminate among, in order to act correctly in the future, must lead M from its initial state to distinct states! Superficially, it may appear that this definition of "state" is of no greater *operational* significance than is the foundational identification of the number *eight* with the infinitely many sets that contain eight elements. This appearance is too simplistic, as we shall see in Chapter 4.

3.2 A Myhill–Nerode-like Theorem for OAs

The formalization of "state" in this section lays the foundation for our development of the powerhouse Myhill–Nerode theorem of finite-automata theory (Theorem 4.1), in Chapter 4. The version of the theorem that we develop now is weakened so that it applies to all OAs, not just finite ones. Even this weakened version of the theorem will expose the mathematical essence of the notion "state." Specifically, we see that a state of an OA M "is" an equivalence class of the relation $\equiv_M$, as defined in

(3.2)—which means, formally, that a state "is" a subset of an equivalence class of the relation $\equiv_{L(M)}$, as defined in (2.2)—which means that the state is an embodiment of the distinctions of past histories that are sufficient to decide membership in the language $L(M)$.

Theorem 3.1. (A Myhill–Nerode-type theorem for OAs)
(a) *If $L = L(M)$ for some OA M, then the right-invariant equivalence relation $\equiv_M$ is a refinement of the right-invariant equivalence relation $\equiv_L$.*
(b) *Every language L is recognized by an OA M_L whose states are the classes of $\equiv_L$.*

Proof. Because our argument will talk explicitly about the words that are in L and those that are not in L, we need to have a name for the alphabet that constitutes those words. Say that $L \subseteq \Sigma^*$.

(a) Let the OA M be given by $M = (Q, \Sigma, \delta, q_0, F)$. We show that for all $x, y \in \Sigma^*$, if $x \equiv_M y$, then $x \equiv_{L(M)} y$. It will follow that each block of relation $\equiv_M$ is a subset of some block of relation $\equiv_{L(M)}$.

By definition, if $x \equiv_M y$, then $\delta(q_0, x) = \delta(q_0, y)$. By the continuation lemma (Lemma 3.1), we then have also that for all $z \in \Sigma^*$, $\delta(q_0, xz) = \delta(q_0, yz)$, or equivalently, $xz \equiv_M yz$. Because all extended strings xz and yz thus share a destination state q (starting from q_0), they are either both accepted by M (if the shared state q belongs to F) or both rejected by M (if the shared state q belongs to $Q \setminus F$). Of course, the phrases "are accepted by M" and "are elements of $L(M)$" are synonymous.

(b) We now present a "construction" of an OA $M_L = (Q_L, \Sigma, \delta_L, q_{0,L}, F_L)$, and we argue that $L(M_L) = L$.

An explanatory note that is somewhat philosophical is called for here.

The "construction" that we are about to present is a mathematical existence proof, not an algorithm that can be followed to actually produce the OA M_L. (That is why we have placed quotation marks around "construction.") This is because the language L is completely arbitrary: we do not have access to any rules that might help us decide of a given word $x \in \Sigma^*$ whether x belongs to L. Indeed, as we shall see in Chapter 9, an arbitrary such set of rules may not be "computable," i.e., may not be able to be translated into a program that runs on an actual computer and makes the required decisions. (I hope that this whets your appetite for Chapter 9, where we adduce conditions that are necessary for a set of rules to be "computable.")

Again to whet your appetite for later material, I note that the reasoning that leads to Corollary 7.2 at the end of this chapter can be used to show that—speaking with an informality that we shall make rigorous via the development in Chapter 7—there are "not enough" algorithms to construct the OAs M_L for all languages L.

To reinforce the present digression, you should refer back to the rest of this proof when you see the real Myhill–Nerode theorem (Theorem 4.1) in Chapter 4.

We specify the four entities that constitute the OA M_L:

1. $Q_L = \{[x]_L \mid x \in \Sigma^*\}$, the set of classes of relation $\equiv_L$.
   The set Q_L is well defined because $\equiv_L$ is an equivalence relation.

2. For all $x \in \Sigma^*$ and all $\sigma \in \Sigma$,
$$\delta_L([x]_L, \sigma) = [x\sigma]_L.$$
You should verify that the right-invariance of relation $\equiv_L$ guarantees that δ_L is a well-defined function—i.e., that there is precisely one equivalence class $[x\sigma]_L$ for each equivalence class $[x]_L$ and each $\sigma \in \Sigma$.

3. $q_{0,L} = [\varepsilon]_L$.
M_L's start state corresponds to its having read nothing.

4. $F_L = \{[x]_L \mid x \in L\}$.
This guarantees that M_L accepts precisely those words that belong to the language L.

The interleaved remarks show that M_L is a well-defined OA. The following simple argument shows that $L(M_L) = L$.

Let $\sigma_1\sigma_2 \cdots \sigma_n$, where $n \geq 0$, be any string in Σ^*. An easy induction, which we leave to the reader, verifies that

$$\delta_L([\varepsilon]_L, \sigma_1\sigma_2 \cdots \sigma_n) = [\sigma_1\sigma_2 \cdots \sigma_n]_L. \tag{3.3}$$

(The required induction repeatedly invokes our remarks about the well-definedness of both the set Q_L and the function δ_L.) By definition, the state $[\sigma_1\sigma_2 \cdots \sigma_n]_L$ belongs to F_L if and only if the string $\sigma_1\sigma_2 \cdots \sigma_n$ belongs to L. It follows that $L(M_L) = L$.
□

3.3 A Concrete OA: The Online Turing Machine

This section develops a variant of the classical *Turing machine* (*TM* for short), the computational model introduced by Alan M. Turing in his monumental study [104] that planted the seeds of the branches of computation theory called computability theory and complexity theory. Our variant of the TM specializes Turing's model to the task of language recognition, i.e., to the computation of a characteristic or a semicharacteristic function, by rendering the model *online,* i.e., by having it receive its input one symbol at a time (through a "funnel"). We call the resulting variant TM an *online TM* (*OTM*, for short).

> You should note how our OTM model represents one particular way of lending structure to the infinite set of states of an OA. This way of achieving that goal is attractive because it translates rather easily into viewing an OTM as a finite program that is endowed with a simple data structure for "scratchwork," with no a priori bound on the size of the data structure.

An OTM can be viewed as an OA that has a finite set of states—the *control structure* of the model—augmented with an auxiliary storage device (the data structure just referred to) that contains only a finite amount of information at any time, but whose capacity has no a priori upper bound; cf. Figure 3.4.

> If we eliminate the auxiliary storage device, then we are left with a finite OA, the model that we study in Chapter 4 under the name *finite automaton* (*FA*, for short).

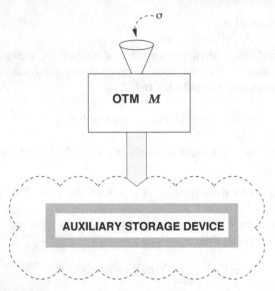

Fig. 3.4 A cartoonish depiction of a generic online Turing machine.

To reiterate a crucial point: An OTM is an OA whose infinitely many states result from appending a simply structured unbounded storage mechanism to a finite-state control. Remarkably, every "reasonable" digitizable proposal for an OTM's auxiliary storage device—or for a TM's auxiliary storage device—has been shown to give no more computing power than the storage device proposed originally by Turing, namely, a *single linear tape*.

> An essential component of a storage device's being "reasonable" is that at every step of a computation, the contents of the device can be represented by a finite sequence of bits. Among other things, this precludes analog devices.

We discuss the importance of the OTM model to the worlds of computation and mathematics at some length in Section 9.8. We develop the model here only to show that the unreasonably abstract—because it cannot be built—OA model can be instantiated in many quite reasonable ways, to produce eminently buildable OAs.

For a computer scientist, the easiest way to view a TM's "tape" is as a linear list of items called *symbols*; we view lists in a manner consistent with the treatment in [53]. Within this traditional model, our earlier comment regarding the storage device's capacity is rendered as follows: Although a tape is always finite in length, there is no bound on its capacity over the course of a computation. The single pointer that a TM uses to access its tape is (for historical reasons) called a *read/write head*.

In preparation for our formal development, we refer the reader to Figure 3.5 during this informal description of the operation of a sample, illustrative, OTM M.

The OTM M can access its tape in the following ways:

- M can "read" the symbol currently pointed to by the read/write head.

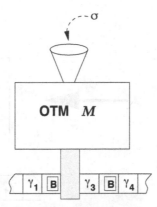

Fig. 3.5 An online Turing machine M whose storage device is a linear tape.

All symbols that appear on the tape belongs to M's *working alphabet* Γ. While Γ may contain some or all of the letters in M's *input alphabet* Σ, Γ differs from Σ at least by containing the designated *blank symbol*, $\boxed{\text{B}}$, which Σ does not contain.

> The blank symbol, $\boxed{\text{B}}$, is an actual symbol that occupies space—namely, a square of the OTM's tape. It should not be confused with the null string ε. What complicates this warning is that when $\boxed{\text{B}}$ occurs either before all nonblank characters on a tape or after all nonblank characters, we usually do not write it. When $\boxed{\text{B}}$ occurs in the midst of nonblank characters on a tape, then we always do write it. Such are the vagaries of convention.

- M can "rewrite" the symbol currently pointed to by its read/write head. This act replaces one symbol by another, but it does not alter the length of the tape.
- M can move its read/write head one unit left or right, thereby:
  - accessing the symbol that currently resides in the square it has moved to, if there is one;
  - appending a new copy of $\boxed{\text{B}}$ to the tape, if necessary to ensure that the read/write head always has at least one tape square to its right and at least one to its left.

This entire process is illustrated in Figure 3.6.

Still staying informal: Each *computation* by M begins with the TM in its designated *initial state* q_0, with the tape "empty." It is convenient when describing M's operation to imagine the initial tape as consisting of three instances of $\boxed{\text{B}}$, rather than being null: M's read/write head sits on (and scans) the middle $\boxed{\text{B}}$ and is flanked by the others. (With this convention, M can never "fall off" its tape by shifting its read/write head. Believe it or not, the problem of modeling whether M should be allowed to "fall off" its tape got a lot of attention not so many decades ago.)

M's set Q of (internal) states—which are the states of M's FA control unit—consists of two disjoint subsets:

1. the set Q_{poll} of *polling* states.
   These states are characterized by two facts:

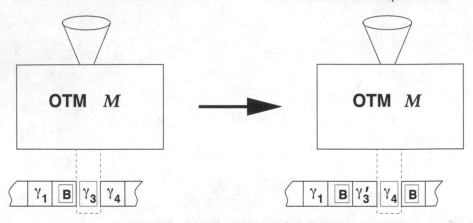

Fig. 3.6 Depicting M's tape's growing because of a move by the read/write head.

   a. M's polling states are the ones that admit input symbols into M's input funnel.

> Let's consider the behavior of the input port in an OTM M's "online" computation, by analogy with OAs. One can view an OA as a device that is passive until a symbol $\sigma \in \Sigma$ is "dropped into" its input port. If the OA M is in a stable configuration at that moment—for an actual electronic machine, this means that all bistable devices (say, flip-flops or transistors) in M's circuitry have stabilized—then M responds to input σ by changing its state. The most interesting aspect of this response is that M indicates whether the entire sequence of input symbols that it has been presented up to that point—i.e., up to and including the last instance of symbol σ—is accepted. Note that M responds to input symbols in an *online* manner—meaning that it makes acceptance/rejection decisions about each prefix of the input string as that prefix has been read. Of course, once M has "digested" the most recent instance of symbol σ, by again reaching a stable configuration, then it is ready to "digest" another input symbol, when and if one is "dropped into" its input port. Thus, all states of an OA are polling states.

   b. M's polling states are the ones that make decisions about the string of inputs that M has "read"—via its funnel—thus far.

   Formally, the fact that M makes decisions only when in a polling state means that Q_{poll} is partitioned into the set $Q_{\mathrm{poll}}^{(\mathrm{acc})}$ of *accepting states* and the set $Q_{\mathrm{poll}}^{(\mathrm{rej})}$ of *rejecting states*.

If M enters a state $q \in Q_{\mathrm{poll}}^{(\mathrm{acc})}$ after having read a word $w \in \Sigma^\star$, then we say that M *accepts* w; if M enters a state $q \in Q_{\mathrm{poll}}^{(\mathrm{rej})}$ after having read a word $w \in \Sigma^\star$, then we say that M *rejects* w.

2. the set Q_{aut} of *autonomous* states.

These states are characterized by the fact that they *do not* admit input symbols via M's funnel.

> Because the structure of an OA M need not exhibit *how* M makes its accept/reject decision—i.e., it need not "justify" its decision via some explicit computation—an OA has no need for autonomous states. In contrast, an OTM (or, for that matter, a TM) must, because of its finite set of states, perform an explicit computation in order to decide how

to react to the most recent instance of symbol σ. (This computation involves the OTM's tape—unless the OTM is actually functioning as an FA.) An OTM's autonomous states provide the formal mechanism for accommodating these inter-input-symbol computations. One can view an OTM as entering a sequence of autonomous states whenever it must "stop to think" before making its decision.

The fact that M makes decisions only in polling states, when it is ready to admit a new input symbol, means that there may be words $w \in \Sigma^\star$ that M neither accepts nor rejects. On such words (if there are any), M enters an infinite sequence of autonomous states, in which case we say that it "fails to halt" on input w! In contrast to an OA, which has no autonomous states, hence always halts when the input sequence is finite, an OTM (or, for that matter, a TM) need never halt. Indeed, the halting problem HP of Chapter 9 originated within the context of TMs.

At every step of a computation—which means "while M is not in a polling state"— M accesses the symbol from Γ that is currently scanned on its tape by its read/write head. (This is the symbol that resides in the square that the read/write head is currently sitting on. If M is currently in a polling state, then it also awaits an input symbol from Σ at its funnel. On the basis of the symbol(s) accessed and the current state of the FA that controls M:

1. M changes the state of its controlling FA, i.e., its *internal* state;
2. M alters its tape by respecifying the accessed symbol and then, possibly, moving its read/write head.

The computation proceeds until M enters the next pollng state—which, as mentioned earlier, may never happen. If a new input symbol now arrives at M's funnel, then the computation proceeds for another stage.

Formally, the single-step operation of M is specified by its *state-transition function*

$$\delta : \big((Q_{\text{poll}} \times \Sigma) \cup Q_{\text{aut}}\big) \times \Gamma \longrightarrow Q \times \Gamma \times \{\text{N}, \text{L}, \text{R}\}. \tag{3.4}$$

The interpretation is as follows. On the basis of the current state q (and the current input symbol $\sigma \in \Sigma$ at its funnel if q is a polling state) and the current symbol $\gamma \in \Gamma$ being scanned on its tape, M makes the "move" specified by δ:

- M enters its next state (which may be the same as the current one);
- it replaces tape-symbol γ by a new one (which can be *any* symbol from the set Γ, including γ or $\boxed{\text{B}}$);
- it moves its read/write head at most one square on the tape: "N" denotes "no move," "L" denotes "move one square to the left," "R" denotes "move one square to the right."

From this point on, we cease using the word "configuration" in its everyday sense of[5] "an arrangement of elements in a particular form, figure, or combination," and we begin to use the word in a technical sense.

The notion of a *computation* by M is formalized via the intermediate notion of a *configuration* (known also as an *instantaneous description*), which is a form of

---

[5] The cited definition comes from the dictionary included in Apple's MacBook computers.

"snapshot" of M's progress in a computation. Letting Γ^+ denote the set $\Gamma^+ \stackrel{\text{def}}{=} \Gamma\Gamma^\star$ of *nonnull* finite strings over Γ, a configuration of M is an ordered pair of strings in the set

$$\mathscr{C}_M \stackrel{\text{def}}{=} \Sigma^\star \times \Gamma^+ Q\Gamma^+ \boxed{\text{B}}. \tag{3.5}$$

The configuration

$$C = \langle w, \gamma_1 \cdots \gamma_m q \gamma_{m+1} \cdots \gamma_n \rangle \in \mathscr{C}_M \tag{3.6}$$

has the following interpretation. At the moment of the snapshot C:

1. M has read the string w at its input port;
2. M is in (internal) state q;
3. M's read/write head is residing on symbol γ_{m+1}; and, M's tape is entirely blank, except possibly for the region delimited by the string $\gamma_1 \cdots \gamma_m \gamma_{m+1} \cdots \gamma_n \in \Gamma^+$.
   We can think of the phrase "entirely blank" in either of the following two ways. Either the portion of the tape referred to is null, or it consists of an infinite string of occurrences of the blank symbol $\boxed{\text{B}}$. The same mathematical formulation describes either of these intuitive views.

We call the portion $\gamma_1 \cdots \gamma_m q \gamma_{m+1} \cdots \gamma_n$ of configuration C in (3.6) the *total state* of M—in contrast to the *internal* state q.

The notion of total state is not just a theoretical construct. It lies at the heart of the *SECD* machine model introduced by Peter J. Landin [55] in his quest for a formal mechanism that could be used both to specify the semantics of a (real) programming language and to guide a person in implementing these semantics.

The *computation by M on input*

$$w = \sigma_1 \sigma_2 \cdots \sigma_\ell \in \Sigma^\star$$

(*if it exists*) is a *finite* sequence of configurations, $C_0^{(M)}(w), C_1^{(M)}(w), \ldots, C_t^{(M)}(w),$[6] that satisfies the following constraints:

- The first configuration, $C_0^{(M)}(w)$, is M's unique *initial configuration*. This means that $C_0^{(M)}(w)$ has the form

$$C_0^{(M)}(w) = \langle \varepsilon, \boxed{\text{B}} q_0 \boxed{\text{B}} \boxed{\text{B}} \rangle,$$

which indicates that M starts out:

  – in its initial state q_0,
  – with an entirely blank worktape,
  – having read none of the input.

---

[6] For obvious reasons, we call the illustrated computation, if it exists, a t-step computation. It is the posited *finiteness* of a computation that may prevent the existence of a computation on certain input strings; cf. the halting problem.

- The final configuration, $C_t^{(M)}(w)$, is a *polling configuration*.

  This means that $C_t^{(M)}(w)$ is a valid configuration having the form

$$C_t^{(M)}(w) = \langle w, xqy \rangle,$$

  where q is a polling state. Note that M must have read precisely the input string w—neither more nor less—by the time it reaches configuration $C_t^{(M)}(w)$.

- Consecutive configurations $C_i^{(M)}(w), C_{i+1}^{(M)}(w)$, where $i \in [1, t-1]$, in the putative computation follow from one another according to M's program.

  This means that for $i \in [1, t-1]$, configuration $C_{i+1}^{(M)}(w)$ is the unique consequent of configuration $C_i^{(M)}(w)$ under M's state-transition function δ. In more detail: Say that $C_i^{(M)}(w)$ has the form

$$\langle \sigma_1 \sigma_2, \cdots \sigma_j, \ x \gamma_1 q \gamma_2 \gamma_3 y \rangle,$$

where $j \geq 0$, $q \in Q$, $\{\gamma_1, \gamma_2, \gamma_3\} \subseteq \Gamma$, and $x, y \in \Gamma^\star$. We consider two main cases, each with three subcases. You should refer back to definition (3.4) of δ as we proceed.

1. State q is an autonomous state.

  (a) If $\delta(q, \gamma_2) = \langle q', \gamma_2', N \rangle$, then

$$C_{i+1}^{(M)}(w) = \langle \sigma_1 \sigma_2, \cdots \sigma_j, \ x \gamma_1 q' \gamma_2' \gamma_3 y, \rangle.$$

  (b) If $\delta(q, \gamma_2) = \langle q', \gamma_2', L \rangle$, then

$$C_{i+1}^{(M)}(w) = \langle \sigma_1 \sigma_2, \cdots \sigma_j, \ x q' \gamma_1 \gamma_2' \gamma_3 y, \rangle.$$

  (c) If $\delta(q, \gamma_2) = \langle q', \gamma_2' R \rangle$, then

$$C_{i+1}^{(M)}(w) = \langle \sigma_1 \sigma_2, \cdots \sigma_j, \ x \gamma_1 \gamma_2' q' \gamma_3 y, \rangle.$$

2. State q is a polling state.

  (a) If $\delta(q, \sigma_{j+1}, \gamma_2) = \langle q', \gamma_2', N \rangle$, then

$$C_{i+1}^{(M)}(w) = \langle \sigma_1 \sigma_2, \cdots \sigma_j \sigma_{j+1}, \ x \gamma_1 q' \gamma_2' \gamma_3 y, \rangle.$$

  (b) If $\delta(q, \sigma_{j+1}, \gamma_2) = \langle q', \gamma_2', L \rangle$, then

$$C_{i+1}^{(M)}(w) = \langle \sigma_1 \sigma_2, \cdots \sigma_j \sigma_{j+1}, \ x q' \gamma_1 \gamma_2' \gamma_3 y, \rangle.$$

  (c) If $\delta(q, \sigma_{j+1}, \gamma_2) = \langle q', \gamma_2', R \rangle$, then

$$C_{i+1}^{(M)}(w) = \langle \sigma_1 \sigma_2, \cdots \sigma_j, \sigma_{j+1}, \ x \gamma_1 \gamma_2' q' \gamma_3 y, \rangle.$$

If one views an OTM and its tape as a hardware construct (as Turing did in [104]—but always keep in mind that programmable digital computers did not exist in those days), then the preceding definition of an OTM and its computations raises myriad thorny questions, such as, Can an OTM fall off its tape? How might one build a tape drive that could handle arbitrarily long tapes (and their arbitrarily large masses)? If one were to expand the model—as many have!—to allow multiple read/write heads on a single tape, how could one design cooperating take-up reels for a thus-endowed tape? If one adopts our recommended software-oriented view of an OTM as a program and of its tape as a data structure, then these questions admit trivial answers, as we all know from our experience programming real digital computers. Read/write heads are just pointers into a list, so there is no physical mass to worry about as a list grows; multiple pointers into a list cause no difficulties that would challenge any competent programmer. Moreover, the "software" view of a TM also gives us access to tractable mathematicizations of a large range of important questions that relate to the relative "powers" and/or "efficiencies" of various types of data structures. We present some such questions in the exercises; we deal with some particularly interesting ones in Section 5.5 and especially in Section 9.8.

Chapter 4
Finite Automata and Regular Languages

4.1 Introduction

A *finite automaton* (*FA*, for short) is an online automaton whose state-set Q is finite. The finite automaton model amply illustrates the three features of computation theory described in Chapter 1. Essentially equivalent models of finite-state systems were developed over a span of three decades, to model a large range of "real-life" systems. In roughly chronological order:

1. Beginning in the 1940s, researchers (e.g., Warren S. McCulloch and Walter H. Pitts [62]) attempting to explain the behavior of *neural systems* (natural and artificial "brains") developed models that were very close to our model of FA. While today's successors to their neurally inspired models have diverged from the standard FA model in many ways, they still share many of its essential features. We shall briefly study an early such model in Section 5.3.

2. Electrical engineers seeking to systematize the design and analysis of *synchronous sequential circuits*—which are clocked circuits that have memory as well as combinational logic—developed a model of finite state machine (FSM, for short) in the 1940s. Edward F. Moore's variant of the FSM model [68] is essentially identical to the FA that we study, the main distinction being that his automata output 0's and 1's, rather than "YES"-es and "NO"-es. George Mealy's variant [64] of the FSM model—which associates 0-1 outputs with state transitions rather than with states—can be translated to our model very easily (with a lag of one-half clock unit, to accommodate the displaced outputs). These models still play an essential role in the design of digital systems—from carry-ripple adders to the control units of digital computers (and often other systems, such as elevator controllers).

3. In the mid-1950s, several *linguists*—Noam Chomsky being the best known and most influential (at least within computer science)—sought formal models that could explain the acquisition of language by children. Chomsky developed a hierarchy of both generative and analytic linguistic models, each augmenting the (linguistic) complexity of its predecessor [12, 13]. Chomsky's lowest-level

A.L. Rosenberg, *The Pillars of Computation Theory*, Universitext,
DOI 10.1007/978-0-387-09639-1_4, © Springer Science+Business Media, LLC 2010

model—his *type-3* grammars and languages models—provide an alternative entry to the world of finite automata and their languages. Chomsky's work was later picked up by compiler designers, who could use his type-3 grammars to generate simple structures such as tokens and his finite automata to check the syntactic integrity of the tokens. (Chomsky's work went far beyond the type-3 artifacts just described, but they are our primary focus in this chapter.)

4. In the late 1950s, researchers (e.g., Michael O. Rabin and Dana Scott) who were disheartened by the resistance of detailed computational models to algorithmic tractability—we'll observe this intractability in Chapter 9, as we study the topic of unsolvability, and in Chapter 13, as we study the topic of hardness—began to study finite automata as a *very coarse, "high-level," model for digital computers*. The appropriateness of FAs as such a model stems from the bistable devices (say, flip-flops or transistors) that implement the hardware of a digital computer—e.g., the CPU and the memory. Being finite—albeit astronomical—in number, these devices can assume only finitely many distinct configurations. The algorithmic tractability of FAs (which we shall observe repeatedly in this chapter) means that in principle—i.e., modulo the astronomical numbers—one can analyze many significant properties of the dynamic behavior of programs, as long as the information one seeks is *very* coarse, so that it can be modeled via FAs.

5. Toward the mid-to-late 1960s (and beyond), as people began to investigate the possibilities for crafting *optimizing compilers* (Frances E. Allen and John Cocke [1]) and *program verifiers* (Robert W. Floyd [26]). To these ends, they developed various families of graphs that abstracted the behavior of programs in a way that facilitates the analysis of the flow of data and the flow of control within the program. The analytical tools that had been developed for studying FAs could be applied—with little or no adaptation—to the analysis of the resulting data- and control-flow graphs for programs.

The pinnacle in this chapter's development of the theory of finite automata is the Myhill–Nerode theorem (Section 4.3), which completely characterizes the power of finite automata by completely characterizing the notion "state." As we develop some of the applications of the theorem, we shall see why we have identified "STATE" as one of the "pillars" of computation theory. We present a second powerhouse result about FAs in Section 11.2: The Kleene–Myhill theorem (Theorem 11.3) completely characterizes the power of finite automata from a totally different perspective from that of the Myhill–Nerode theorem. We do not view the Kleene–Myhill theorem as exposing a "pillar" of computation theory because we view its applications and implications as narrower, within the context of the general theory, than those of the Myhill–Nerode theorem. That said, the Kleene–Myhill theorem is exceedingly important, and it has broad-ranging application within the language-theoretic and software-engineering related applications of finite automata theory. After dealing with the preceding two blockbuster results, we close this chapter with a theorem, known widely as the "pumping lemma (for regular languages)" (Section 6.1.2). This result exploits a simple aspect of the structure of FAs—their finiteness. In my estimation, the pumping lemma has been promoted in textbooks far beyond its intrinsic importance. We attempt to put the lemma into perspective by expounding, at greater length than is customary in computation theory texts, on its strengths and its limitations within the context of FA theory. We close the chapter with an "enrichment" section devoted to extensions of the central concepts we have developed.

4.2 Preliminaries

Much of the groundwork for our tour of finite automata theory was laid in Chapter 3, since, as we noted there, an FA is just an OA whose set of states is finite. The one point that we want to repeat here for emphasis is that despite the fact that our abstract development of FA theory treats input symbols (i.e., elements of Σ) as atomic (i.e., indivisible) entities, when we design specific FAs, the symbols usually have intrinsic structure, which is endowed by their meanings. We shall see this immediately as we design our first concrete sample FA of this chapter. The following concrete FAs are intended to add some flesh to the abstract model we began to develop in Chapter 3; we hope that these examples will hone both the reader's intuition for and appreciation of this simple, elegant computational model. We begin gently, with the simple example of a sequential carry-ripple adder—call it M. The input to M will be a sequence/string of *pairs* of bits.

Thus, the input alphabet Σ_M for our first concrete FA *has structure*, being the set $\{0,1\} \times \{0,1\}$ of ordered pairs of bits.

The adder M interprets each input string as a pair of binary numerals that are being fed in simultaneously, low-order bit to high-order bit. M produces an output bit immediately after reading each input symbol: the aggregate output after M has read n input symbols is the numeral comprising the n lowest-order bits of the sum of the pair of numerals represented by the input string. Symbolically, the behavior of M after having read n input symbols can be depicted as follows:

$$
\begin{array}{llll}
\text{INPUT:} & \langle \alpha_{n-1}, \beta_{n-1} \rangle \; \langle \alpha_{n-2}, \beta_{n-2} \rangle \; \cdots \; \langle \alpha_1, \beta_1 \rangle & \langle \alpha_0, \beta_0 \rangle \\
\text{OUTPUT:} & \gamma_n\; _1 \qquad\quad \gamma_{n-2} \qquad \cdots \quad \gamma_1 & \gamma_0
\end{array}
$$

where all $\alpha_i, \beta_i, \gamma_i \in \{0,1\}$ and where

$$
\begin{array}{rll}
& \cdots \; \alpha_{n-1} \; \alpha_{n-2} \; \cdots \; \alpha_1 \; \alpha_0 \\
+ & \cdots \; \beta_{n-1} \; \beta_{n-2} \; \cdots \; \beta_1 \; \beta_0 \\
= & \cdots \; \gamma_{n-1} \; \gamma_{n-2} \; \cdots \; \gamma_1 \; \gamma_0.
\end{array}
$$

Note that M requires both its input numerals to have the same length. This is no real restriction, because a short numeral can be "padded out" with leading 0's without changing its numerical value.

How can we design the circuitry necessary to implement M? It suffices to implement M as an FA that has four states. We produce M in a sequence of steps that move us gradually from the preceding functional specification to a formal specification. The tabular representation of M in Table 4.1 specifies the four states and their roles in the addition process; the rightward arrow indicates M's initial state. The next step

State-name	State "meaning"
→ 0, no-C	output 0 is produced no carry is propagated
0, C	output 0 is produced a carry is propagated
1, no-C	output 1 is produced no carry is propagated
1, C	output 1 is produced a carry is propagated

Table 4.1 The states of our carry-ripple adder M.

in designing M is to convert the representation of Table 4.1 to that of Table 4.2. One

State	$\langle 0,0 \rangle$	$\langle 0,1 \rangle$	$\langle 1,0 \rangle$	$\langle 1,1 \rangle$
→ 0, no-C	0, no-C	1, no-C	1, no-C	0, C
1, no-C	0, no-C	1, no-C	1, no-C	0, C
0, C	1, no-C	0, C	0, C	1, C
1, C	1, no-C	0, C	0, C	1, C

Table 4.2 A tabular representation of the 4-state carry-ripple adder M.

can directly read off from the latter table the graph-theoretic representation of M as a Moore-style finite state machine, i.e., one whose outputs are associated with states, as depicted in Figure 4.1.

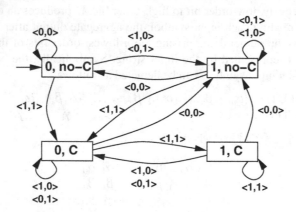

Fig. 4.1 A graph-theoretic representation of the 4-state carry-ripple adder M.

Let us pause in our process of translating one representation of M to another, to follow M's behavior on a small sample input string: let us add 15 $(= 1111_2)$ and 6 $(= 0110_2)$:

Time $\longrightarrow$

State:	[0, no-C]		[1, no-C]		[0, C]		[1, C]		[1, no-C]

State: [0, no-C]   [1, no-C]   [0, C]   [1, C]   [1, no-C]
Input: $\langle 1,0 \rangle$ $\langle 1,1 \rangle$ $\langle 1,1 \rangle$ $\langle 1,0 \rangle$
Output: (ignored)   1   0   1   1

The reader may note that it would be more intuitive to use the Mealy variant of FSM for our adder, rather than the Moore model, for the former associates outputs with *arcs*, or *state transitions*, rather than with states. We stick with the Moore model because it is the standard that appears in all computation-theoretic studies based on finite automata, save, perhaps, studies that use FSMs to design synchronous sequential circuits.

Our final translational step somewhat simplifies our representation of FSMs—and turns our FSMs into the finite automata that dominate our study in this chapter. This translation consists in partitioning M's set of states into those that produce output 1 and those that produce output 0. Having done this, we can suppress explicit mention of the output in our representation of M, relying on the partition to tell us what output M is emitting at each step. We illustrate this abbreviated notation in Table 4.3, where the states that give output 1 are boxed, and those that give output 0 are not. (Of course, we shall have a more elegant mechanism for distinguishing these classes of states when we turn to the formal development of the model.)

State	$\langle 0,0 \rangle$	$\langle 0,1 \rangle$	$\langle 1,0 \rangle$	$\langle 1,1 \rangle$
$\rightarrow$ no-C	no-C	no-C	no-C	C
no-C	no-C	no-C	no-C	C
C	no-C	C	C	$\boxed{C}$
$\boxed{C}$	no-C	C	C	$\boxed{C}$

Table 4.3 A tabular representation of the 4-state carry-ripple adder M, with "implicit" outputs.

Each of the two abstract formulations of FAs we have presented—the algebraic formulation and the graph-theoretic one—lends powerful intuition to the capabilities and limitations of the devices, as we shall see in Sections 4.3 and 11.2.

We now shift gears from the engineering antecedents of FA theory to the language-theoretic antecedents that continue to dominate the terminology and, largely, the focus of the theory. A language L is *regular* (is a *regular set* or *regular language*) precisely if there is an FA M such that $L = L(M)$. This terminology reflects the fact that the structure of the strings in L can be delimited via strict rules ("regulae" in Latin).

Using only the background material we have developed thus far, we can already prove a simple, yet significant, result about the class of regular languages.

Lemma 4.1. *The class of regular languages over any alphabet Σ is closed under complementation; that is, a language $L \subseteq \Sigma^\star$ is regular if and only if L's complement $\overline{L} = \Sigma^\star \setminus L$ is regular.*

Proof. Focus on a language $L \subseteq \Sigma^*$ whose regularity is witnessed by the fact that $L = L(M)$ for the FA

$$M = (Q, \Sigma, \delta, q_0, F).$$

To show that $\overline{L}$ is regular, we need only have the FA M "complement" (or "flip") its answer regarding the acceptance or nonacceptance of all input words. This is achieved formally by switching the roles of M's accepting and nonaccepting states. Thus, $\overline{L} = L(\overline{M})$, for the FA

$$\overline{M} = (Q, \Sigma, \delta, q_0, Q \setminus F).$$

In more detail: We know that the computations by M and $\overline{M}$ under any word $w \in \Sigma^*$ comprise the same sequences of states—because M and $\overline{M}$ share the same state-transition function δ. In particular, we know that $\delta(q_0, w)$ within M is the same state as $\delta(q_0, w)$ within $\overline{M}$. But this state is an accepting state within one of these FAs and a rejecting state within the other. It follows that $L(\overline{M}) = \overline{L(M)}$. $\square$

We shall have many opportunities to invoke Lemma 4.1 as we develop the theory of FAs and their languages.

One final remark before we turn to the first "big" result about FAs, the Myhill–Nerode theorem. The study of FAs and their languages is interesting only when these languages are infinite, for only then do the limitations of the FA as a computational model impose limitations on the structures of the languages. The following lemma, whose proof is left as an exercise, states the preceding assertion formally.

Lemma 4.2. *Every finite language over a finite alphabet[1] is regular.*

The two theorems that form the high points of FA theory—the Myhill–Nerode theorem (Theorem 4.1) and the Kleene–myhill theorem (Theorem 11.3)—expose quite distinct aspects of the influence of an FA's finiteness on the language that it accepts.

4.3 The Myhill–Nerode Theorem for FAs

This section is devoted to establishing the theorem that tells us what the notion "state" really means within our formal framework. This demonstration will allow us to examine the importance of the intuitive notion of "state" as enunciated in Chapter 3:

> The state of a system comprises that fragment of its history that allows it to behave correctly in the future.

Decades of experience with state-based systems have taught that all but the simplest such systems display a level of complexity that makes them hard—conceptually

---

[1] All alphabets that we consider are finite. We mention this finiteness in the statement of the Lemma only for emphasis.

and/or computationally—to design and analyze. One brilliant candle in this gloomy scenario is the Myhill–Nerode theorem, which supplies a *rigorous mathematical* analogue of the preceding informal characterization of "state." The theorem turns out to be a conceptual and technical powerhouse in analyzing a surprising range of problems concerning the state-transition systems that occur in so many guises within the world of computation. Indeed, although the theorem resides most naturally within the theory of finite automata—it first appeared in [72]; an earlier, weaker version appeared in [70]; the most accessible presentation appeared in [79]—it has manifold lessons for the analysis of many problems associated with *any* state-transition system, even those that have infinitely many states (as we shall see in the next chapter).

4.3.1 The Theorem: States Are Equivalence Classes

We refine the development in Section 3.2 as we specialize it to *finite* OAs. Recall that the *index* of an equivalence relation $\equiv$ on Σ^* is the number of classes into which $\equiv$ partitions Σ^*.

Theorem 4.1 ([70, 72, 79]). (The Myhill–Nerode theorem) *The following statements about a language $L \subseteq \Sigma^*$ are equivalent.*

1. L is regular.
2. L is the union of some of the equivalence classes of a right-invariant equivalence relation over Σ^* of finite index.
3. The right-invariant equivalence relation $\equiv_L$ of (2.2) has finite index.

> The earliest version of the theorem, in [70], uses *congruences*—i.e., equivalence relations that are both right- and left-invariant.

Proof. We prove the (logical) equivalence of the theorem's three statements by verifying the three cyclic implications: statement 1 implies statement 2, which implies statement 3, which implies statement 1.

> You should verify that the proposed proof strategy is valid. Here is the challenge.
>
> A literal reading of the Theorem asserts the *(logical) equivalence* of statements 1, 2, and 3. Using a version of logical notation in which
>
> $$\text{``}A \Rightarrow B\text{'' means ``}A \text{ implies } B\text{,'' i.e., ``if } A \text{ then } B\text{,''}$$
>
> and
>
> $$\text{``}A \Leftrightarrow B\text{'' means ``}A \text{ and } B \text{ are logically equivalent,'' i.e., ``}A \text{ if and only if } B\text{,''}$$
>
> the theorem asserts that $[(1) \Leftrightarrow (2)]$, $[(1) \Leftrightarrow (3)]$, and $[(2) \Leftrightarrow (3)]$. We are instead proposing to prove $[(1) \Rightarrow (2)]$, $[(2) \Rightarrow (3)]$, and $[(3) \Rightarrow (1)]$.
>
> Your challenge is to verify that our strategy does, indeed, provide a legitimate proof of the theorem.
>
> Hint: You should first verify that logical implication, $\Rightarrow$, is a *transitive* relation on statements.

On with the proof!

(1) ⇒ (2). Say that the language L is regular. There is, then, an FA $M = (Q, \Sigma, \delta, q_0, F)$ such that $L = L(M)$. Note that the right-invariant equivalence relation $\equiv_M$ of (3.2) has, by definition, index no greater than $|Q|$.

The relation could have index smaller than $|Q|$ if some of M's states were isolated, i.e., not accessible from q_0.

(One would likely never intentionally design M with isolated states, but we have all inadvertently written programs that had inaccessible regions.)

Moreover, L is the union of some of the classes of relation $\equiv_M$; specifically:

$$L = \{x \in \Sigma^\star \mid \delta(q_0, x) \in F\} = \bigcup_{f \in F} \{x \in \Sigma^\star \mid \delta(q_0, x) = f\}.$$

(2) ⇒ (3). We claim that if L is "defined" via some (read: any) finite-index right-invariant equivalence relation $\equiv$ on $\Sigma^\star$, in the sense of statement 2, then the specific right-invariant equivalence relation $\equiv_L$ has finite index. We verify the claim by showing that the relation $\equiv$ that "defines" L must *refine* relation $\equiv_L$, in the sense that every equivalence class of $\equiv$ is a subset of—i.e., is totally contained in—some equivalence class of $\equiv_L$. To see this, consider any strings $x, y \in \Sigma^\star$ such that $x \equiv y$. By the right-invariance of relation $\equiv$, for all $z \in \Sigma^\star$, we have $xz \equiv yz$. Because L is, by assumption, the union of entire classes of $\equiv$, we must have

$$[xz \in L] \quad \text{if and only if} \quad [yz \in L].$$

But—cf. (2.2)—this logical equivalence means that $x \equiv_L y$. Because x and y were arbitrary strings from $\Sigma^\star$, we conclude that

$$[x \equiv y] \;\Rightarrow\; [x \equiv_L y].$$

Because relation $\equiv$ has only finitely many classes, and because each class of relation $\equiv_L$ is the union of some of the classes of relation $\equiv$, it follows that relation $\equiv_L$ has finite index.

(3) ⇒ (1). Say finally that the specific right-invariant equivalence relation $\equiv_L$ on $\Sigma^\star$ has finite index and that L is the union of some of the classes of $\equiv_L$. Let the distinct classes of $\equiv_L$ be $[x_1], [x_2], \ldots, [x_n]$, for some n strings $x_i \in \Sigma^\star$.

Note that because of the transitivity of relation $\equiv_L$, we can identify a class uniquely via any one of its constituent strings. This works, of course, for any equivalence relation.

We claim that these n classes form the states of an FA $M = (Q, \Sigma, \delta, q_0, F)$ that accepts L. We identify the various components of M as follows.

This is essentially the same construction as in the proof of Theorem 3.1.

1. $Q = \{[x_1], [x_2], \ldots, [x_n]\}$.
   This set is finite because $\equiv_L$ has finite index.
2. For all $x \in \Sigma^\star$ and all $\sigma \in \Sigma$, define $\delta([x], \sigma) = [x\sigma]$.
   The right-invariance of relation $\equiv_L$ guarantees that δ is a well-defined function.

3. $q_0 = [\varepsilon]$.

   M's start state corresponds to its having read nothing.

4. $F = \{[x] \mid x \in L\}$.

   We are guaranteed that L is the union of some of the classes of $\equiv_L$.

We leave as an exercise the inductive argument that M is a well-defined FA that accepts L. $\square$

4.3.2 What Do Equivalence Classes Look Like?

While it is often easy to argue abstractly about states "being" equivalence classes of $\equiv_L$, in the sense of Theorems 3.1 and 4.1, it is just as often a daunting exercise to identify these equivalence classes explicitly when presented with even a moderately simple language. Put somewhat differently, but equivalently: precisely what distinctions have to be made in order to correctly decide membership in a language L? In this section, we analyze relation $\equiv_L$ for a few sample languages L, in an attempt to lend readers some intuition that will help them gain operational command over this important component of FA theory.

$L_1 = \{a^i \mid i \equiv 0 \bmod 3\}$:

—As we argued informally in Section 3.1, the language L_1 is accepted by the first FA depicted in Figure 3.2, hence is regular. Hopefully, this example is simple enough that the reader will recognize the following equivalence classes of $\equiv_{L_1}$ as "operational names" of the states of the depicted FA.

> I want to digress momentarily to distinguish names that are *operationally useful* from those that are not. Anyone who has tried to do arithmetic with Roman numerals will certainly appreciate the algorithmic manipulability of our conventional positional (e.g., binary or decimal) numbering systems. Some of the ways we "name" numbers are much clumsier operationally than Roman numerals. One example that comes readily to mind is the number $e = 2.718281828\ldots$, the base of the natural logarithm. Indeed, calling this number "e" leaves one with no idea of how to perform *any* arithmetic operation on it—even an operation as simple as identifying the number's first 100 decimal digits. In contrast, a name such as $\sum_{k=0}^{\infty}(1/k!)$ allows one—albeit with some difficulty—to perform any variety of arithmetic operations on it. Of course, what we have just said about e is just as true about $\pi = 3.14159\ldots$, the ratio of the circumference of a circle to its diameter.
>
> Our point here is that just endowing entities with arbitrary names need not afford one any "power" over the entities. One should always strive for *operational* names, such as we describe here for the states of an FA, and in Section 11.2 for regular languages.

1. $[\varepsilon]$

   This class consists of all strings whose length is divisible by 3.

   All of these strings are in L_1; in fact, $L_1 = [\varepsilon]$.

   For any word x that belongs to this class, an extension of x by a word z is in L_1 if and only if $z \in [\varepsilon]$. Symbolically:

   $$(\forall x \in [\varepsilon])(\forall z \in \Sigma^\star)\Big[[xz \in L_1] \Leftrightarrow [z \in [\varepsilon]]\Big].$$

2. $[a]$
   This class consists of every string whose length leaves a remainder of 1 when divided by 3.
   None of these words is in L_1.
   For any word x that belongs to this class, an extension of x by a word z—i.e., the word xz—is in L_1 if and only if $z \in [aa]$. Symbolically:

$$(\forall x \in [\varepsilon])(\forall z \in \Sigma^\star)\Big[[xz \in L_1] \Leftrightarrow [z \in [aa]]\Big].$$

3. $[aa]$
   This class consists of every string whose length leaves a remainder of 2 when divided by 3.
   None of these words is in L_1.
   For any word x that belongs to this class, an extension of x by a word z—i.e., the word xz—is in L_1 if and only if $z \in [a]$. Symbolically:

$$(\forall x \in [\varepsilon])(\forall z \in \Sigma^\star)\Big[[xz \in L_1] \Leftrightarrow [z \in [a]]\Big].$$

Because all positive integers are congruent either to 0 or to 1 or to 2 modulo 3, we have thus partitioned $\{a\}^\star$. The annotations that accompany our descriptions of the blocks of the partition demonstrate that the partition is the sought relation $\equiv_{L_1}$, and that L_1 is the union of some of the classes of this relation—in this case, the "union" of one of the classes.

$L_2 = \{x1111y \mid x, y \in \{0,1\}^\star\}$:
—This language is accepted by the third FA depicted in Figure 3.2, hence is regular. One should recognize the following equivalence classes of $\equiv_{L_2}$ as "operational names" of the states of the depicted FA.

The conditions listed (a), (b), … in the following descriptions of sets of binary words are the raw materials for the arguments needed to establish that each of the described sets is a separate equivalence class of the relation $\equiv_{L_2}$.

1. ε, plus all words ending in 0 that do not contain a run of four 1's.
   This class does not intersect any of the other classes we enumerate because: (a) none of the words in this class belong to L_2; (b) no $z \in \{1, 11, 111\}$ extends any of these words to a word in L_2; (c) every word $z \in L_2$ does extend any of these words to a word in L_2.
2. All words ending in 01 that do not contain a run of four 1's.
   This class does not intersect any of the others because: (a) none of these words belongs to L_2; (b) no $z \in \{1, 11\}$ extends any of these words to a word in L_2; (c) every word[2] $z \in \{111\} \cdot \{0,1\}^\star \cup L_2$ does extend any of these words to a word in L_2.
3. All words ending in 011 that do not contain a run of four 1's.

---

[2] Recall that the set $\{111\} \cdot \{0,1\}^\star$ comprises all strings that start with three 1's.

This class does not intersect any of the others because: (*a*) none of these words belongs to L_2; (*b*) the string $z = 1$ does not extend any of these words to a word in L_2; (*c*) every word $z \in \{11, 111\} \cdot \{0, 1\}^* \cup L_2$ does extend any of these words to a word in L_2.

4. All words ending in 0111 that do not contain a run of four 1'zas.

This class does not intersect any of the others because: (*a*) none of these words belongs to L_2; (*b*) every word $z \in \{1, 11, 111\} \cdot \{0, 1\}^* \cup L_2$ does extend any of these words to a word in L_2.

5. All words in L_2.

This class does not intersect any of the others because every word $z \in \{0, 1\}^*$ extends any of these words to a word in L_2.

We have thus partitioned $\{0, 1\}^*$ into five blocks. The conditions accompanying each of the blocks can be used to show that the words in each block share precisely the same set of extensions into L_2. Therefore, our partition is, in fact, relation $\equiv_{L_2}$.

$L_3 = \{a^n b^n \mid n \in \mathbb{N}\}$:[3]
—We shall see in Section 5.1 that this language—which is $L(M_4)$ for the OA M_4 in Figure 3.3—is *not* regular, so that by Theorem 4.1, $\equiv_{L_3}$ has infinitely many classes. (This fact should not intimidate, because the infinitely many classes are quite easy to describe.) One should recognize the following list of classes as corresponding to the (infinitely many) states of the first OA depicted in Figure 3.3.

1. $[\varepsilon]$

This is a singleton class, because ε is the unique word in L_3 that has a nonnull continuation that is also in L_3. (In fact, it has infinitely many: any $x \in L_3 \setminus \{\varepsilon\}$ will work.)

2. $[ab]$

This class contains all elements of $L_3 \setminus \{\varepsilon\}$, hence is infinite. All words in this class belong to L_3, but none of them admits any nonnull continuation that is also in L_3.

3. For each integer $i > 0$, there is the class $[a^i]$.

For each integer $i > 0$, the class $[a^i]$ contains all words that are completed to elements of L_3 by the word b^i. The few shortest words in the class $[a^i]$ are a^i, $a^{i+1}b$, $a^{i+2}b^2$. No word in any class $[a^i]$ belongs to L_3.

4. There is a single "dead" class that contains all words that are not in L_3 and that have no completion into L_3. These words have one of the following forms: $a^i b^j$ with $j > i > 0$, or some word that is not a block of a's followed by a block of b's. The latter contingency includes all words that start with a b (symbolically, all words in $b\Sigma^*$) and all words that contain three or more nonnull blocks of a's and b's, such as aba and $aabaab$.

The Palindromes:
—Our final example in this section is the language

---

[3] for any word x, the notation x^n denotes a string that is a concatenation of n occurrences of x.

$$L_4 = \{x \in \{0,1\}^\star \mid x \text{ reads the same forward and backward;}$$

$$\text{symbolically, } x = x^R\},$$

where x^R, the *reversal* of string $x \in \{0,1\}^\star$, denotes string x written backward

$$(\sigma_1 \sigma_2 \cdots \sigma_{n-1} \sigma_n)^R = \sigma_n \sigma_{n-1} \cdots \sigma_2 \sigma_1.$$

The words in L_4 are called *"palindromes."*

We shall see in Section 5.1 that this language, too, is not regular, so that, by Theorem 4.1, $\equiv_{L_4}$ has infinitely many classes. We organize our analysis of L_4 a bit differently from our analyses of L_1, L_2, and L_3, because the classes of L_4 have a rather dramatic structure: each is a singleton!

We prove that every $x \in \{0,1\}^\star$ resides in its own class of $\equiv_{L_4}$ using a proof by contradiction. To this end, we assume that there exist distinct words $x, y \in \{0,1\}^\star$ such that $x \equiv_{L_4} y$. We distinguish two cases.

     Case 1: $\ell(x) = \ell(y)$.[4]

In this case, the fact that $xx^R \in L_4$, while $yx^R \notin L_4$, contradicts the assumed $\equiv_{L_4}$-equivalence of x and y.

     Case 2: $\ell(x) \neq \ell(y)$.

Say, with no loss of generality, that $\ell(x) < \ell(y)$. Specifically, let

$$x = \alpha_1 \alpha_2 \cdots \alpha_m \quad \text{and} \quad y = \beta_1 \beta_2 \cdots \beta_m \beta_{m+1} \cdots \beta_n, \tag{4.1}$$

where each $\alpha_i, \beta_j \in \{0,1\}$. Consider the binary word

$$z = \overline{\beta}_{m+1} \alpha_m \cdots \alpha_2 \alpha_1.$$

(Recall that $\overline{\beta}_{m+1} = 1 - \beta_{m+1}$.) On the one hand, the word xz is a palindrome:

$$xz = \alpha_1 \alpha_2 \cdots \alpha_m \overline{\beta}_{m+1} \alpha_m \cdots \alpha_2 \alpha_1.$$

On the other hand, the word yz is *not* a palindrome:

$$yz = \beta_1 \beta_2 \cdots \beta_m \beta_{m+1} \cdots \beta_n \overline{\beta}_{m+1} \alpha_m \cdots \alpha_2 \alpha_1;$$

as one reads yz forward and backward, one is certain to encounter a mismatch no later than step $m+1$. Because the word z thus extends x into L_4 but extends y into $\overline{L}_4$, we conclude that $x \not\equiv_{L_4} y$, contradicting the assumed $\equiv_{L_4}$-equivalence of x and y.

Because x and y were arbitrary binary words, we have thus verified that every word in $\{0,1\}^\star$ occupies its own class of $\equiv_{L_4}$.

---

[4] Recall from Section 2.4.1 that $\ell(w)$ is the length of word w.

Chapter 5
Applications of the Myhill–Nerode Theorem

This chapter is devoted to justifying our praise for the Myhill–Nerode theorem, by developing a few of its applications. We strive to display both the usefulness of the theorem and its versatility.

The usefulness of the Myhill–Nerode theorem. In Section 5.1, we show how to use the theorem to *prove that certain languages are not regular*. Indeed, we argue, both in that section and in Section 6.1.2 that the Myhill–Nerode theorem is always the preferred tool for this purpose.

In Section 5.2, we use the theorem to develop an algorithm for *minimizing the number of states in an FA*. The input to the algorithm we develop is a FA M; the output is a FA M' that is *equivalent* to M, in the sense that $L(M') = L(M)$, and that has the smallest number of states of any FA that is equivalent to M. This algorithm is usually part of every college curriculum in computer or electrical engineering, but the theoretical underpinnings of the algorithm—namely, the theorem—are, regrettably, seldom taught.

The far-reaching implications of the Myhill–Nerode theorem. In Section 5.3, we use the theorem to analyze aspects of a probabilistic FA-like model from [77], which, despite its 1963 vintage, shares many of the characteristics of models that are used in modern studies of machine (computer) learning. The main result of the section shows that a certain class of these models accept only regular languages—so that *their computational power is not enhanced by allowing probabilistic state transitions*. In Section 5.4, we use the theorem to *derive a lower bound*, first observed in [47], *on the amount of memory that is needed to decide membership in any nonregular language*. The bound assumes the following form. Let L be a nonregular language. Then for infinitely many integers n, any FA that correctly decides membership/nonmembership in L of all words of length $\leq n$ must have no fewer than $f(n)$ states. The bound $f(n)$ is derived via an analogue of the equivalence relation $\equiv_L$ of the Myhill–Nerode theorem. Finally, in Section 5.5, we use the theorem to *derive a lower bound*, first observed in [38], *on the time needed by an online TM that has t tapes, each of dimensionality d, to solve a simplified database problem*. By considering algorithms for solving this problem on other models, one obtains significant information about the computational implications of the "online" regimen for computing and about the relative powers of a variety of families of data structures.

5.1 Proving that Languages Are *Not* Regular

Finite automata are very limited in their computing power due to the finiteness of their memories, i.e., of their sets of states. Indeed, as one might infer from the Myhill–Nerode theorem, the standard way to expose the limitations of FAs—by proving that a language L is not regular—is to establish somehow that the structure of L requires distinguishing among infinitely many distinct situations.

The finite-index lemma and fooling sets. Given the conceptual parsimony and power of Theorem 4.1, it is not surprising that the theorem affords one a simple, yet powerful, tool for proving that a language is not regular. This tool is encapsulated in the following lemma, which is an immediate corollary of the equivalence of statements (1) and (3) in the theorem, and which can be viewed as a strengthening of the continuation lemma for OAs (Lemma 3.1). For reasons that we hope will become suggestive imminently, we refer to the upcoming lemma as "the finite-index lemma."

We maintain that the ensuing development should be viewed as the primary tool *for proving that a language is not regular.*

Lemma 5.1. (The finite-index lemma) *Let $L \subseteq \Sigma^\star$ be an infinite regular language. Every sufficiently large set of words over Σ contains at least two distinct words, x and $y \neq x$, such that $x \equiv_L y$.*

Proof. Let us say that the infinite regular language L is accepted by the FA $M = (Q, \Sigma, \delta, q_0, F)$. Because the set Q is finite, any infinite set of words from $\Sigma^\star$ must—by the *pigeonhole principle*[1]—always contain two distinct words, x and $y \neq x$, that are indistinguishable to M, in the sense that $x \equiv_M y$. Clearly, then, $x \equiv_{L(M)} y$; cf. Lemma 3.2. (The validating argument proceeds as follows. Let us enumerate $\Sigma^\star$ in some way—the specific way is not relevant to the argument—and note, for each word $w \in \Sigma^\star$ in our enumeration, the state $\delta(q_0, w)$ to which w leads M. Because Q is finite, we must eventually find distinct words, x and y, such that $\delta(q_0, x) = \delta(q_0, y)$. By definition, $x \equiv_M y$.)

A direct calculation based on the way we extended the state-transition function δ to the domain $Q \times \Sigma^\star$ now verifies that

$$(\forall z \in \Sigma^\star) \Big[\delta(q_0, xz) = \delta(\delta(q_0, x), z) = \delta(\delta(q_0, y), z) = \delta(q_0, yz) \Big].$$

(We proceed from expression #2 in this chain to expression #3 via the universal algebraic operation of substituting equals for equals.) This system of equalities means that $x \equiv_L y$. □

---

[1] The "pigeonhole principle" asserts: *If you put $n + 1$ balls into n bins, some bin must receive more than one ball.* It is sometime called "Dirichlet's box principle," after Johann P.G. Lejeune Dirichlet

The finite-index lemma has a natural interpretation in terms of FAs, namely, that an FA M has no "memory of the past" other than its current state. Specifically, if strings x and y lead M to the same state (from its initial state)—i.e., if $\delta(q_0, x) = \delta(q_0, y)$, or, in our shorthand, $x \equiv_M y$—then no continuation/extension of the input string will ever allow M to determine which of x and y it actually read. (Note how this reasoning builds on the right-invariance of every FA-based equivalence relation $\equiv_M$.)

One applies the finite-index lemma, Lemma 5.1, to the problem of showing that an infinite[2] language $L \subseteq \Sigma^*$ is not regular by constructing a *fooling set* for L, i.e., an infinite set of words no two of which are equivalent with respect to L. Said another way: *An infinite set $S \subseteq \Sigma^*$ is a fooling set for L if for every pair of words $x, y \in S$, there exists a word $z \in \Sigma^*$ such that precisely one of xz and yz belongs to L.*

The fooling-set technique also has a natural interpretation in terms of FAs. As noted in the proof of Lemma 5.1, there must be distinct words, x and y, such that $x \equiv_M y$ (so that $x \equiv_{L(M)} y$). Given such words, the continuation lemma (Lemma 3.1) tells us that no continuation word z can ever allow M to distinguish between having read x and having read y.

We now look at a few sample proofs of the nonregularity of languages, which are based on the finite-index lemma and fooling sets. We shall observe how direct and simple such proofs can be.

Application 1. *The language $L_1 = \{a^n b^n \mid n \in \mathbb{N}\} \subset \{a, b\}^*$ is not regular. ($L_1 = L(M_4)$ for the OA M_4 of Figure 3.3.)*
We claim that the set $S_1 = \{a^k \mid k \in \mathbb{N}\}$ is a fooling set for L_1. To see this, note that for any distinct words $a^i, a^j \in S_1$, we have $a^i b^i \in L_1$, while $a^j b^i \notin L_1$; hence $a^i \not\equiv_{L_1} a^j$. By Lemma 5.1, L_1 is not regular. $\square$

Application 2. *The language $L_2 = \{a^k \mid k \text{ is a perfect square}\}$ is not regular.*
This application requires a bit of subtlety. We claim that L_2 is a fooling set for itself! To see this, consider any distinct words $a^{i^2}, a^{j^2} \in L_2$, where $j > i$. On the one hand, $a^{i^2} a^{2i+1} = a^{i^2+2i+1} = a^{(i+1)^2} \in L_2$; on the other hand, $a^{j^2} a^{2i+1} = a^{j^2+2i+1} \notin L_2$, because $j^2 < j^2 + 2i + 1 < (j+1)^2$; hence $a^{i^2} \not\equiv_{L_2} a^{j^2}$. By Lemma 5.1, L_2 is not regular. $\square$

The "subtlety" in Application 2 resides in knowing that we must use the smaller of i and j in order to construct the fooling continuation a^{2i+1}. You should try to complete the argument using a^{2j+1} in place of a^{2i+1}. You will run into trouble if you attempt an argument as simple as ours, because for some $j > i$, $i^2 + 2j + 1$ *could* be a perfect square—so your "punch line" would not work. (Consider, for instance, the possibility that $j = 3i + 4$, so that $2j + 1 = 6i + 9$, so that $i^2 + 2j + 1 = (i+3)^2$.)

Applications 3 and 4. *The language L_3 comprising all palindromes over $\{0, 1\}^*$ and the language*

$$L_4 = \{x \in \{0, 1\}^* \mid (\exists y \in \{0, 1\}^*)[x = yy]\}$$

(whose words are often called "squares") are not regular.

---

[2] You will be asked in the exercises to show that every finite language L is regular.

We present two proofs for these two languages, so that the reader will not assume that there is only one road to proofs of nonregularity.

1. We claim that the set $S_3 = \{10^k 1 \mid k \in \mathbb{N}\}$ is a fooling set for both L_3 and L_4. To see this, consider any pair of distinct words $10^i 1$ and $10^j 1$ from S_3. On the one hand, $10^i 1 10^i 1 \in L_3 \cap L_4$: the word is both a palindrome and a square. On the other hand, $10^i 1 10^i 1 \notin L_3 \cup L_4$: this word is neither a palindrome nor a square. We thus see that $10^i 1$ and $10^j 1$ are not equivalent with respect to either L_3 or L_4: $10^i 1 \not\equiv_{L_3} 10^j 1$, and $10^i 1 \not\equiv_{L_4} 10^j 1$. By Lemma 5.1, neither L_3 nor L_4 is regular.

2. We claim that the set $\{0,1\}^\star$ is a fooling set for both L_3 and L_4. To see this, consider any pair of distinct binary words x and y. Assume first that $\ell(x) = \ell(y)$. In this case, xx^R is a palindrome, while yx^R is not, and xx is a square, while yx is not. Alternatively, say that $\ell(x) < \ell(y)$. In this case, both $x1^{\ell(y)}x^R$ and $x0^{\ell(y)}x^R$ are palindromes, but at least one of $y1^{\ell(y)}x^R$ and $y0^{\ell(y)}x^R$ is not a palindrome. Similarly, both $x1^{\ell(y)}x1^{\ell(y)}$ and $x0^{\ell(y)}x0^{\ell(y)}$ are squares, but at least one of $y1^{\ell(y)}x1^{\ell(y)}$ and $y0^{\ell(y)}x0^{\ell(y)}$ is not a square. We again conclude that $10^i 1 \not\equiv_{L_3} 10^j 1$, and $10^i 1 \not\equiv_{L_4} 10^j 1$, so by Lemma 5.1, neither L_3 nor L_4 is regular. □

5.2 On Minimizing Finite Automata

Theorem 4.1 and its proof tell us several important things regarding the design of finite automata.

1. The notion of "state" that underlies the FA model is embodied in the relations $\equiv_M$ for FAs M. In more detail, a state of an FA M "is" the set of input strings that M "identifies"—in the sense of "does not distinguish among." This identification is permissible, indeed desirable, because—and so that—any two strings in the set are treated identically as histories, when M makes future decisions about membership in $L(M)$.
2. The *coarsest*—i.e., smallest-index—equivalence relation $\equiv_M$ that "works correctly" for the (regular) language L—in the sense of allowing M to make precisely the correct distinctions among input strings—is, by definition, the relation $\equiv_L$. This means that the *smallest* FA that accepts language L has a state-set whose cardinality is the index of the relation $\equiv_L$.
3. The *structure* of the smallest FA that accepts language L is determined *uniquely* by the construction in the "(3) $\Rightarrow$ (1)" step of the proof of Theorem 4.1.

We can turn the preceding intuition into an algorithm for minimizing the state-set of a given FA. You can look at this algorithm as starting with any given equivalence relation that "defines" L (usually presented via an FA M such that $L(M) = L$) and iteratively "coarsifying" the relation as much as possible, thereby "sneaking up" on the relation $\equiv_{L(M)}$ (which by hypothesis is identical to the relation $\equiv_L$).

The resulting algorithm for minimizing an FA $M = (Q, \Sigma, \delta, q_0, F)$ proceeds by iteratively computing the following equivalence relation on M's state-set Q. (The iteration embodies the process of "sneaking up" on the desired relation.) For $p, q \in Q$,

$$[p \equiv_\delta q] \text{ if and only if } (\forall x \in \Sigma^\star)\Big[[\delta(p,x) \in F] \Leftrightarrow [\delta(q,x) \in F]\Big]$$

(You should verify that $\equiv_\delta$ is indeed an equivalence relation.) This relation says that no input string will allow one to distinguish M's being in state p from M's being in state q. When $p \equiv_\delta q$, one can therefore *merge* (or *identify*) states p and q, in order to obtain an FA that is smaller than M but that also accepts $L(M)$. Therefore, if one can compute the entire equivalence relation $\equiv_\delta$, then the equivalence classes of the relation, i.e., the sets

$$\{[p]_{\equiv_\delta} \mid p \in Q\},$$

are the states of the smallest FA—call it $\widehat{M}$—that accepts $L(M)$. The state-transition function $\widehat{\delta}$ of $\widehat{M}$ is given by

$$\widehat{\delta}([p]_{\equiv_\delta}, \sigma) = [\delta(p,\sigma)]_{\equiv_\delta}. \tag{5.1}$$

Finally, the initial state of $\widehat{M}$ is $[q_0]_{\equiv_\delta}$, and the accepting states are $\{[p]_{\equiv_\delta} \mid p \in F\}$. (Why is $\widehat{M}$ a well-defined FA? In other words, why is $\widehat{\delta}$ well defined, and why does the indicated choice of initial state and final states guarantee that $L(\widehat{M}) = L(M)$? You will be asked to answer these basic questions as an exercise. As you ponder the questions, keep in mind that $\equiv_\delta$ is an equivalence relation.)

The FA state-minimization algorithm. We simplify our explanation of how to compute the relation $\equiv_\delta$ by describing an example concurrently with our description of the algorithm. We start with a *very coarse* approximation to $\equiv_\delta$ and iteratively improve the approximation. Figure 5.1 presents, in tabular form, the FA

$$M = (\{a,b,c,d,f,g,h\}, \{0,1\}, \delta, a, \{c\})$$

that we use as our running example.

M	q	$\delta(q,0)$	$\delta(q,1)$	$q \in F?$
(start state) $\rightarrow$	a	b	f	$\notin F$
	b	g	c	$\notin F$
(final state) $\rightarrow$	c	a	c	$\in F$
	d	c	g	$\notin F$
	e	h	f	$\notin F$
	f	c	g	$\notin F$
	g	g	e	$\notin F$
	h	g	c	$\notin F$

Fig. 5.1 The FA M that we minimize.

Our initial partition[3] of Q is $\langle Q \setminus F, F \rangle$. This partition acknowledges that the null string ε, as an input to M, witnesses the fact that no accepting state of M is $\equiv_\delta$-equivalent to any nonaccepting state. We thus have the following initial, stage-0, partition of the states of our running example FA M:

$$[a,b,d,e,f,g,h]_0, \ [c]_0.$$

The subscript "0" here indicates that this is the first discriminatory stage of our algorithm. The general notation is that the "stage-t" partition of Q is obtained by considering the discriminatory impact of all input strings of length $\leq t$, i.e., of the set $\bigcup_{i=0}^{t} \Sigma^i$.

We observe that state c, being M's unique final state, is not $\equiv_\delta$-equivalent to any other state. This means that we can henceforth ignore it as we refine the initial partition, because its $\equiv_\delta$-class will always remain a singleton. Had M possessed more than one accepting state, then state c would not have ended up isolated at this stage of the algorithm.

Inductively—meaning "at a general stage of the algorithm"—we now look at the current, stage-t, partition and try to "break apart" stage-t blocks. We do this by feeding single input symbols to pairs of states, say p and q, that reside in the same stage-t block. If any symbol $\sigma \in \Sigma$, as an input to M, leads states p and q to different stage-t blocks, then, by induction, we will have found a string x that discriminates between p and q—so that they must reside in distinct stage-$(t+1)$ blocks.

The preceding sentence, being crucial to the development of the algorithm, deserves elaboration. Say that there exist states r and s such that $\delta(p, \sigma) = r$ and $\delta(q, \sigma) = s$. Say further that there is a string x that discriminates between r and s—by showing them not to be equivalent under $\equiv_\delta$. In this case, *the string σx discriminates between states p and q*. This is because saying that x "discriminates between r and s" means that one of $\delta(r,x)$ and $\delta(s,x)$ belongs to F, while the other does not. If this is the case, though, then clearly, one of

$$\delta(p, \sigma x) = \delta(r, x) \quad \text{and} \quad \delta(q, \sigma x) = \delta(s, x)$$

belongs to F, while the other does not. This means, as stated, that the string σx discriminates between p and q—so that the stage-t block containing these states must be split by relegating p and q to distinct stage-$(t+1)$ blocks.

In our example, we find that input "0" breaks the big stage-0 block, so that we get the "stage-0.5" partition

$$[a,b,e,g,h]_{0.5}, \ [d,f]_{0.5}, \ [c]_{0.5}.$$

(We call this the "stage-0.5" partition because we still have another input symbol, namely, input "1," to apply to M, before we will have considered the impact of all input strings of length $t+1 = 1$.) We find that input "1" further breaks the block down. We end up with the stage-1 partition

$$[a,e]_1, \ [b,h]_1, \ [g]_1, \ [d,f]_1, \ [c]_1.$$

---

[3] Recalling that partitions and equivalence relations are (operationally) just different ways of looking at the same concept, we continue to use notation "$[a,b,\ldots,z]$" to denote the set $\{a,b,\ldots,z\}$ viewed as a block of a partition (= class of an equivalence relation).

Let's see how this happens. First, we find that

$$\delta(d,0) = \delta(f,0) = c \in F,$$

while

$$\delta(q,0) \notin F$$

for $q \in \{a,b,e,g,h\}$. This leads to our "stage-0.5" partition. At this point, input "1" leads states a and e to block $\{d,f\}$, symbolically

$$\{\delta(a,1),\delta(e,1)\} \subseteq [d,f]_{0.5};$$

it leads states b and h to block $\{c\}$, symbolically

$$\{\delta(b,1),\delta(h,1)\} = [c]_{0.5};$$

and it leaves state g in its present block, symbolically

$$\delta(g,1) \in [a,b,e,g,h]_{0.5}.$$

The main point of this analysis is that states a and e are broken away, as a pair, from the rest of the class $[a,b,e,g,h]_{0.5}$, and the same is true for states b and h, as a pair, and, finally, of the state g by itself. We thus end up with the indicated stage-1 partition.

One now determines that any further application of single inputs to M leaves the stage-1 partition unchanged! This means that the stage-1 partition must be the coarsest partition that preserves $L(M)$.

The preceding sentence is critical, in that it embodies the halting condition for our algorithm. The halting criterion is justified by a simple inductive argument that establishes the following fact. If at some stage of the described algorithm, a partition persists under—i.e., is unchanged by—all single-letter inputs, then the partition in fact persists under all input strings. We claim that such a "stable" partition embodies the relation $\equiv_M$, hence, by Lemma 3.2, the relation $\equiv_{L(M)}$.

Lemma 5.2. *Let* $M = (Q,\Sigma,\delta,q_0,F)$ *be an FA. Consider states p and q such that* $\delta(q_0,x) = p$ *and* $\delta(q_0,y) = q$, *for some $x,y \in \Sigma^\star$. If $p \equiv_\delta q$, then $x \equiv_M y$.*

Proof. If $p \equiv_\delta q$, then the state-minimization algorithm places p and q into the same block of a partition that persists under all input strings. The stability of the partition means that for all $z \in \Sigma^\star$, the states $r \stackrel{\text{def}}{=} \delta(p,z)$ and $s \stackrel{\text{def}}{=} \delta(q,z)$ belong to the same block of the partition; hence either both states belong to F or neither does. Recalling that $\delta(q_0,x) = p$ and $\delta(q_0,y) = q$, we have the following dichotomy. Because $\delta(q_0,x) = p$ and $\delta(q_0,y) = q$, so that $\delta(p,z) = \delta(q_0,xz)$ and $\delta(q,z) = \delta(q_0,yz)$, we have:

1. If r and s both belong to F (i.e., both are accepting states), then both xz and yz belong to $L(M)$.
2. If r and s both belong to $Q \setminus F$ (i.e., neither is an accepting state), then both xz and yz belong to $\Sigma^\star \setminus L(M)$ (i.e., neither xz nor y belongs to $L(M)$).

By definition, then, $x \equiv_{L(M)} y$. Since our argument applies to arbitrary states p and q, it shows that the relation $\equiv_\delta$ is just an encoding of the relation $\equiv_{L(M)}$. $\square$

Returning for the final time to our running example, our algorithm has identified the FA $\widehat{M}$ of Figure 5.2 as the minimum-state version of M.

$\widehat{M}$	q	$\widehat{\delta}(q,0)$	$\widehat{\delta}(q,1)$	$q \in F?$
(start state) $\rightarrow$	$[ae]$	$[bh]$	$[df]$	$\notin F$
	$[bh]$	$[g]$	$[c]$	$\notin F$
(final state) $\rightarrow$	$[c]$	$[ae]$	$[c]$	$\in F$
	$[df]$	$[c]$	$[g]$	$\notin F$
	$[g]$	$[g]$	$[ae]$	$\notin F$

Fig. 5.2 The minimum-state FA $\widehat{M}$ that minimizes the FA M of Figure 5.1.

We turn now to a series of three studies that cast a somewhat wider net in their search for applications of the Myhill–Nerode theorem.

5.3 Finite Automata with Probabilistic Transitions

This section focuses on a computational model that differs significantly from FAs, or even from OAs, by allowing state-transitions to be *probabilistic*: Such an FA, M, moves from state p to state q (in response to an input symbol) only with a designated probability. In accord with this new scenario, we view M as accepting an input string x only if the probability that M ends up in an accepting state after reading string x exceeds a preassigned threshold. The main result that we develop in this section comes from a 1963 paper by Michael O. Rabin [77]. The result exhibits a nontrivial, rather surprising, situation in which probabilistic state-transitions add no power to the FA model: The restricted class of "probabilistic" FAs accept only regular sets.

As of the time of the writing of this book, "probabilistic" FAs are a very timely model to study. The utility of probabilistic state-transition systems as conceptual tools is being amply demonstrated in several areas of artificial intelligence, notably the steadily growing area of machine learning.

5.3.1 PFAs and Their Languages

We start with an FA $M = (Q, \Sigma, \delta, q_0, F)$ and make its state-transitions and acceptance criterion *probabilistic*. We call the resulting model a *probabilistic finite automaton* (*PFA*, for short). We begin by fleshing out the various features of the PFA model.

States. We simplify the exposition in our formal development by positing that the state-set of the PFA M is $Q = \{1, 2, \ldots, n\}$, with $q_0 = 1$, and $F = \{m, m+1, \ldots, n\}$ for some $m \in Q$.

The names we assign to the state of an FA are clearly irrelevant to our analyses of any PFA's properties. Can you see why? (This question really relates to mathematical models and systems in general.)

State-transitions. We replace the state-transition function δ of an FA with *a set of state-transition tables* for a PFA, with one table for each symbol in Σ. The table associated with $\sigma \in \Sigma$ indicates, for each pair of states $q, q' \in Q$, the probability— call it $\rho^{(\sigma)}(q, q')$—that M ends up in state q' when we start M in state q and "feed" it input symbol σ. It is convenient—for our subsequent manipulations and analyses of M and its behavior—to present the state-transition tables as matrices, instead of tables.

Observe how we make use of the basic arithmetic operations on matrices as this section evolves, in order to simplify our work with PFAs. Try to imagine how awkward it would be to replace these familiar operations with little programs involving the entries of tables.

To be specific, we represent the table associated with $\sigma \in \Sigma$ via the σ-state-transition matrix

$$\Delta_\sigma = \begin{pmatrix} \rho^{(\sigma)}(1,1) & \rho^{(\sigma)}(1,2) & \cdots & \rho^{(\sigma)}(1,n) \\ \rho^{(\sigma)}(2,1) & \rho^{(\sigma)}(2,2) & \cdots & \rho^{(\sigma)}(2,n) \\ \vdots & \vdots & \ddots & \vdots \\ \rho^{(\sigma)}(n,1) & \rho^{(\sigma)}(n,2) & \cdots & \rho^{(\sigma)}(n,n) \end{pmatrix}.$$

Within each matrix Δ_σ each[4] $\rho^{(\sigma)}(i, j)$ resides in the interval $[[0, 1]]$, and for each $i = 1, 2, \ldots, n$,

$$\rho^{(\sigma)}(i, 1) + \rho^{(\sigma)}(i, j) + \cdots + \rho^{(\sigma)}(i, n) = 1.$$

The preceding sum reflects the fact that M must end up in *some* state on input σ, and states $1, 2, \ldots, n$ are the only choices.

PFA states, revisited. The probabilistic nature of M's state-transitions forces us to distinguish between M's set of states—the set Q—and the "state" that reflects M's situation at any point in M's "computation" on an input string x—which is a probability distribution over Q. We therefore define the *state-distribution vector* of M at each step of a computation to be a vector of probabilities $\mathbf{q} = \langle \pi_1, \pi_2, \ldots, \pi_n \rangle$, where each π_i is the probability that M is in state i at the step we are looking at. M's *initial state-distribution vector* is $\mathbf{q}_0 = \langle 1, 0, \ldots, 0 \rangle$, reflecting the fact that M begins each computation in state 1 (with certainty, i.e., probability 1).

PFA state-transitions, revisited. Under the preceding formalism, the PFA analogue of an FA's single-symbol state-transition $\delta(q, \sigma)$ is the vector–matrix product

$$\widehat{\Delta}(\mathbf{q}, \sigma) = \mathbf{q} \times \Delta_\sigma.$$

---

[4] In this section, we use the notation $[[0, 1]]$ for the closed *real* interval $\{x \mid 0 \leq x \leq 1\}$.

By extension, the PFA analogue of the FA string state-transition $\delta(q, \sigma_1 \sigma_2 \cdots \sigma_k)$, where each $\sigma_i \in \Sigma$, is

$$\widehat{\Delta}(\mathbf{q}, \sigma_1 \sigma_2 \cdots \sigma_k) \stackrel{\text{def}}{=} \mathbf{q} \times \Delta_{\sigma_1} \times \Delta_{\sigma_2} \times \cdots \times \Delta_{\sigma_n}. \tag{5.2}$$

The language accepted by a PFA. An FA accepts a string x by ending up in a final state after reading x—which is encapsulated by the condition $\delta(q_0, x) \in F$. The probabilistic analogue of accepting via a final state builds on the notion of an *(acceptance) threshold*, which is a probability: $\theta \in [[0, 1]]$. We say that the string $x \in \Sigma^\star$ is *accepted* by the PFA M just when

$$p_M(x) \stackrel{\text{def}}{=} \sum_{i=m}^{n} \widehat{\Delta}(\mathbf{q}_0, x)_i > \theta, \tag{5.3}$$

where $\widehat{\Delta}(\mathbf{q}, x)_i$ denotes the ith coordinate of the tuple $\widehat{\Delta}(\mathbf{q}, x)$.

> What does condition (5.3) really say? Recall first that M's final states are those whose integer-names are no smaller than m. Because of this convention, the formal analogue of the assertion that M accepts the string x with a probability that exceeds the threshold θ takes the mathematically simple form of (5.3). We need only sum the last $n - m + 1$ terms in the first row of the state-transition matrix $\widehat{\Delta}(\mathbf{q}, x)$ and see how the sum compares with θ. This is because that sum is the probability that string x leads M's initial state, 1, to one of the states $m, m+1, \ldots n$, *which are all of M's accepting states.* Notice how seemingly innocuous conventions that we have built up come back to reward us by simplifying the formal development. (Conversely, a poor choice of conventions could come back to hurt us.)

Thus, M accepts x if and only if the probability that it leads M from its initial "state" to a final state exceeds θ. As with all FAs, the *language accepted by M* is the set of all strings that M accepts. To acknowledge the crucial role of the acceptance threshold θ in defining the language accepted by the PFA M, we denote this language by $L(M, \theta)$:

$$L(M, \theta) \stackrel{\text{def}}{=} \{x \in \Sigma^\star \mid p_M(x) > \theta\}. \tag{5.4}$$

5.3.2 PFA Languages and Regular Languages

Nonregular PFA languages. It is not difficult to show that there exist simple—e.g., two-state—PFAs M, with associated thresholds θ, such that $L(M, \theta)$ is not regular. Consider the following two-state PFA M, whose design is attributed in [77] to Edward F. Moore.

M's states: M has two states, denoted by 1 and 2 by our convention; also by our convention, state 1 is M's initial state; we choose state 2 as M's unique accepting state.

M's input alphabet is $\Sigma = \{0, 1\}$.

M's state-transitions are specified by the following state-transition matrices:

$$\Delta_0 = \begin{pmatrix} 1 & 0 \\ 1/2 & 1/2 \end{pmatrix} \quad \text{and} \quad \Delta_1 = \begin{pmatrix} 1/2 & 1/2 \\ 0 & 1 \end{pmatrix}$$

One can prove by an induction that we leave as an exercise that for every string $x = \beta_1 \beta_2 \cdots \beta_{n-1} \beta_n \in \Sigma^\star$, the probability that x takes M from its initial state, state 1, to its accepting state, state 2, is, when written as a binary numeral,

$$p_M(x) = p_M(\beta_1 \beta_2 \cdots \beta_{n-1} \beta_n) = 0.\beta_n \beta_{n-1} \cdots \beta_2 \beta_1. \tag{5.5}$$

(You may want to refer back to (5.3) to verify this.)

The relationship between x and $p_M(x)$ that we just revealed leads to a proof that there exist acceptance thresholds θ for which the language $L(M, \theta)$ is not regular. While this proof relies only on quite general principles, these principles are not developed in this book until Chapter 7. With ample apologies to the reader for the following forward reference, we now sketch this proof—because this is where the result belongs. As partial penance, we insert a backward pointer at the end of Chapter 7, urging the reader to return to the following lemma after studying the material in that chapter. We here urge the reader either to persist through the following lemma, using prior background to get at least an intuitive understanding of the lemma's proof, or to skip *just this proof* and to rejoin us after it. It is important to skip no more than this lemma and its proof, because we return immediately after them to material that is accessible with the current flow of the text.

Lemma 5.3 ([77]). *Focusing on the PFA M just specified: There exist acceptance thresholds θ for which the language $L(M, \theta)$ is not regular.*

Proof. Recall that an acceptance threshold can be any real number $\theta \in [[0, 1]]$. Consider, therefore, two arbitrary *positive* real numbers, θ_1 and $\theta_2 > \theta_1$, in this range; say that θ_1 and θ_2 have the respective binary numerals

$$\theta_1 = 0.\alpha_1 \alpha_2 \cdots \alpha_k 10 \cdots \quad \text{and} \quad \theta_2 = 0.\alpha_1 \alpha_2 \cdots \alpha_k 11 \cdots$$

Given these (possibly infinite) binary numerals, we see that the real number $\xi \in [[0, 1]]$ whose (finite!) binary numeral is

$$\xi = 0.\alpha_1 \alpha_2 \cdots \alpha_k 11$$

satisfies

$$\theta_1 < \xi \leq \theta_2.$$

Now define x to be the (finite) binary string

$$x = 11\alpha_k \cdots \alpha_2 \alpha_1.$$

As noted in our earlier discussion of the PFA M, $p_M(x) = \xi$. By definition (5.4) of acceptance by a PFA, we have $x \in L(M, \theta_2) \setminus L(M, \theta_1)$.

What have we learned thus far? We have shown that *every two distinct positive acceptance thresholds θ_1 and $\theta_2 > \theta_1$ define distinct languages when associated with* *the PFA M*. In fact, given such θ_1 and θ_2, we see that $L(M, \theta_1)$ is a proper subset of $L(M, \theta_2)$. It follows that there are *uncountably many* distinct languages $L(M, \theta)$, as θ ranges over all possible positive acceptance thresholds.

> Uncountability is the first of the two notions that we use in this proof but do not develop until Chapter 7; countability is the other.

A straightforward application of techniques developed in Chapter 7 shows that the set of regular languages is countable!

The existence of uncountably many PFA languages $L(M, \theta)$, but only countably many regular languages, shows that some—indeed most—of the former languages are not regular. □

One quite unsatisfying aspect of the proof of Lemma 5.3 is its *nonconstructive* nature: The lemma establishes that some of the languages $L(M, \theta)$ are nonregular, but it does not explicitly identify even a single such language. Some of these recalcitrant languages are identified explicitly in [77], but without proof. We cite these examples here and supply a proof. You may want to cover the page and try to derive your own proof before looking at ours.

An *enumeration* of $\{0, 1\}^*$ is an infinite list of finite binary words that contains each such word at least once. You could—as but one example—form an enumeration by listing the finite binary words in *lexicographic order*: 0, 1, 00, 01, 10, 11, and so on. An *enumerative* real number is a positive real number $\xi \in [[0, 1]]$ whose binary numeral is formed by concatenating all of the words in an enumeration of $\{0, 1\}^*$. For definiteness, you could think of $\xi = 0.0100011011 \cdots$, which uses the lexicographic enumeration.

Lemma 5.4. *Focusing on our two-state PFA M: For any acceptance threshold θ that is an enumerative real number, the language $L(M, \theta)$ is not regular.*

Proof. Assume, for the sake of contradiction, that there is an enumerative real number $\widehat{\theta}$ for which the language $L(M, \widehat{\theta})$ is regular. Say specifically that $L(M, \widehat{\theta}) = L(M')$ for the FA $M' = (Q', \{0, 1\}, \delta', q'_0, F')$.

Denote the infinite sequence of binary words—i.e., the enumeration of $\{0, 1\}^*$— that underlies $\widehat{\theta}$ by $S = w_1, w_2, \ldots$; thus, as a binary numeral, $\widehat{\theta} = 0.w_1 w_2 \cdots$.

Because sequence S contains *all* binary words, we can identify an infinite sequence of prefixes of S of increasing lengths, each having an associated binary numeral formed by concatenating its elements,

$$
\begin{aligned}
S_1 &= w_1, w_2, \ldots, w_{k_1} & N_1 &= 0.w_1 w_2 \cdots w_{k_1} \\
S_2 &= w_1, w_2, \ldots, w_{k_1}, \ldots, w_{k_2} & N_2 &= 0.w_1 w_2 \cdots w_{k_1} \cdots w_{k_2} \\
&\;\;\vdots & &\;\;\vdots \\
S_i &= w_1, w_2, \ldots, w_{k_1}, \ldots, w_{k_2}, \ldots, w_{k_i} & N_i &= 0.w_1 w_2 \cdots w_{k_1} \cdots w_{k_2} \cdots w_{k_i} \\
&\;\;\vdots & &\;\;\vdots
\end{aligned}
\tag{5.6}
$$

that is chosen for the following property. For each integer j, if we were to continue the enumeration S beyond the prefix S_j, then the next word, $w_{k_j + 1}$, would be a string of n_j 1's, where n_j is strictly bigger than the length of any string of 1's that occurs in the numeral $N_j = 0.w_1w_2\cdots w_{k_j}$.

Why do we care about the two sequences in (5.6)? Precisely because of the following inequalities:

$$\text{For all } i < j, \quad 0.w_1w_2\cdots w_{k_j}w_{k_j+1} < \widehat{\theta} < 0.w_1w_2\cdots w_{k_i}w_{k_i+1}. \qquad (5.7)$$

We leave it to the reader to validate the two inequalities in (5.7). As a hint, we remind the reader that because $\widehat{\theta}$ is enumerative, one can find both a 0 and a 1 to the right of any finite bit-position of the numeral $\widehat{\theta}$.

In order to continue the proof, we must now digress to develop some technical machinery. We focus, for convenience, on the FA M' such that $L(M') = L(M, \widehat{\theta})$, but note that the following development holds for any FA and any input alphabet Σ. Define the following binary relation on $\{0,1\}^*$: For any strings $x, y \in \{0,1\}^*$, say that x is *totally equivalent* to y for FA M', denoted $x \cong y$, just when

$$(\forall q \in Q')\Big[\delta(q,x) = \delta(q,y)\Big].$$

Proposition 5.1 *For any FA M', the relation $\cong$ is an equivalence relation of finite index. Moreover, if $x \cong y$ for strings $x, y \in \{0,1\}^*$, then M' either accepts both of x and y or it accepts neither of them.*

Proof (of Proposition). Relation $\cong$ is clearly reflexive, symmetric, and transitive because of its definition in terms of equality. The finite-index property is a little subtler. One can view each string $x \in \{0,1\}^*$ as a total function from Q' to Q', defined by

$$x(q) = \delta(q,x).$$

There are clearly no more than $|Q'|^{|Q'|}$ such functions, because each $q \in Q'$ has no more than $|Q'|$ "places to go" under any such function.

The assertion about $\cong$-equivalent strings and $L(M')$ is immediate by definition of the relation. $\square$

Proposition 5.2 *For any FA M', the equivalence relation $\cong$ is left-invariant, in the following sense. If $x \cong y$ for strings $x, y \in \{0,1\}^*$, then for all $z \in \{0,1\}^*$, $zx \cong zy$. Moreover, the relation $\cong$ is also right-invariant.*

Proof (of Proposition). The proof that relation $\cong$ is left-invariant can be viewed as a backward version of the continuation lemma (Lemma 3.1). Say that $x \cong y$. Then $\delta(q,x) = \delta(q,y)$ for all $q \in Q'$. Hence, in particular, given any $z \in \{0,1\}^*$ and $q \in Q'$,

$$\delta(q,zx) = \delta(\delta(q,z),x) = \delta(\delta(q,z),y) = \delta(q,zy).$$

The middle equation here follows by instantiating the "q" in the definition of $\cong$ with "$\delta(q,z)$."

The proof that relation $\cong$ is right-invariant is left to the reader, since it follows from the same argumentation as does the continuation lemma (Lemma 3.1). $\square$

Back to proving the lemma. Because there are infinitely many paired sequences in (5.6), there must be two prefixes of the enumeration S of $\{0,1\}^*$, say S_a and S_{a+b}, such that[5]

$$(w_1 w_2 \cdots w_a)^R \cong (w_1 w_2 \cdots w_{a+b})^R.$$

(This follows by the pigeonhole principle.) By Proposition 5.2, therefore,

$$w_{a+b+1}(w_1 w_2 \cdots w_a)^R \cong w_{a+b+1}(w_1 w_2 \cdots w_{a+b})^R. \qquad (5.8)$$

We designed the sequence of S_i's to ensure that w_{a+b+1} is a string of 1's, so that we can rewrite (5.8) as

$$(w_1 w_2 \cdots w_a w_{a+b+1})^R \cong (w_1 w_2 \cdots w_{a+b} w_{a+b+1})^R.$$

Now, however, we hark back to the special nature of the PFA M, specifically, the fact that the probability that M accepts a string x is given by $p_M(x)$, as defined in (5.5). This means that the probability that M accepts $(w_1 w_2 \cdots w_a w_{a+b+1})^R$ is

$$p_M((w_1 w_2 \cdots w_a w_{a+b+1})^R) = 0.w_1 w_2 \cdots w_a w_{a+b+1},$$

and

$$p_M((w_1 w_2 \cdots w_{a+b} w_{a+b+1})^R) = 0.w_1 w_2 \cdots w_{a+b} w_{a+b+1}.$$

But the system of inequalities (5.7) now tells us that

$$(w_1 w_2 \cdots w_a w_{a+b+1})^R \in L(M, \widehat{\theta}),$$

while

$$(w_1 w_2 \cdots w_{a+b} w_{a+b+1})^R \notin L(M, \widehat{\theta}).$$

This pair of assertions means, however, that $L(M, \widehat{\theta}) \neq L(M')$, because Proposition 5.1 tells us that either both strings

$$(w_1 w_2 \cdots w_a w_{a+b+1})^R \text{ and } (w_1 w_2 \cdots w_{a+b} w_{a+b+1})^R$$

belong to $L(M')$ or neither does.

Since we assumed nothing about the FA M' (except that it accepted $L(M, \widehat{\theta})$), we conclude that M' does not exist, because $L(M, \widehat{\theta})$ is not regular. $\square$

$L(M, \theta)$ **is regular when** θ **is "isolated."** In view of the preceding demonstration that even a simple—e.g., a two-state—PFA can accept a nonregular language when coupled with an "unfavorable" acceptance threshold, it is a bit surprising that there can exist PFAs M and associated acceptance thresholds θ that are "favorable" for M, in the sense that the language $L(M, \theta)$ is regular! The Myhill–Nerode theorem provides the tools necessary to show that "favorable" thresholds do exist. We begin with the formal notion of an *isolated threshold* for a PFA M.

---

[5] Recall that x^R denotes the reversal of string x.

The threshold $\theta \in [[0,1]]$ is *isolated* for the PFA M just when there exists a real *constant of isolation* $\kappa > 0$ such that for all $x \in \Sigma^*$,

$$|p_M(x) - \theta| \geq \kappa. \tag{5.9}$$

Theorem 5.1 ([77]). *For any PFA M and associated* isolated *acceptance threshold θ, the language $L(M, \theta)$ is regular.*

Proof. The proof is a direct application of Theorem 4.1. Specifically, we show that if M has n states, a of which are accepting states, and if κ is the constant of isolation from (5.9), then the index $I_{L(M,\theta)}$ of the equivalence relation $\equiv_{L(M,\theta)}$ does not exceed

$$I_{L(M,\theta)} = [1 + (a/\kappa)]^{n-1}. \tag{5.10}$$

We establish the bound of (5.10) by considering a set of k words—call them x_1, $x_2, \ldots, x_k \in \Sigma^*$—that are *mutually inequivalent* under $\equiv_{L(M,\theta)}$. This inequivalence means that for each pair of distinct such words x_i, x_j, there must exist a word $y \in \Sigma^*$ such that $x_i y \in L(M, \theta)$ while $x_j y \notin L(M, \theta)$ (or vice versa). We now show that k cannot exceed the bound of (5.10). The theorem will then follow by Theorem 4.1.

Our technical development begins by our converting M's language-related problem to a geometric setting. For any string $w = \sigma_1 \sigma_2 \cdots \sigma_h \in \Sigma^*$, let $\Delta(w)$ denote the matrix

$$\Delta(w) \stackrel{\text{def}}{=} \Delta_{\sigma_1} \times \Delta_{\sigma_2} \times \cdots \times \Delta_{\sigma_h}.$$

Then—cf. (5.2)—$\widehat{\Delta}(q_0, w)$, the state distribution of M after reading w, is just the first row of $\Delta(w)$; and the sum of the last a entries of this row is the probability that M accepts w.

Referring back to our designated triple of words, x_i, x_j, y, we consider the following three points, two in n-dimensional space and one in a-dimensional space:

Corresponding to x_i: $\langle \xi_1^{(i)}, \xi_2^{(i)}, \ldots, \xi_n^{(i)} \rangle$
(the first row of $\Delta(x_i)$, i.e., $\widehat{\Delta}(q_0, x_i)$);
Corresponding to x_j: $\langle \xi_1^{(j)}, \xi_2^{(j)}, \ldots, \xi_n^{(j)} \rangle$
(the first row of $\Delta(x_j)$, i.e., $\widehat{\Delta}(q_0, x_j)$);
Corresponding to y: $\langle \eta_1, \eta_2, \ldots, \eta_n \rangle$
(the coordinatewise sum of the last a *columns* of $\Delta(y)$).
Easily (cf. (5.3)),

$$p_M(x_i y) = \xi_1^{(i)} \eta_1 + \xi_2^{(i)} \eta_2 + \cdots + \xi_n^{(i)} \eta_n;$$
$$p_M(x_j y) = \xi_1^{(j)} \eta_1 + \xi_2^{(j)} \eta_2 + \cdots + \xi_n^{(j)} \eta.$$

We have focused on the strings x_i, x_j, and y because M accepts $x_i y$ but does not accept $x_j y$. Therefore, since the acceptance threshold θ is isolated and has associated constant of isolation κ, we must have

$$\theta + \kappa \leq \xi_1^{(i)} \eta_1 + \xi_2^{(i)} \eta_2 + \cdots + \xi_n^{(i)} \eta_n;$$
$$\theta - \kappa \geq \xi_1^{(j)} \eta_1 + \xi_2^{(j)} \eta_2 + \cdots + \xi_n^{(j)} \eta_n.$$

It follows by subtraction that

$$2\kappa \leq (\xi_1^{(i)} - \xi_1^{(j)})\eta_1 + (\xi_2^{(i)} - \xi_2^{(j)})\eta_2 + \cdots + (\xi_n^{(i)} - \xi_n^{(j)})\eta_n.$$

Since each entry of $\Delta(y)$, being a probability, cannot exceed 1, we have each $\eta_l \leq a$. Exploiting this fact, we find that

$$2(\kappa/a) \leq |\xi_1^{(i)} - \xi_1^{(j)}| + |\xi_2^{(i)} - \xi_2^{(j)}| + \cdots + |\xi_n^{(i)} - \xi_n^{(j)}|. \tag{5.11}$$

The preceding reasoning has transported our automata/language-theoretic problem to a geometric setting. Let us, accordingly, view each tuple $\langle \xi_1, \xi_2, \ldots \xi_n \rangle$ as a point in n-dimensional Euclidean space. Consider, for each $i \in \{1, 2, \ldots, k\}$ (recall that k is the number of mutually inequivalent words), the set Λ_i comprising all points $\langle \xi_1, \xi_2, \ldots, \xi_n \rangle$ such that

- $\xi_l \geq \xi_l^{(i)}$ for all $l \in \{1, 2, \ldots, n\}$.
- $\sum_{l=1}^{n} (\xi_l - \xi_l^{(i)}) = (\kappa/a)$.

Easily, each Λ_i is a translate of the set

$$\Lambda = \left\{ \langle \xi_1, \xi_2, \ldots, \xi_n \rangle \mid \text{ all } \xi_l \geq 0 \text{ and } \sum_{l=1}^{n} \xi_l = (\kappa/a) \right\},$$

which is an $(n-1)$-dimensional simplex that is a subset of the hyperplane $\sum_{l=1}^{n} \xi_l = (\kappa/a)$. The volume of Λ as a function of κ is readily seen to be $c(\kappa/a)^{n-1}$ for some absolute constant $c > 0$.

Now, because $\sum_{l=1}^{n} \xi_l^{(i)} = 1$, it follows that $\sum_{l=1}^{n} \xi_l = 1 + (\kappa/a)$ for every point $\langle \xi_1, \xi_2, \ldots, \xi_n \rangle \in \Lambda_i$. Therefore, Λ_i is a subset of the locus of points

$$\widehat{\Lambda} \stackrel{\text{def}}{=} \left\{ \langle \xi_1, \xi_2, \ldots, \xi_n \rangle \mid \text{ all } \xi_l \geq 0 \text{ and } \sum_{l=1}^{n} \xi_l = 1 + (\kappa/a) \right\}.$$

An elementary argument shows that the k sets Λ_i share no *interior* points, i.e., points $\langle \xi_1, \xi_2, \ldots, \xi_n \rangle$ for which each $\xi_l < \xi_l^{(i)}$. This means that the volumes of the sets Λ_i satisfy

$$kc(\kappa/a)^{n-1} = \sum_{l=1}^{n} \text{Vol}(\Lambda_l) \leq \text{Vol}(\widehat{\Lambda}) = c(1 + (\kappa/a))^{n-1},$$

so that $k \leq [1 + (a/\kappa)]^{n-1}$, as was claimed. $\square$

5.4 State as a Memory-Constraining Resource

We turn now to a topic that is part of the study of general, possibly infinite-state, OAs but that we deferred to this point in order to have access to the powerful technical machinery of Theorem 4.1. A literal reading of the theorem tells us that any OA that accepts a nonregular language must have infinitely many states. A closer reading, though—which is enhanced by Theorem 3.1—tells us that the states of an OA M are, in fact, the equivalence classes of the relation $\equiv_M$, which, in turn, is a refinement of the language-defining relation $\equiv_{L(M)}$. Under certain circumstances, one can therefore glean detailed quantitative information about the number of states of M that must be accessible from its initial state when M processes words of given lengths. The reader should recognize from earlier developments in this chapter that the states of an OA contain memory as well as control logic. With this in mind, the reader will recognize that the analyses of this subsection are really using characteristics of the relation $\equiv_{L(M)}$ to bound the *memory requirements* of computations that decide nonregular languages.

> The reader will find a less abstract approach to studying the memory requirements of computations, one that is tailored to the Turing machine model, in Section 13.2.2.

This subsection builds on the technical machinery developed in [47], which will allow us to derive *lower bounds* on memory requirements. Section 5.4.2, in particular, follows [47] in developing the strongest possible *general* lower bounds on the memory requirements of OAs that accept nonregular languages.

We begin with an OA $M = (Q, \Sigma, \delta, q_0, F)$ that accepts a nonregular language L (so $L = L(M)$). Our goal is to determine how hard it is to construct "regular approximations" of L, in the following sense. For any integer $n > 0$, consider the set $L^{(n)}$ that consists of every word in $L(M)$ whose length does not exceed n. Of course, $L^{(n)}$ is a finite set, so by Lemma 4.2, there is an FA $M^{(n)}$ such that $L(M^{(n)}) = L^{(n)}$ (i.e., $M^{(n)}$ accepts $L^{(n)}$). We call $M^{(n)}$ an *order-n approximation* of M. The formal specification of $M^{(n)}$'s behavior—which should be redundant for you at this point—is as follows:

$$L(M^{(n)}) \ = \ \{x \in L(M) \mid \ell(x) \leq n\}.$$

We denote by $\mathscr{A}_M(n)$ the (obviously monotonically nondecreasing) number of states in the smallest order-n approximation of M, as a function of n. The quantity $\mathscr{A}_M(n)$ can be viewed as measuring $L(M)$'s memory requirements (or "space complexity") because one needs $\lceil \log_2 \mathscr{A}_M(n) \rceil$ bistable devices in order to implement an order-n approximation of M in circuitry. The main result of this subsection, which is due to Richard M. Karp in [47], is an "infinitely often" lower bound on $\mathscr{A}_M(n)$, i.e., a lower bound that holds for infinitely many n.

> Bounds—both upper and lower—come in at least three flavors. The most satisfying, in some sense, are the "universal" bounds. An easily proved "universal" bound is the following: *For all positive integers n, $2\lfloor n/2 \rfloor \leq n$.* (The highlighted phrase is the source of the qualifier "universal.") Perhaps next along the satisfaction line are the "eventual" bounds. An easily

proved "eventual" upper bound is the following: *For all sufficiently large positive integers n,*
$n^3 > 2^{15493} n^2$. (Again, the highlighted phrase is the source of the qualifier.) Finally, there are
the "infinitely often" bounds, such as the one we focus on here. An easily proved "infinitely
often" bound is the following: *For every positive integer n, there is an integer m > n such that*
if one represents both *m* and *n* in binary, then one must perform more carries when adding
1 to *m* than when adding 1 to *n*. (I have purposely not used the phrase "infinitely often"
in the preceding example, to show you how such bounds are frequently expressed.) While
"infinitely often" bounds may not be as emotionally satisfying as the two stronger types of
bounds, they are often the strongest bounds that hold. Moreover, in many circumstances—the
result of this subsection being an example—it is amazing that any nontrivial bound can be
proved!

Quite surprisingly, the bound we prove for $\mathscr{A}_M(n)$, which is *tight* for many
languages—meaning that it cannot be replaced by a larger general lower bound—
assumes nothing about M other than that L(M) is not regular. (Indeed, *M*'s state-
transition function δ need not even be computable!)

5.4.1 $\mathscr{A}_M(n)$ for Two Specific Infinite OAs

In order to better appreciate the lower bound we derive for arbitrary nonregular OAs
and their languages, let us determine $\mathscr{A}_M(n)$ for two OAs that accept nonregular
languages.

Example 1: $L(M) = \{a^k b^k \mid k \in \mathbb{N}\}$.

Focus first on the OA *M* that we called M_4 in Section 4. We enumerated all of the
classes of $\equiv_{L(M)}$ in Section 4.3.2, so we can use this list as a guide in our analysis
of $L(M)$; see Figure 3.3. An order-*n* approximation $M^{(n)}$ of *M* may safely make the
following identifications. (In other words, the following classes of words may safely
lead $M^{(n)}$ from its initial state to the same state.)

1. All words that are *not* of the form $a^i b^j$, where $[i \geq j]$ *and* $[2i \leq n]$, can be identified
   via a nonaccepting "dead" state. None of these strings is in $L(M)$, and no extension
   will bring them into $L(M)$. This category accounts for *one state* of $M^{(n)}$.
2. For each $h \in [0, \lfloor n/2 \rfloor - 1]$, there is a state that identifies all strings of the form
   $a^i b^{i-h}$, where $[h < i]$ *and* $[2i \leq n]$. Each of these states has a unique continuation,
   b^h, into $L(M)$. This category accounts for $\lfloor n/2 \rfloor$ *states* of $M^{(n)}$, one for each indicated value of *h*.
3. For each $i \in [0, \lfloor n/2 \rfloor]$, there is a state that is dedicated to the single string a^i.
   This string can be continued into $L(M)$ by any string of the form $a^j b^{i+j}$, where
   $2(i + j) \leq n$. This category accounts for $\lfloor n/2 \rfloor + 1$ *states* of $M^{(n)}$, one for each
   indicated value of *i*.

Since no two of the thus-enumerated states can be identified (or merged), we see that
$\mathscr{A}_M(n) = 2 \lfloor n/2 \rfloor + 2$.

We have thus proved the following bound on *L*'s memory requirements.

Lemma 5.5. *For any FA M that accepts the language* $L(M) = \{a^k b^k \mid k \in \mathbb{N}\}$, *we
have* $\mathscr{A}_M(n) = 2 \lfloor n/2 \rfloor + 2$.

Example 2: $L(M)$ comprises the palindromes over the alphabet $\{0,1\}$. Once again, we can be guided by our enumeration in Section 4.3.2 of all of the classes of $\equiv_{L(M)}$.

1. Since we do not care about any words of length $> n$, we can relegate all such words to a single "dead" state. This category gives us *one state*.
2. We allocate all words of length n to two new states, an accepting state for the words that are palindromes and a nonaccepting state for the words that are not. This category gives us *two states*.
3. We allocate each word x of length $\ell(x) = m \le n/2$ to a state that it shares with no other word; this state is an accepting state if x is a palindrome and a nonaccepting state otherwise. The string x is the sole occupant of its state because no other word $y \in \{0,1\}^m$ shares the property that yx^R is a palindrome of length $2m \le n$. This category gives us $2^{n/2+1} - 1 = 2^{\Omega(n)}$ *states*.
4. Each word x of length $n/2 < \ell(x) < n$ is allocated to a state based on the subset S_x of

$$\bigcup_{k=0}^{n-\ell(x)} \{0,1\}^k$$

all of whose elements extend x to a palindrome (perforce of length $\le n$). Words that share the same subset are allocated to the same state; words that have different subsets are allocated to different states. Note that the $(k=0)$ component of the union guarantees that if x is a palindrome, then so also are all strings y such that $S_y = S_x$; therefore, we are safe in mandating that x be allocated to an accepting state iff $\varepsilon \in S_x$.

As noted, the words of category #3 already show that $\mathscr{A}_M(n) = 2^{\Omega(n)}$.

We have thus proved the following bound on the memory requirements of the palindromes.

Lemma 5.6. *For any FA M that accepts the palindromes over the alphabet $\{0,1\}$, we have $\mathscr{A}_M(n) = 2^{\Omega(n)}$.*

5.4.2 A Bound on $\mathscr{A}_M(n)$ for Any OA M with Nonregular L(M)

We now continue developing the general bounding technique from [47]. Building on general principles that derive from the conceptual framework of Theorem 4.1, we shall derive an "infinitely often" lower bound on $\mathscr{A}_M(n)$ that works with *any* OA M that accepts a nonregular language—and that is within a factor of 2 of the bound that we just derived by analyzing the detailed structure of $L(M_4) = \{a^k b^k\}$. Of course, our bound for $L(M_4)$ is a "universal" bound—it holds *for all n*. That said, it is still remarkable that the upcoming general lower bound is—for those n to which it applies—just a factor of 2 smaller than the bound that holds for a specific language that we can analyze in complete detail. (It is less surprising that the general lower bound is far too small—in fact exponentially so—in the case of the palindromes.)

Theorem 5.2 ([47]). *If M is an OA that accepts a nonregular language, then for infinitely many n,*

$$\mathscr{A}_M(n) > \frac{1}{2}n + 1. \tag{5.12}$$

Proof. Let M_1 and M_2 be OAs. For $n \in \mathbb{N}$, we say that M_1 and M_2 are *n-equivalent*, denoted $M_1 \equiv_n M_2$, just when

$$\{x \in L(M_1) \mid \ell(x) \le n\} = \{x \in L(M_2) \mid \ell(x) \le n\}.$$

Saying that $M_1 \equiv_n M_2$ is clearly (logically) equivalent to saying that each of M_1 and M_2 is an *n*-approximation of the other. Moreover, the relation $\equiv_n$ between M_1 and M_2 can be viewed as an approximation to the relations $\equiv_{M_1}$ and $\equiv_{M_2}$; hence, the relation $\equiv_n$ allows us to bring the conceptual power of Theorem 4.1 to bear on the problem of bounding $\alpha_{M_1}(n)$ and $\alpha_{M_2}(n)$.

Our analysis of *n*-approximations of OAs builds on the following bound on the "degree" of equivalence of pairs of *FA*s.

Lemma 5.7 ([68]). *Let M_1 and M_2 be FAs having s_1 and s_2 states, respectively. If $L(M_1) \ne L(M_2)$, then $M_1 \not\equiv_{s_1+s_2-2} M_2$.*

Proof (of Lemma 5.7). We establish the result by bounding from above the number of partition-refinements that the state-minimization algorithm of Section 5.2 must perform in order to distinguish the initial states of M_1 and M_2. (Because FAs M_1 and M_2 are, by hypothesis, not equivalent, their initial states must be distinguishable: there must be at least one word that one of M_1 and M_2 accepts while the other doesn't.)

Because the state-minimization algorithm is actually a "state-equivalence tester," we can apply it to state-transition systems that are not legal FAs, as long as we are careful to keep final and nonfinal states segregated from one another. We can therefore apply the algorithm to the following "disconnected" FA *M*. Say that for $i = 1, 2$, $M_i = (Q_i, \Sigma, \delta_i, q_{i,0}, F_i)$, where $Q_1 \cap Q_2 = \emptyset$. Then $M = (Q, \Sigma, \delta, \{q_{1,0}, q_{2,0}\}, F)$, where

- $Q = Q_1 \cup Q_2$;
- for $q \in Q$ and $\sigma \in \Sigma$: $\delta(q, \sigma) = \begin{cases} \delta_1(q, \sigma) \text{ if } q \in Q_1, \\ \delta_2(q, \sigma) \text{ if } q \in Q_2; \end{cases}$
- $F = F_1 \cup F_2$.

Now, the fact that $L(M_1) \ne L(M_2)$ implies (*a*) that $q_{1,0} \not\equiv_\delta q_{2,0}$, and (*b*) that neither $Q \setminus F$ nor F is empty.

Recall that the algorithm proceeds in stages, where each stage applies every letter from Σ to all states within each block of the then-current partition of Q, to determine whether there exists a letter that will drive one state in a block to an accepting state and another state in the same block to a nonaccepting state (thereby establishing the inequivalence of those states). How many stages of the algorithm could be needed, in the worst case, to distinguish states $q_{1,0}$ and $q_{2,0}$ within *M*, when the algorithm starts with the initial partition $\{Q \setminus F, F\}$? (This initial partition is created by applying the null string ε to $q_{1,0}$ and $q_{2,0}$.) Each stage of the algorithm, save the last, must "split" some block of the partition into two nonempty subblocks—or else no further "splits" will ever occur. Because one "split," namely, the separation of $Q \setminus F$ from F, occurs before the algorithm starts applying input symbols, and because $|Q| = s_1 + s_2$, the algorithm can proceed for no more than $s_1 + s_2 - 2$ stages, because after that many stages, all blocks would be singletons! In other words, if $p \not\equiv_\delta q$, for states $p, q \in Q$, then there is a string of length $\leq s_1 + s_2 - 2$ that witnesses the inequivalence. Because we know that $q_{1,0} \not\equiv_\delta q_{2,0}$, this completes the proof. $\square$

Back to the theorem. For each $k \in \mathbb{N}$, Theorem 4.1 guarantees—by its guarantee that $\equiv_{L(M)}$ has infinite index—that there is a smallest integer $n > k$ such that $\mathscr{A}_M(k) = \mathscr{A}_M(n-1) < \mathscr{A}_M(n)$. The preceding inequality implies the existence of FAs M_1 and M_2 such that:

1. M_1 has $\mathscr{A}_M(n-1)$ states and is an $(n-1)$-approximation of M;
2. M_2 has $\mathscr{A}_M(n)$ states and is an n-approximation of M.

By statement 1, $M_1 \equiv_{n-1} M_2$; by statements 1 and 2, $M_1 \not\equiv_n M_2$. By Lemma 5.7, then, $M_1 \not\equiv_{\mathscr{A}_L(n-1)+\mathscr{A}_L(n)-2} M_2$. Because $M_1 \equiv_{n-1} M_2$, we therefore have $\mathscr{A}_M(n-1) + \mathscr{A}_M(n) > n+1$, which yields inequality (5.12), because $\mathscr{A}_M(n-1) \leq \mathscr{A}_M(n) - 1$. $\square$

It is shown in [47] that Theorem 5.2 is as strong as possible, in two senses: (1) the constants $\frac{1}{2}$ and 1 in (5.12) cannot be improved; (2) the phrase "infinitely many" cannot be strengthened to "for all but finitely many."

> The phrase "for all but finitely many" is one of many equivalent linguistic devices for specifying an "eventual" bound on integers.

Because our focus here is only on illustrating the power of Theorem 4.1—and not on proceeding deeper into the subject of approximations to nonregular languages—we refer the interested reader to [47] for these embellishments of Theorem 5.2.

5.5 State as a Time-Constraining Resource

This section complements the previous section's analysis of state as a *memory-constraining* resource. We now develop lower bounds on the *time* that online Turing machines with varying numbers of tapes of varying structures (cf. Section 3.3) require in order to perform certain specific computations that relate to the encoding and retrieval of information. There is actually a lesson to be learned from how we represent our information-retrieval problem as a (nonregular) language L. (As we shall see, each word of L represents a rather simple database, plus a series of queries to the database.) As in Section 5.4, we develop the bound of this section by adapting the Myhill–Nerode theorem (Theorem 4.1) to a broad class of OTMs. In this section, the required adaptation of Theorem 4.1 is achieved by parameterizing the word-relating equivalence relation $\equiv_M$: for each integer $t > 0$, the parameter-t relation $\equiv_M^{(t)}$ behaves like $\equiv_M$, but *it exposes only discriminations that M can make in t or fewer steps*. The development here is based on the pioneering paper [38] by Fred C. Hennie.

Because the study in [38] focuses on the impact of an OTM's tape topology on its efficiency in retrieving sets of words, the bounds we develop here can be viewed as an early contribution to the theory of data structures. This perspective underlies both the *data graph* model of this book's author [81] and the *storage modification machine* model of Arnold Schönhage [94].

The interesting features in this section are the formulation of an information-retrieval problem as a formal language, and the use of the concepts underlying Theorem 4.1 to analyze the problem. For completeness, we rephrase here some of the material from Section 3.3, with an eye to the study we present.

5.5.1 Online TMs with Multiple Complex Tapes

A *d-dimensional tape* is a linked data structure with a meshlike topology; it is thus essentially identical to the *orthogonal list* data structure discussed in [53]. A tape is accessed via a *read/write head*—the OTM-oriented name for a pointer. Each cell of a tape holds one symbol from the OTM's *work alphabet* Γ, which always contains the designated *blank* symbol $\boxed{\text{B}}$; e.g., in a 32-bit computer, Γ could be the set of 32-bit binary words, and the *blank* symbol could be the word of all 1's. Access to cells within a tape is sequential: at each step, the read/write head either remains stationary, or it moves from its current cell to a neighboring one in any of the $2d$ permissible directions.

An *OTM M with t d-dimensional work tapes* can be viewed as an FA that has access to t d-dimensional orthogonal lists (cf. [53]). As with any FA, M has an *input port*, which it uses in the manner described in Section 3.3 in order to receive a sequence of input symbols that come from M's input alphabet Σ; M has a designated initial state and a designated set of final states. We shall see concretely in this section why OTMs need both polling and autonomous states: During the "passive" periods in which an OTM does not accept new input symbols at its input port, the OTM may be doing quite valuable subcomputations using its work tapes. Indeed, the study in this section can be viewed as bounding (from below) the cumulative time that must be devoted to these "introspective" subcomputations as the OTM performs certain computations. With this intuitive background in place, we note that, formally, a computational step by M depends on:

- its current state,
- the current input symbol, *if M's program reads the input at this step*,
- the t symbols (elements of Γ) currently scanned by the pointers on M's t work-tapes.

On the basis of these, M:

- enters a new state (which may be the same as the current one),

- independently rewrites the symbols currently scanned on its t worktapes (possibly with the same symbol as the current one),
- independently moves the read/write head on each tape at most[6] one square in one of the $2d$ allowable directions.

A few clarifying comments are in order at this point.

First, I want to ensure that we are "on the same page" regarding this new model. There are twice as many *directions* to move as there are dimensions to the tape, because there is an analogue of both UP and DOWN (or LEFT and RIGHT) in each dimension.

Next, I hope that it is clear to the reader that when $d = 1$, the OTM M has t *linear* (i.e., one-dimensional) tapes. Hence, when $d = t = 1$, M is precisely an OTM as defined in Section 3.3.

The final comment is more in the way of food for thought than clarification. We have chosen to discuss tapes with topologies that are meshlike mainly because meshlike data structures are useful in many computational scenarios—see, e.g., the discussion of orthogonal lists in [53]; and the analyses that we are about to embark on require us to know the topologies of M's tapes. We shall try to use exercises to show how the upcoming analyses can accommodate tapes with a broad range of regular topologies that are quite "unmeshlike," for instance, tapes with the topologies of trees of various arities. The details of the analyses—especially the quantitative details—change with tape topology. But the "flow" of the arguments adapts quite readily.

One extends M's one-step computation to a multistep computation (whose goal is language recognition, as usual) as follows. To determine whether a word $w = \sigma_1 \sigma_2 \cdots \sigma_n \in \Sigma^*$ is accepted by M—i.e., is in the language $L(M)$—one makes w's n symbols available, in sequence, at M's input port. If M starts in its initial state with all cells of all tapes containing the *blank* symbol $\boxed{\text{B}}$, and it proceeds through a sequence of N steps that:

- includes n steps during which M "reads" an input symbol,
- ends with a step in which M is programmed to "read" an input symbol,

then M is said to *decide w in N steps*; if, moreover, M's state at step N is an accepting state, then M is said to *accept w in N steps*. (Note that N can be much larger than n, because of the "introspective" subcomputations alluded to earlier.) See Section 3.3 for details.

With the current model, as with all online automata, we need the just-defined double condition for acceptance ("includes ..." and "ends with ..."). This somewhat complicated condition ensures that if M accepts a word w, then it does so unambiguously. Specifically, after M reads the last symbol of w, it does not "give its answer" until it is prepared to read a new input symbol (if that ever happens). This means that M cannot oscillate between accepting and nonaccepting autonomous states after reading the last symbol of w.

---

[6] The qualifier "at most" indicates that a read/write head is allowed to remain stationary.

5.5.2 An Information-Retrieval Problem as a Language

Following [38], we shall use a database-inspired language L_{DB} to expose the potential effect of tape structure/topology on the time necessary for a OTM to perform certain computations.

We simplify the description of L_{DB} by first describing the language's words in the computation-oriented terms. Each word $w \in L_{DB}$ specifies—in a manner that will be clarified imminently—a sequence of like-length, not necessarily distinct, binary words; let k denote the common length of these words. We view the *set* consisting of all the distinct words in this sequence as a (rather primitive) *database*. After the portion of w that specifies the database comes another sequence of length-k (again, not necessarily distinct) binary words. We view each of these latter length-k words as a *query* into the database. M's role in this scenario is as follows. M begins by reading in the database and somehow storing it on its worktapes. Once M reaches the sequence of queries, it reads these query-words in order. After reading each query-word, M responds "YES" if the query-word occurs in the database, and "NO" if doesn't. Wait! There is a technical problem here. By definition, M must emit a "YES"–"NO" output before reading each new input symbol: It cannot do so just after reading the special query-words. (This is because every one of M's polling states is either an accepting state (that emits "YES") or a rejecting state (that emits "NO").) In order to accommodate this requirement of the OTM model, we shall have M produce the output "NO" before reading each input symbol *unless* the input string it has read thus far represents a database, followed by a string of query-words the last of which occurs in the database.

Let us now rewrite the preceding scenario in language-theoretic terms, by formalizing the *database language L_{DB}*.

L_{DB} is a language over the 3-letter alphabet $\Sigma = \{0, 1, :\}$, wherein ":" is a symbol distinct from "0" and "1." Each word in L_{DB} has the form

$$\xi_1 : \xi_2 : \cdots : \xi_m :: \eta_1 : \eta_2 : \cdots : \eta_n,$$

where for some $k \in \mathbb{N}$,

- each ξ_i ($1 \leq i \leq m$) and each η_j ($1 \leq j \leq n$) is a length-k binary string;
- $m = 2^k$;
- $\eta_n \in \{\xi_1, \xi_2, \ldots, \xi_m\}$.

The *set* of ξ_i's, namely $\{\xi_1, \xi_2, \ldots, \xi_m\}$, is our *database*. The *database string*

$$\xi_1 : \xi_2 : \cdots : \xi_m$$

is just the mechanism we use to present the database to M. While the database must contain at least one word, it could have many *fewer* than m words, because of possible repetitions. Each word η_i in the string

$$\eta_1 : \eta_2 : \cdots : \eta_n$$

(which, as a sequence of length-k words, could also contain many repetitions) is a *query*. In each word $x \in L_{DB}$, the double colon "::" separates the database from the queries, while the single colon ":" separates consecutive binary words.

The fact that we are interested only in whether *the last* query appears in the database reflects the *online* nature of the computation: M must respond to each query as it appears, with no knowledge of which one is the last, i.e., the important one. (This is essentially the challenge faced by all online algorithms.)

Note how an OTM M that accepts the language L_{DB} can be used to solve the motivating database problem. Say that one has a set S of length-k binary words, and one wants to determine whether a given length-k binary word x belongs to S. One can present the OTM M that accepts L_{DB} with any string

$$\xi_1 : \xi_2 : \cdots : \xi_m :: x,$$

where $\xi_1 : \xi_2 : \cdots : \xi_m$ encodes (as a string) *any* length-2^k sequence of binary words formed using all and only words from the set S. By definition of the language L_{DB}, if M accepts this string, then $x \in S$; if M rejects this string, then $x \notin S$.

5.5.3 The Impact of Tape Structure on Memory Locality

The *configuration* of an OTM M having t d-dimensional tapes, at any step of a computation by M, is the $(t+1)$-tuple

$$\langle q, \tau_1, \tau_2, \ldots, \tau_t \rangle$$

defined as follows. (More details appear in Section 3.3 for the case $t = d = 1$.)

- q is the state of M's finite-state control (its associated FA);
- each τ_i is the d-dimensional configuration of symbols from Γ that comprises the non-"blank" portion of tape i, with one symbol highlighted (in some way) to indicate the current position of M's read/write head on tape i.

The sometimes-encountered term "total state" as an alternative for "configuration" presages the time-parameterized variant of the equivalence relation $\equiv_M$ that we introduce now; cf. (3.2).

Say that for $i = 1, 2$, the database-string $x_i \in \Sigma^\star$ leads M to configuration $C_M(x_i) = \langle q_i, \tau_{i1}, \tau_{i2}, \ldots, \tau_{it} \rangle$. If:

- $q_1 = q_2$; i.e., the configurations share the same state;
- for some integer $r \geq 1$, and all $i \in \{1, 2, \ldots, t\}$, tape configurations τ_{1i} and τ_{2i} are identical within r symbols of their highlighted symbols (which indicate where M's read/write heads reside),

then we say that the configurations $C_M(x_1)$ and $C_M(x_2)$ are *r-equivalent to M* and that the databases specified by x_1 and x_2 are *r-indistinguishable by M*. We denote these synonymous relations by the following notation: $x_1 \equiv_M^{(r)} x_2$.

Consider what the relation "r-indistinguishable" on databases (or synonymously, the relation "r-equivalent" on configurations) means. Say that two distinct databases, x_1 and x_2, leave M in configurations that are r-equivalent. The distinctness of the databases means that there is a length-k word η that belongs to one of the databases but not to the other. Say that we feed one of these databases to M, and then apply the query η. Of course, M must give the answer "YES" if the database we supplied contains η, and the answer "NO" otherwise. By hypothesis, though, it will take M at least r steps to determine which of these databases it has read. It follows that M must perform an "introspective computation" of at least r steps' duration in order to respond correctly to the length-k query η.

The preceding somewhat lengthy story can be told more compactly using mathematical terminology. The following lemma can be viewed as a time-parameterized version of the continuation lemma (Lemma 3.1), just as the relation $\equiv_M^{(r)}$ is a time-parameterized version of the relation $\equiv_M$.

Lemma 5.8. *Say that $x_1 \equiv_M^{(r)} x_2$ and that there exists a $y \in \Sigma^\star$ such that one of $x_1 y$ and $x_2 y$ belongs to $L(M)$, while the other does not. If M has read either x_1 or x_2, then it must compute for more than r steps while reading y.*

The reader may want to prove Lemma 5.8 formally, in order to get practice with a slightly more complicated version of such an argument.

5.5.4 Tape Dimensionality and the Time-Complexity of L_{DB}

To simplify notation in what follows, we focus now on certain sublanguages of L_{DB} that are defined by the common length of the binary words in their databases and query sets. For each $k \in \mathbb{N}$, the language $L_{DB}^{(k)}$ consists of all words from L_{DB} whose binary subwords all have length k. In the notation of the preceding subsection, these "binary subwords" are the ξ_i that make up the databases and the η_j that are the queries. Note that *each database-string in $L_{DB}^{(k)}$ has length $(k+1)2^k - 1$*.

Focus on any fixed (but arbitrary) language $L_{DB}^{(k)}$, and let x_1 and x_2 be two database-strings whose constituent binary words all have length k. If x_1 and x_2 specify *distinct* databases, then there exists a query η that appears in the database specified by one of the x_i but not the other—so, precisely one of the strings $x_1 :: \eta$ and $x_2 :: \eta$ belongs to $L_{DB}^{(k)}$. Database-strings that specify distinct databases must therefore lead M to distinct configurations. We now consider how "big" these configurations must be, in terms of the necessary "radius of indistinguishability."

On the one hand, the database-strings that occur within the words of $L_{DB}^{(k)}$ can specify $2^{2^k} - 1$ distinct databases (corresponding to that number of nonempty sets of length-k ξ_i's). This means that if M is to distinguish all possible length-k databases—which it must do in order to correctly decide membership in $L_{DB}^{(k)}$—then the configurations that M uses to encode length-k databases must have a "radius" r that is big

enough so that the database-indistinguishability relation $\equiv_M^{(r)}$ has index $\geq 2^{2^k} - 1$.
(There must be $\geq 2^{2^k} - 1$ equivalence classes.) This is because for this maximum-r
"radius," the relation $\equiv_M^{(r)}$ must, in fact, be the relation $\equiv_M$.

On the other hand, for any M with t d-dimensional tapes, there is a constant $\alpha_M > 0$
that depends only on M's structure such that M has $\leq \alpha_M^{tr^d}$ distinct configurations of
"radius" r—meaning that all nonblank symbols on all tapes reside within r cells of
the read/write heads. Thus, in order for each database to get a distinct configuration
(so that $\equiv_M^{(r)}$ has index $\geq 2^{2^k} - 1$), the "radius" r must exceed $\beta_M \cdot 2^{k/d}$, for some
constant $\beta_M > 0$ that depends only on M's structure.

> I have slipped some actually simple arithmetic past you in the preceding paragraph. Here's a
> hint at how it goes.
>
> (1) M has $|Q|$ states.
>
> (2) If the number of "radius"-r configurations that can occur on each of M's t tapes is c_r, then
> the total number of tape configurations that M could conceivably reach is $\leq c_r^t$. To wit, each
> configuration can occur independently on each tape.
>
> (3) The number c_r can be bounded via the following overestimate. Make believe that you
> put some grease on one of M's read/write heads and then repeat the following experiment as
> often as you want. Have the read/write head move r steps and then return to its starting place.
> On a one-dimensional tape, $2r + 1$ tape squares will get greasy. On a two-dimensional tape,
> the number is $2r^2 + 2r + 1$. The exact number gets harder to compute as the dimensionality
> d grows, but it is not too hard to show that it is always proportional to r^d. The upper bound
> $(2r + 1)^d$ is easy to derive by just imagining a side-$(2r + 1)$ d-dimensional chess board.
>
> (4) Finally, since each of the roughly r^d tape squares must hold a symbol from the alphabet
> Γ, we end up with roughly $|\Gamma|^{r^d}$ possible configurations for each of M's tapes, hence with
> roughly $|Q| \cdot |\Gamma|^{tr^d}$ possible configurations in all. The "roughly" in this paragraph covers up a
> lot of sins that become just constant factors in the next paragraph.
>
> We now have the desired upper bound on the number of potential configurations that a
> database string can leave M in, and this bound has the form $\alpha_M^{tr^d}$. In order for this number
> to exceed 2^{2^k}, we must have, roughly (hiding another constant), $tr^d \geq 2^k$ (by taking loga-
> rithms of both sides), so that finally, we get the desired rough (hiding yet another constant)
> bound on r, namely, $r \geq \beta^{k/d}$ for some constant $\beta > 0$.
>
> It is worth doing the calculations here carefully, for practice. But the result we are seeking
> really needs just the rough estimates that we have outlined here.

Combining the preceding reasoning with Lemma 5.8, we arrive at the following
time bound.

Lemma 5.9. *If* $L(M) = L_{DB}^{(k)}$, *then for some length-k query* η, *M must take*[7] *more
than* $\beta_M \cdot (2^{1/d})^k$ *steps while reading* η, *for some* $\beta_M > 0$ *that depends only on M's
structure.*

The reasoning behind Lemma 5.9 is *information-theoretic*. Specifically, the bound
depends only on the fact that the number of distinct databases specified by database-
strings in $L_{DB}^{(k)}$ is *doubly exponential* in k, while the number of bounded-"radius"

---

[7] We write $2^{k/d}$ in the unusual form $(2^{1/d})^k$ to emphasize that the dimensionality of M's tapes (which
is a *fixed* constant) appears only in the base of the exponential.

OTM configurations is *singly exponential*. (This is why we could be so cavalier with our calculations; we needed just this gross result.) The ultimate message is this: *No matter how M reorganizes its tape contents while responding to one bad query, there must always be a query that is bad for the new configuration!* By focusing on strings with 2^k bad queries, we thus obtain the following result:

Theorem 5.3 ([38]). *Any OTM M with d-dimensional tapes that recognizes the language L_{DB} must, for infinitely many N, take time $> \beta_M \cdot (N/\log N)^{1+1/d}$ to process inputs of length N, for some constant $\beta_M > 0$ that depends only on M's structure.*

Proof. Consider some fixed, but arbitrary, sublanguage of L_{DB}, all of whose words consist of binary words of length k *and* have 2^k queries. (Thus, this sublanguage of L_{DB} is also a sublanguage of $L_{DB}^{(k)}$.) Call this sublanguage $L_{DB}^{\langle k \rangle}$.

Now, every word in $L_{DB}^{\langle k \rangle}$ has length $(k+1)2^{k+1} - 1$. What is important for the bound of the theorem is that this common length is roughly $k2^k$; the constant factors we thereby ignore will get "absorbed" into the constant β_M. Repeated invocation of Lemma 5.9 tells us that no matter how M organizes—and reorganizes—its databases, at least one of these strings will require M to compute for a number of steps that is proportional to roughly $2^{k/d}$ *for every query*. This "bad" string thus causes M to compute for roughly $2^{k(1+1/d)}$ steps on an input of length roughly $N = k2^k$.

The remainder of the proof is the exercise of expressing the quantity $2^{k(1+1/d)}$ as a function of $N = k2^k$. To accomplish this, we note that $\log N$ is roughly k. (In fact, of course, $\log N = k + \log k$, but given any positive fraction φ, for all sufficiently large k, this sum is less than $(1 + \varphi)k$.) This means that 2^k is roughly $N/\log N$, so that $2^{k(1+1/d)}$ is roughly $(N/\log N)^{1+1/d}$.

By doing the calculations more carefully, one shows finally that as a function of $N = k2^k$, the quantity $2^{k(1+1/d)}$ deviates from $(N/\log N)^{1+1/d}$ by only a constant factor. $\square$

One finds in [38] a companion upper bound of $O(N^{1+1/d})$ for the problem of recognizing L_{DB}. Hence, Theorem 5.3 does, indeed, expose the potential of nontrivial impact of data-structure topology on computational efficiency.

In its era (the late 1960s), the theorem also exposed one of the earliest examples of the cost of requiring a computation to be *online*. Specifically, L_{DB} can clearly be accepted *in linear time* by an OTM M that has just a single linear work tape, but that operates in an *offline* manner—meaning that M gets to see the entire input string before it must give an answer (so that it knows which query is important before it starts computing).

Chapter 6
Enrichment Topics

6.1 Pumping in Formal Languages

This section is devoted to discussing a phenomenon called *"pumping,"* which is a characteristic of any finite closed mathematical system. We introduce this notion in order to explain the so-called *pumping lemma* for regular languages (and, briefly, its analogue for so-called *context-free languages*). Most textbooks introduce the pumping lemma as the primary tool for exposing the limitations of FAs. The reader will see from the discussion throughout this section that we disagree with this point of view, on both conceptual and methodological grounds. We hope that the reader will agree with us after reading this section's rather thorough explanation of the pumping lemma, its origins, its strengths, and its weaknesses.

6.1.1 The Phenomenon of Pumping in Finite, Closed Systems

Example 1: Semigroups. A *semigroup* is a set of elements that is closed under an associative binary operation that, by convention, is called "multiplication."

> Each of the following number systems, the integers, the rational numbers, and the real numbers, forms a semigroup under both the operation of addition and that of multiplication. None of these systems forms a semigroup under the operation of division: the integers are not closed under this operation, and while the rationals and the reals are closed under division, this operation is not associative, as one sees from endpoints of the following chain:
>
> $$a/(b/c) = ac/b \neq a/(bc) = (a/b)/c.$$

A semigroup is one of the simplest algebraic systems, yet also one that is rich in applications. And finite semigroups are among the simplest examplars of the phenomenon of *pumping*.

Consider any finite *semigroup* formed by the set of elements $S = \{\alpha_1, \alpha_2, \ldots, \alpha_n\}$ and some associative binary multiplication (which we shall denote here by juxtaposi-

A.L. Rosenberg, *The Pillars of Computation Theory*, Universitext,   91
DOI 10.1007/978-0-387-09639-1_6, © Springer Science+Business Media, LLC 2010

tion). Consider an arbitrary sequence of products of elements of S (with repetitions), where each sequential item is obtained from its predecessor via post-multiplication—i.e., multiplication on the right—with some element of S:

$$\alpha_{i_1}$$
$$\alpha_{i_1}\alpha_{i_2}$$
$$\alpha_{i_1}\alpha_{i_2}\alpha_{i_3}$$
$$\vdots$$

Because S is finite, every sufficiently long such sequence—specifically, every sequence of length $> n = |S|$—must contain two distinct products,

$$\alpha_{i_1}\alpha_{i_2}\cdots\alpha_{i_k}$$

and

$$\alpha_{i_1}\alpha_{i_2}\cdots\alpha_{i_k}\alpha_{i_{k+1}}\cdots\alpha_{i_{k+\ell}},$$

that are equal within the semigroup; i.e., they denote the same element of S:

$$\alpha_{i_1}\alpha_{i_2}\cdots\alpha_{i_k} = \alpha_{i_1}\alpha_{i_2}\cdots\alpha_{i_k}\alpha_{i_{k+1}}\cdots\alpha_{i_{k+\ell}}. \tag{6.1}$$

This is an instance of the pigeonhole principle. Do you see how it applies here?

Any such pair of equal products is the seed of an instance of the phenomenon of *pumping* within semigroup S. To wit, the associativity of the semigroup multiplication allows us to iterate the "absorption" in (6.1) to prove that for all $h \in \mathbb{N}$,[1]

$$\alpha_{i_1}\alpha_{i_2}\cdots\alpha_{i_k}\left(\alpha_{i_{k+1}}\cdots\alpha_{i_{k+\ell}}\right)^h = \left(\alpha_{i_1}\alpha_{i_2}\cdots\alpha_{i_k}\alpha_{i_{k+1}}\cdots\alpha_{i_{k+\ell}}\right)\left(\alpha_{i_{k+1}}\cdots\alpha_{i_{k+\ell}}\right)^{h-1}$$

$$= \left(\alpha_{i_1}\alpha_{i_2}\cdots\alpha_{i_k}\right)\left(\alpha_{i_{k+1}}\cdots\alpha_{i_{k+\ell}}\right)^{h-1}$$

$$\vdots$$

$$= \left(\alpha_{i_1}\alpha_{i_2}\cdots\alpha_{i_k}\right)\left(\alpha_{i_{k+1}}\cdots\alpha_{i_{k+\ell}}\right)$$

$$= \alpha_{i_1}\alpha_{i_2}\cdots\alpha_{i_k}.$$

Example 2: Edge-labeled directed graphs. Consider next the following little tale, which suggests how the phenomenon of pumping manifests itself in finite graphs. Say that you are in a park in Paris (*lucky you!*) that is organized as a set of n statues interconnected by one-way paths. (Think of the statues as the nodes of a directed graph and of the paths as its arcs.) Assume that, as is common in parks, the pattern of paths is sufficiently complex that every statue marks the end of one one-way path and the beginning of another. Say that you take a *long* walk in the park—specifically, long enough for you to traverse n interstatue paths. Since the graph/park is finite—note another application of the pigeonhole principle here—you must encounter some specific statue at least twice in your walk. Moreover, you can keep repeating the

---

[1] The power notation implies iterated multiplication within the semigroup.

portion of your path that led you to the same statue twice—as many times as you want—and it will always return you to that statue.

Because one of our ways of visualizing a finite automaton M is as a directed graph whose nodes are states and whose arcs and their labels represent M's state-transition function δ_M, I am sure that you can already see some form of the pumping lemma for regular languages just below the surface of Example 2. If you are willing to look at the FA M just a bit differently from how we have been looking at them, then you will see a version of the pumping lemma emerge from Example 1 also. Specifically, the fact that δ_M is a function with domain $Q_M \times \Sigma$ means that *we can view each letter $\sigma \in \Sigma$ as a function that maps Q_M into Q_M*. The process of investigating M's behavior under finite input strings from Σ^* is equivalent to studying the semigroup generated by the letters in Σ under *functional composition*—which is easily shown to be an associative "multiplication." This semigroup of letters-as-functions is clearly finite, since there are only n^n distinct total functions that map an n-element set to itself. ($n = |Q_M|$ in this case.) The development in the rest of this section thus consists in just adding FA-specific details to Examples 1 and 2.

6.1.2 Pumping in Regular Languages

Now let's talk like automata theorists and translate the basic elements of the phenomenon exposed in Examples 1 and 2 into an FA-theoretic framework.

Focus on an FA M. In Example 1, semigroup elements become input symbols, viewed as functions from Q_M to Q_M, and sequences of such elements become input strings. In Example 2, statues become states, and interstatue directed paths become input strings. Within the contexts of both examples, the phenomenon of pumping ensures the following. Say that the language $L(M)$ that M accepts contains infinitely many strings. Among other things, we know that given any integer m, $L(M)$ contains (infinitely many) strings that are longer than m. In particular, no matter how many states M has, there is a string $w \in L(M)$—in fact, infinitely many of them—whose length is $\geq |Q_M|$. When we feed any such string w to M (starting from the initial state q_0, of course), the sequence of states that we pass through must contain some state— call it q—at least twice. To analyze this situation in more detail, let's "parse" w into the form $w = xyz$, where:

- x is the prefix of w that leads us from state q_0 to state q for the *first* time;
- y is the maximal-length internal portion of w that takes us from state q back to q;
- z is the suffix of w that leads us from state q to an accepting state $\widehat{q}$ of M.

Clearly, for all integers $k = 0, 1, \ldots$, the string $xy^k z$ acts essentially like w, in the sense that it takes us from q_0 to q (using the prefix x), loops around to q k times (using the internal portion y^k), and then leads from q to $\widehat{q}$ (using the suffix z). If we recast this description of "pumping" into the formalism of FAs, then we can describe it as follows.

Any word $w \in \Sigma^\star$ of length $\ell(w) \geq |Q_M|$ can be parsed into the form $w = xy$, where $y \neq \varepsilon$,[2] in such a way that $\delta(q_0, x) = \delta(q_0, xy)$.

Because M is deterministic—so that δ is a function—we find that for all $h \in \mathbb{N}$,

$$\delta(q_0, x) = \delta(q_0, xy^h). \tag{6.2}$$

Because the "pumping" depicted in (6.2) occurs also with words $w \in \Sigma^\star$ that admit a continuation $z \in \Sigma^\star$ that places them into $L(M)$—i.e., $wz \in L(M)$—we arrive finally at a formal statement of the pumping lemma for regular languages. (As you read along, note the implicit invocation of Lemma 5.1 in the argument that we have been developing.)

Lemma 6.1. (The pumping lemma for regular languages) *For every infinite regular language L, there exists an integer $n \in \mathbb{N}$ such that every word $w \in L$ of length $\ell(w) \geq n$ can be parsed into the form $w = xyz$, where $\ell(xy) \leq n$ and $\ell(y) > 0$, in such a way that for all $h \in \mathbb{N}$, $xy^h z \in L$.*

The proper way to look at Lemma 6.1 is as a strengthened version of the continuation lemma for OAs (Lemma 3.1) when the latter is applied to FAs. The technique of using Lemma 6.1 to prove that sets are not regular differs from the fooling set/finite-index lemma technique of Section 5.1 mainly in the new (and nonintrinsic!) requirement that one of the "fooling" words must be a prefix of the other. Indeed (inexplicably to me), most standard texts actually build all of their proofs of nonregularity of languages on the exposure of undesired "pumping" activity. This method of argumentation violates the *principle of parsimony*, by leading to proofs that are longer than necessary and that focus on restrictions that are extraneous (mainly, demanding that one "fooling" word be a prefix of the other). Note that we are not suggesting that the problems based on pumping are wrong—only that they unnecessarily complicate the proof process and the proofs themselves.

It is worth spending a moment to contemplate the *principle of parsimony* (*lex parsimoniae*), which is attributed to the fourteenth-century logician, William of Occam,

The principle, which is also known as Occam's razor, mandates that one always use the simplest possible setting that is sufficient to achieve one's goals. The extraneous condition on the "fooling" words that we have just discussed is a clear violation of this principle.

For the Latin lovers among you, Occam's razor is often stated in the following form, which seems never to have been established as the actual terminology in which William of Occam promulgated his principle:

Entia non sunt multiplicanda praeter necessitatem.

One simple example will illustrate my reasons for recommending that proofs of nonregularity *not* be based on Lemma 6.1. Consider the following pumping-based proof of the nonregularity of the language L_1 of Application 1 (Section 5.1). One notes that the "pumped" word y of Lemma 6.1:

1. cannot consist solely of a's, or else the block of a's becomes longer than the block of b's;

---

[2] Of course, we could countenance the case $y = \varepsilon$, but this would get us (and M) nowhere.

2. cannot consist solely of b's, or else the block of b's becomes longer than the block of a's;
3. cannot contain both an a and a b, or else the pumped word no longer has the form "a block of a's followed by a block of b's."

Even when one judiciously avoids this three-case argument by invoking the lemma's length limit on the prefix xy, one is inviting/risking excessive complication by seeking a string that pumps. For instance, when proving the nonregularity of the language L_3 of palindromes, one must cope with the fact that any palindrome *does* pump about its center! (That is, for any palindrome w and any integer ℓ, if one parses w into $w = xyz$, where x and z both have length ℓ, then indeed, for all $h \in \mathbb{N}$, the word xy^hz *is* indeed a palindrome.)

The danger inherent in using Lemma 6.1 to prove that a language is not regular is mentioned explicitly in [60]:

The pumping lemma is difficult for several reasons. Its statement is complicated, and it is easy to go astray in applying it.

We show now that the condition for a language to be regular that is provided in Lemma 6.1 is *necessary* but not sufficient. This contrasts with the *necessary and sufficient* condition provided by Theorem 4.1.

Lemma 6.2. *Every string of length > 4 in the nonregular language*

$$L_5 = \{uu^Rv \mid u,v \in \{0,1\}^*;\ \ell(u), \ell(v) \geq 1\}$$

pumps in the sense of Lemma 6.1.

I have been unable to trace the source of the example in Lemma 6.2; the example is discussed (anonymously) in [106], but it certainly had been in circulation decades before 2005.

Proof. Each string in L_5 consists of a nonempty even palindrome followed by another nonempty string. Say first that $w = uu^Rv$ and that $\ell(w) \geq 4$. If $\ell(u) = 1$, then we can choose the first character of v as the nonnull "pumping" substring of Lemma 6.1. (Of course, the "pumped" strings are uninteresting in this case.) Alternatively, if $\ell(u) > 1$, then consider the first character of u, call it a. Because a^k is a palindrome for every $k > 1$, we can let this first letter be the nonnull "pumping" substring of Lemma 6.1. In either case, the lemma holds. $\square$

Notably, the discussion of Lemma 6.2 in [106] ends with the following comment.

For a practical test that exactly characterizes regular languages, see the Myhill—Nerode theorem.

For the record, we note that the Myhill–Nerode theorem (Theorem 4.1) provides the following simple proof that L_5 is not regular. Let x and y be distinct strings from the infinite language $L = (01)(01)^*$, with $\ell(y) > \ell(x)$. (Strings in L consist of a sequence of one or more instances of 01.) Easily, xx^R is an even-length palindrome, hence belongs to L_5 (with $v = \varepsilon$). However, one verifies easily that yx^R does not begin with an even-length palindrome, so that $yx^R \notin L_5$. To wit, if one could write yx^R in the form uu^Rv, then:

- u could not end with a 0, because the "center" substring 00 does not occur in yx^R;
- u could not end with a 1, because the unique occurrence of 11 in yx^R occurs to the right of the center of the string.

It now follows by Lemma 5.1 that L_5 is not regular.   □

For completeness, we end this section by citing, without proof, a version of Lemma 6.1 whose underlying condition is both necessary and sufficient for a language to be regular. This version is rather nonperspicuous and a bit cumbersome, hence is only of academic interest: it shows that there is a version of pumping that actually characterizes the property of being a regular language. That said, I am standing by my assertion that Theorem 4.1 should be your main tool when proving that a language is not regular.

Theorem 6.1 ([46]). (The necessary-and-sufficient pumping lemma for regular languages) *A language* $L \subseteq \Sigma^*$ *is regular if and only if there exists an integer* $n \in \mathbb{N}$ *such that every word* $w \in \Sigma^*$ *of length* $\ell(w) \geq n$ *can be parsed into the form* $w = xyz$, *where* $\ell(y) > 0$, *in such a way that for all* $z \in \Sigma^*$:

- *if* $wz \in L$, *then for all* $h \in \mathbb{N}$, $xy^h z \in L$;
- *if* $wz \notin L$, *then for all* $h \in \mathbb{N}$, $xy^h z \notin L$.

6.1.3 Pumping in Nonregular Languages

This section exposes another enrichment topic that centers around the phenomenon of pumping. We illustrate a language-theoretic setting—the theory of *context-free* languages—where pumping plays a more fundamental role than with regular languages, because here it is the primary tool for negative proofs. The material in this section has its roots in Chomsky's "type-2" grammars and languages of [12, 13].

A *context-free grammar* (*CFG*, for short) is specified as follows:

$$G = (V, \Sigma, S, \mathscr{P}),$$

where

- V is a finite *vocabulary* of *nonterminal symbols* that play the roles of "syntactic categories";
- Σ is a finite alphabet of *terminal symbols*;
- $S \in V$ is the *sentence symbol*, the most general "syntactic category";
- $\mathscr{P} \subseteq V \times (V \cup \Sigma)^*$ is a relation whose elements are called *productions*.

Conventionally, a production (A, w), where $A \in V$ and $w \in (V \cup \Sigma)^*$, is written in the form "$A \rightarrow w$." (The arrow notation suggests, evocatively, that productions will be used to rewrite letters as strings.)

Informally, one starts with the sentence symbol S and begins generating strings by rewriting nonterminal symbols in manners allowed by the productions.

The just-described process can be viewed as "nondeterministic"—in contrast to procedural—or as "descriptive"—in contrast to prescriptive—in the following sense. While rewriting a string (or "sentential form"), you can always employ any production that applies to the "sentential form" that you have at the moment. You will generally have many options at every step in the process.

One continues to rewrite the symbols of the current "sentential form" until one has a string of terminal symbols: a *sentence*. Here is one simple example to provide intuition.

- $V = \{$Mathematical-Expression, Sum, Product, Var$\}$
- $\Sigma = \{(,),+,-,\times,\div,X,Y,Z\}$
- $S =$ Mathematical-Expression
- $\mathscr{P}$ consists of the following ten productions:
  Mathematical-Expression $\rightarrow$ Sum
  Mathematical-Expression $\rightarrow$ Product
  Mathematical-Expression $\rightarrow$ Var
  Sum $\rightarrow$ (Mathematical-Expression $+$ Mathematical-Expression)
  Sum $\rightarrow$ (Mathematical-Expression $-$ Mathematical-Expression)
  Product $\rightarrow$ (Mathematical-Expression $*$ Mathematical-Expression)
  Product $\rightarrow$ (Mathematical-Expression $/$ Mathematical-Expression)
  Var $\rightarrow$ X
  Var $\rightarrow$ Y
  Var $\rightarrow$ Z

Hopefully, you can intuit how this CFG specifies a "Mathematical-Expression" as either a variable or a fully parenthesized sum/difference or product/ratio of "Mathematical-Expression"s. Figure 6.1 contains a simple sample derivation of a specific mathematical expression from the sentence symbol Mathematical-Expression; we shall return to this example formally. In order to fit within margins, we abbreviate "Mathematical-Expression" by "M-E" in the figure. (In this example, we always expanded the leftmost eligible nonterminal symbol; as the preceding comment indicates, we need not have done this.)

We now formalize our intuition about how CFGs work, by supplying the "semantics" of CFGs.

Let us be given a CFG $G = (V, \Sigma, S, \mathscr{P})$. Consider any string uAv, where $A \in V$ and $u, v \in (V \cup \Sigma)^{\star}$. If there is a production $(A, w) \in \mathscr{P}$, then we write

$$uAv \Rightarrow_G uwv,$$

meaning that string uAv can be rewritten as string uwv under G. This defines $\Rightarrow_G$ as a new "rewriting" relation on strings:

$$\Rightarrow_G \subseteq (V \cup \Sigma)^{\star} V (V \cup \Sigma)^{\star} \times (V \cup \Sigma)^{\star}.$$

We are really interested in the *reflexive, transitive closure* of relation $\Rightarrow_G$, which we denote by $\Rightarrow_G^{\star}$ and define as follows. For strings $u, v \in (V \cup \Sigma)^{\star}$, we write $u \Rightarrow_G^{\star} v$, articulated as "$v$ is derivable from u under G," just when the following holds:

$$\begin{aligned}
\text{M-E} &\to \text{Sum} \\
&\to (\text{M-E} + \text{M-E}) \\
&\to (\text{Var} + \text{M-E}) \\
&\to (X + \text{M-E}) \\
&\to (X + \text{Sum}) \\
&\to (X + (\text{M-E} - \text{M-E})) \\
&\to (X + (\text{Prod} - \text{M-E})) \\
&\to (X + ((\text{M-E} * \text{M-E}) - \text{M-E})) \\
&\to (X + ((\text{Var} * \text{M-E}) - \text{M-E})) \\
&\to (X + ((X * \text{M-E}) - \text{M-E})) \\
&\to (X + ((X * \text{Var-E}) - \text{M-E})) \\
&\to (X + ((X * Y) - \text{M-E})) \\
&\to (X + ((X * Y) - \text{Prod})) \\
&\to (X + ((X * Y) - (\text{M-E/M-E}))) \\
&\to (X + ((X * Y) - (\text{Var/M-E}))) \\
&\to (X + ((X * Y) - (Y/\text{M-E}))) \\
&\to (X + ((X * Y) - (Y/\text{Var}))) \\
&\to (X + ((X * Y) - (Y/Z)))
\end{aligned}$$

Fig. 6.1 A sample derivation of the expression $(X + ((X * Y) - (Y/Z)))$ under the CFG G.

$$u \Rightarrow_G^\star v \text{ means } \begin{cases} \text{either } u = v \\ \text{or there exist strings } w_1, w_2, \ldots, w_n \text{ such that} \\ \quad u \Rightarrow_G w_1 \Rightarrow_G w_2 \Rightarrow_G \cdots \Rightarrow_G w_n \Rightarrow_G v. \end{cases}$$

We can now, finally, define the *context-free language* (*CFL*, for short) $L(G)$ that is *generated* by the CFG G:

$$L(G) = \{x \in \Sigma^\star \mid S \Rightarrow_G^\star x\}.$$

Each *derivation*

$$S \Rightarrow_G y_1 \Rightarrow_G y_2 \Rightarrow_G \cdots \Rightarrow_G y_n \Rightarrow_G x \tag{6.3}$$

of a string $x \in L(G)$ under the CFG G can be depicted in a natural way as a rooted, oriented tree—called the *derivation tree* of x under G—where:

- S is the root of the tree;
- for each rewriting $uAv \Rightarrow_G uwv$ in the sequence of rewritings that constitute derivation (6.3), all of the letters of strings w are, from left to right, the children of node A;
- the left-to-right sequence of leaves of the tree constitute string x.

Figure 6.2 illustrates these concepts by means of the derivation tree of the expression $(X + ((X * Y) - (Y/Z)))$ under the CFG for mathematical expressions presented earlier. Note how one can read off the derivation in Figure 6.1 from this tree by always rewriting the leftmost possible variable.

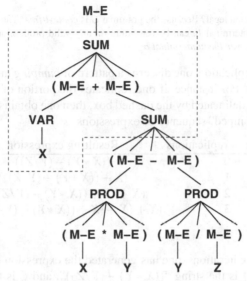

Fig. 6.2 A tree-structured depiction of the derivation of the expression $(X + ((X * Y) - (Y/Z)))$ under the CFG G; cf. Figure 6.1.

On to the pumping lemma for the CFG G! We derive this lemma by analyzing the ramifying (branching) structure in G's derivation trees, so let us focus on an arbitrary such tree T. Note that every root-to-leaf path in the derivation tree T is a string $\beta_1\beta_2 \cdots \beta_m\sigma$, where each $\beta_k \in V$, and $\sigma \in \Sigma$. Let us compactify this path to eliminate "nonproductive" productions—those of the form $\beta_i \rightarrow \beta_j$. "Nonproductive" productions add no interesting structure to the tree T; they merely rename syntactic categories. While such renaming may be quite significant linguistically, it has no significance in the structural analysis we are engaged in. After we have thus compactified T, the root-to-leaf path we started with has become a string $\beta_1'\beta_2' \cdots \beta_\ell'\sigma$, where each β_k' either produces a leaf of tree T (i.e., a terminal symbol) or has more than one child in tree T. We make two simple, yet important, observations:

- If the path is long enough, then some nonterminal along the path must repeat. This follows from the pigeonhole principle, because the set V is finite.
- If $L(G)$ is infinite, then the compactified root-to-leaf paths in derivation trees get arbitrarily long. This is verified using the same reasoning as in the famous "infinity lemma" of Dénes König [52], because there is an upper bound on the number of children a node can have in a derivation tree under G—namely, the length of the longest right-hand string in a production—and because there is no upper bound

on the number of leaves that a derivation tree for G can have (because $L(G)$ is infinite).

Consider now the effect of taking the portion of a derivation tree consisting of two occurrences of the same β_i on a path, plus the subtree subtended by these occurrences, and replicating it.

> Why is such replication legal? Because the grammar G is *context-free!* The context-free property means that nonterminal β_i can be rewritten via any valid production *at any step in a derivation* and *wherever the nonterminal occurs.*

If one repeats this replication, one discerns a pattern of *pumping* in the terminal string derived via the tree! For instance, if one replicates the portion of the derivation tree of Figure 6.2 that is delineated by the dashed box, then one obtains, via repeated such replications, the "pumped" sequence of expressions

Number of replications	Resulting expression
0	$((X*Y)-(Y/Z))$
1	$(X+((X*Y)-(Y/Z)))$
2	$(X+(X+((X*Y)-(Y/Z))))$
3	$(X+(X+(X+((X*Y)-(Y/Z)))))$
$\vdots$	$\vdots$

Symbolically, after k iterations, one has generated the expression $\xi^k \eta \zeta^k$, where ξ is the string "$(X+$", η is the string "$((X*Y)-(Y/Z))$", and ζ is the string "$)$". The situation we have described illustrates a special case of the general phenomenon of pumping in CFLs, as described in the following lemma. The lemma's detailed proof is left as an exercise, but all of the raw material appears in the preceding paragraphs.

Lemma 6.3. (The pumping lemma for context-free languages) *For every infinite context-free language L, there exists an integer $m \in \mathbb{N}$ such that every word $z \in L$ of length $\ell(z) \geq m$ can be parsed into the form $z = uvwxy$, where $\ell(uv) \leq m$ and $\ell(vx) > 0$, in such a way that for all $h \in \mathbb{N}$, $uv^h wx^h y \in L$.*

As suggested earlier, Lemma 6.3 is the primary tool for proving that languages are not context-free. We present just one simple example; others appear as exercises.

Application 5. *The language $L = \{a^n b^n c^n \mid n \in \mathbb{N}\}$ is not context-free.*

The proof consists of a case-by-case analysis of where the pumping pair of strings, v, x, of Lemma 6.3 can reside, relative to the blocks of a's, b's, and c's in each string in L.

1. If the strings v and x exist, then each contains instances of only one letter. To wit, if either of these strings, say v, contained instances of two or more letters, then after a single application of pumping, the resulting string would no longer consist of a block of a's followed by a block of b's followed by a block of c's; hence the string would not belong to the language L. (As a simple illustration, after a single application of pumping, the string ab would become $abab$.)

2. By item 1, each of v and x is contained within a single one of the three blocks of letters that make up string of L. This means that after a single application of pumping, at most two of the three blocks will have increased in length. Once this happens, the three blocks will no longer share the same length, so the pumped string will not belong to L.

The language L is particularly simple prey for the pumping lemma. The reader is invited to prove (the true, but harder to verify, fact) that the language of "squares" considered in Application 4 of Section 5.1 is not context-free.

We now leave the subject of pumping in formal languages having given the reader the rudiments necessary for delving further into the topic.

6.2 Closure Properties of the Regular Languages

So-called *closure properties* of classes of languages have attracted much attention since the earliest days of studying formal languages. This interest has several antecedents. From a mathematical perspective, closure properties can expose important aspects of the intrinsic nature of a class of languages. Indeed, the Kleene–Myhill theorem (Theorem 11.3) that we study in Section 11.2 actually *characterizes* the regular languages in terms of closure properties.

> By "characterizes," we mean that the theorem employs closure properties to tell us precisely which languages are regular and which not. This is a mathematical, rather than algorithmic, description of the class: it does not yield a programmable test for regularity. In fact, the theorem can be stated in the following form; cf. [51].
>
> *The family of regular languages over an alphabet Σ is the smallest class of subsets of Σ^* that contains all finite subsets of Σ^* and that is closed under a finite number of applications of the operations of . . .*

From a linguistic perspective, closure properties can expose significant structural features of a class of languages.

> Perhaps the simplest example of this use of closure properties occurs with CFLs. One can intuit from Lemma 6.3 that the hallmark of CFLs is that they can "match" pairs of sites in a string, as those sites develop (say, via pumping). Pairs can be matched, but not larger groupings. One can observe this phenomenon in action from proofs that the CFLs are *not* closed under intersection. To wit:
>
> (a) We leave as an exercise the verification that each of the following languages is a CFL:
>
> $$L_1 = \{a^i b^j c^k \mid i = j\} \quad \text{and} \quad L_2 = \{a^i b^j c^k \mid i = k\}.$$
>
> We proved via Application 5 of the preceding section that
>
> $$L_1 \cap L_2 = \{a^n b^n c^n \mid n \in \mathbb{N}\}$$
>
> is *not* a CFL.
>
> Intersection thus forces two pair-matching languages to become a triple-matching language. We can easily force the matching of arbitrarily high groupings, as the reader can verify from the following example.

(*b*) We leave as an exercise the verification that each of the following languages is a CFL:

$$L_3 = \{a^i b^j c^k d^\ell \mid i = j \text{ and } k = \ell\} \quad \text{and} \quad L_4 = \{a^i b^j c^k d^\ell \mid j = k\}.$$

A slightly more complicated analogue of the argument in Application 5 of the preceding section proves, however, that

$$L_3 \cap L_4 = = \{a^n b^n c^n d^n \mid n \in \mathbb{N}\}$$

is *not* a CFL.

From a computational perspective—the most important one, given our focus on developing computation theory via formal languages—closure properties afford one a high-level mechanism for discussing often-intricate manipulations of the language-recognizing algorithms that we call "automata."

We shall imminently see the *direct-product* construction for automata, which is an important example of these "intricate manipulations." Other significant examples appear in Section 11.2.

Finally, many significant interlanguage relationships can be expressed and studied by means of sentences involving closure properties. Here are some important examples.

Focus on two languages, L_1 and L_2:

$$(L_1 \subset L_2) \text{ iff } ((L_1 \setminus L_2 = \emptyset) \text{ and } (L_2 \setminus L_1 \neq \emptyset));$$
$$(L_1 \subseteq L_2) \text{ iff } (L_1 \cup L_2 = L_2);$$
$$(L_1 = L_2) \text{ iff } ((L_1 \setminus L_2) \cup (L_2 \setminus L_1) = \emptyset).$$

These examples illustrate but certainly do not exhaust the point here.

Our development of language theory thus far has given us access to a large repertoire of closure properties of the class of regular languages. We enumerate several of these (redundantly, for emphasis) in the following definitions and the subsequent theorem.

Let L_1 and L_2 be languages over the alphabet Σ.

The **union** of L_1 and L_2 is the set-theoretic union $L_1 \cup L_2$, as defined in Section 2.1.

The **intersection** of L_1 and L_2 is the set-theoretic intersection $L_1 \cap L_2$, as defined in Section 2.1.

The **complement** of L_1 is the set-theoretic complement $\Sigma^\star \setminus L_1$, as defined in Section 2.1.

The **concatenation** of L_1 and L_2 (in that order!) is:

$$L_1 \cdot L_2 \stackrel{\text{def}}{=} \{xy \in \Sigma^\star \mid [x \in L_1] \text{ and } [y \in L_2]\}.$$

Thus, $L_1 \cdot L_2$ consists of all strings that have a prefix in L_1 whose removal leaves a string in L_2.

"Powers" of a language. For any integer $k \geq 0$,

$$L_1^0 \stackrel{\text{def}}{=} \{\varepsilon\} \quad \text{and, inductively,} \quad L_1^{k+1} \stackrel{\text{def}}{=} L_1 \cdot L_1^k.$$

Clearly, $L_1^1 = L_1$. We call each language L^k the kth *power* of language L—using the word "power" as with exponentiation.

The **star-closure** of L_1, denoted by $L_1^\star$, is, informally, the language comprising all finite concatenations of strings from L_1. Formally,

$$L_1^\star \stackrel{\text{def}}{=} \bigcup_{i=0}^{\infty} L_1^i = \{\varepsilon\} \cup L_1 \cup L_1^2 \cup L_1^3 \cup \cdots.$$

Note the somewhat unintuitive fact that $\emptyset^\star = \{\varepsilon\}$. (In fact, $\emptyset^\star$ is the only finite star-closure language.)

The **reversal** of L_1, denoted L_1^R, is the language obtained by reversing all strings in L_1; i.e.,

$$L_1^R \stackrel{\text{def}}{=} \{\sigma_n \sigma_{n-1} \cdots \sigma_2 \sigma_1 \mid \sigma_1 \sigma_2 \cdots \sigma_{n-1} \sigma_n \in L_1\}.$$

All σ_i here belong to Σ.

The three operations union, intersection, and complementation are known collectively as the *Boolean* operations; less universally, but quite commonly, the three operations union, concatenation, and star-closure are known collectively as the *Kleene* operations.

At this point, we only suggest, via pointers, how to prove the following theorem; details are available by following the pointers.

Theorem 6.2. (Closure properties of the regular languages) *The regular languages are closed under the following operations.*

1. *the Kleene operations: if L_1 and L_2 are regular, then so also are $(L_1)^\star$, $L_1 \cup L_2$, and $L_1 \cdot L_2$.*
2. *the operation of language-reversal: if L is regular, then so also is its reversal L^R.*
3. *the Boolean operations: if L_1 and L_2 are regular, then so also are $\overline{L}_1$, $L_1 \cup L_2$, and $L_1 \cap L_2$.*

Proof Sketch. We defer the proof of the closure of the regular languages under the Kleene operations until we cover Lemma 11.1, which embodies the portion of the Kleene–Myhill theorem (Theorem 11.3) that converts a given *regular expression* R to an FA M_R that accepts the language denoted by R. (The lemma's companion, Lemma 11.2, embodies the portion of the theorem that converts a given FA M to a regular expression R_M that denotes $L(M)$.) The detailed definition of regular expressions must be deferred until Section 11.2, because its technical details build so heavily on the "pillar" topic NONDETERMINISM, but we note here that each such expression is a finite string that *denotes* a regular language.

> We use the word "denotes" here to indicate that regular expressions have the same descriptive power as FAs, in the sense that one can translate either the regular-expression "name" of a regular language or the FA "name" of the language into the other "name."

The closure of the regular languages under the Boolean operations follows from their closure under union and complementation (Lemma 4.1), because of De Morgan's laws (Section 2.1). □

There is another, direct, proof that the regular languages are closed under the Boolean operations. This proof—or really, its easy generalizations to OAs and their languages—is important in many applications of OA theory, both to other components of computation theory and to application areas such as logic design. We now present this alternative proof, via the *direct-product construction* for FAs. (The reader can easily extend the technique to general OAs.)

The direct-product construction. We illustrate both the construction and the use of direct products of FAs by focusing on an arbitrary "generalized Boolean operation" $\otimes$, i.e., a binary operation that can be formed as a finite composition of one or more of the three basic set-theoretic Boolean operations, union, intersection, and complementation. (Some of the more important such operations are mentioned in Section 2.1.) We prove, via the direct-product construction, that the regular languages are closed under the operation $\otimes$, obtaining thereby a rather strong, "one size fits all," proof of the closure of the regular languages under Boolean operations.

The intuitive strategy that underlies the direct-product construction of FAs is to "run two FAs in parallel," state-transition by state-transition, and then combine their answers to the current input string via the *logical* Boolean operation that corresponds to the target *set-theoretic* Boolean operation $\otimes$. One achieves this effect formally as follows. Focus on regular languages L_1 and L_2, where each L_i, for $i = 1, 2$, is accepted by the FA

$$M_i = (Q_i, \Sigma, \delta_i, q_{i0}, F_i).$$

We construct the following direct-product FA:

$$M_{1,2}^{\otimes} = (Q_1 \times Q_2, \Sigma, \delta_{1,2}, \langle q_{10}, q_{20} \rangle, F_{1,2}^{\otimes}),$$

where

- For all $q_1 \in Q_1$, $q_2 \in Q_2$, and $\sigma \in \Sigma$, define

$$\delta_{1,2}(\langle q_1, q_2 \rangle, \sigma) = \langle \delta_1(q_1, \sigma), \delta_2(q_2, \sigma) \rangle.$$

Note how the definition of $\delta_{1,2}$ can be viewed as "running the FAs M_1 and M_2 in parallel."

- $F_{1,2}^{\otimes} = \{\langle q_1, q_2 \rangle \in Q_1 \times Q_2 \mid ([q_1 \in F_1] \otimes [q_2 \in F_2] = 1)\}$.

To explain the definition of $F_{1,2}^{\otimes}$ in more detail: Note that the predicate "$q_i \in F_i$" can be viewed as evaluating to 0 (if the predicate is false) or to 1 (if the predicate is true). Thus intepreted, it is meaningful to combine predicates using the logical version of the set-theoretic Boolean operation $\otimes$.[3] The condition that delimits the pairs of states of $M_{1,2}^{\otimes}$ that belong to $F_{1,2}^{\otimes}$ can, therefore, be translated as follows:

1. Take the truth values of the statements $[q_1 \in F_1]$ and $[q_2 \in F_2]$, and combine them using the *logical* Boolean operation $\otimes$.
2. Add the pair $\langle q_1, q_2 \rangle$ to $F_{1,2}^{\otimes}$ if and only if the expression $[q_1 \in F_1] \otimes [q_2 \in F_2]$ evaluates to 1.

---

[3] The semantic overloading of the operation symbol "$\otimes$" should cause no problems, because each use of the symbol inherits its "type" (set-theoretic or logical) from its arguments.

We claim that $L(M_{1,2}^{\otimes}) = L_1 \otimes L_2$, whence the latter language is regular. To see this, note that when we extend the state-transition function $\delta_{1,2}$ of the product FA $M_{1,2}^{\otimes}$ to act on strings, rather than single letters (in a way that should be quite clear by this point in our study) we find that for all strings $x \in \Sigma^\star$,

$$\delta_{1,2}(\langle q_{10}, q_{20}\rangle, x) = \langle \delta_1(q_{10}, x), \delta_2(q_{20}, x)\rangle.$$

By definition, then, the state $\langle \delta_1(q_{10}, x), \delta_2(q_{20}, x)\rangle$ belongs to $F_{1,2}^{\otimes}$—or equivalently, $x \in L(M_{1,2}^{\otimes})$—precisely when

$$[\delta_1(q_{10}, x) \in F_1] \otimes [\delta_2(q_{20}, x) \in F_2] = 1,$$

using the logical version of operation $\otimes$. But this is equivalent to saying that

$$L(M_{1,2}^{\otimes}) = L(M_1) \otimes L(M_2) = L_1 \otimes L_2,$$

as was claimed.

You should keep the direct-product construction in an easily accessed place in your computation-theoretic toolbox. It is an invaluable tool for formally introducing parallel operation in a succinct manner.

6.3 Systems of Linear Equations with Languages as Coefficients

We have already referred several times to the importance of the Kleene–Myhill theorem (Theorem 11.3) within the theory of finite automata and regular languages. This importance is attested to by the large number of algorithms that have been developed for producing a regular expression R_M that denotes the language $L(M)$ accepted by a given FA M. One of the mathematically most interesting of these algorithms, which appears in [2], produces from any FA M a system of linear equations whose coefficients are sets (in fact, regular languages in the intended application in [2]) such that any solution to the system can be translated into a regular expression that denotes $L(M)$.

> The previous sentence is worded so craftily that the reader may suspect that some "sleight of pen" is being perpetrated here: "*any* solution to the system *can be translated* as *a* regular expression that denotes $L(M)$" (highlights added). The careful (but not "crafty") wording of the sentence is needed because—as we shall see imminently—*many* quite distinct regular expressions denote the same regular language. Although every solution to a given linear system denotes the same language, the expressions that different solutions yield may look totally different. Keep this in mind as we proceed.

A complete treatment of the FA-to-regular expression algorithm of [2] is beyond the scope of this book. Section 11.2 contains a complete treatment of one such algorithm, and none of these algorithms has a dramatic advantage over any other in efficiency. From a conceptual vantage point, though, the algorithm of [2] is quite interesting, in that it helps one develop a deeper understanding of some familiar linear-system-solving algorithms from linear algebra, by seeing the basic ideas of the algorithms—pivoting, unknown elimination, back-substitution—at work with algebras that look quite different from rings and their kin. (See a standard text on modern algebra, such as [5].) Therefore, we spend some time now exposing the reader to the underlying mathematical topic of systems of linear equations whose coefficients are sets. Specifically, we focus (for definiteness) on a fixed but arbitrary alphabet Σ, and we consider systems of linear equations of the form

$$\begin{aligned}
X_1 &= A_{11} \cdot X_1 \cup A_{12} \cdot X_2 \cup \cdots \cup A_{1n} \cdot X_n, \\
X_2 &= A_{21} \cdot X_1 \cup A_{22} \cdot X_2 \cup \cdots \cup A_{2n} \cdot X_n, \\
&\vdots \quad \vdots \qquad \vdots \\
X_n &= A_{n1} \cdot X_1 \cup A_{n2} \cdot X_2 \cup \cdots \cup A_{nn} \cdot X_n,
\end{aligned} \tag{6.4}$$

where:

- the "multiplication" "$\cdot$" denotes the operation of concatenation on languages;
- the "addition" "$\cup$" denotes the operation of union on languages;
- $X_1, X_2, \ldots, X_n$ are unknowns that range over subsets of $\Sigma^\star$ (i.e., languages);
- each coefficient A_{ij} is an ε-free language over Σ, i.e., a subset $A_{ij} \subseteq \Sigma^\star \setminus \{\varepsilon\}$.[4]

We now show how to solve a system of the form (6.4) for the unknowns $X_1, X_2, \ldots,$ X_n, and we show that the resulting solution is unique.

> We emphasize that the solution is unique *as a sequence of n languages*. That is to say, for each $i \in [1,n]$, in every solution, X_i will denote *the same language*. However, the *expressions* that one generates to denote this unique language will possibly be very different depending on factors such as the order in which one performs the various operations called for in the algorithm.

Theorem 6.3 ([2]). *Every linear system of the form (6.4) can be solved uniquely for the unknowns $X_1, X_2, \ldots, X_n$. If each coefficient set A_{ij} is a* regular *language, then all n solution languages[5] are regular languages.*

Proof. We proceed by induction on the size n of the system (6.4).

Case: $n = 1$. When the system (6.4) has one equation in one unknown, then the theorem can be strengthened in the following way, to specify the unique solution.

Lemma 6.4. *Every linear equation of the form*

$$X_1 = A_{11} \cdot X_1 \cup B_1, \tag{6.5}$$

where $A_{11} \subseteq \Sigma^\star \setminus \{\varepsilon\}$, has the unique solution

$$X_1 = A_{11}^\star \cdot B_1.$$

---

[4] We consider only ε-free languages for technical reasons. The coefficient languages that occur in the algorithms of [2] are indeed ε-free.

[5] The "n solution languages" are the solutions for $X_1, X_2, \ldots, X_n$.

We defer until Section 11.2 the proof that the solution language X_1 is regular whenever both A_{11} and B_1 are; this fact is a simple corollary of Theorem 11.3.

Proof. We begin our proof of Lemma 6.4 by noting that the set $A_{11}^\star \cdot B_1$ is a solution of equation (6.5), as the following chain of equalities demonstrates:

$$
\begin{aligned}
A_{11}^\star \cdot B_1 &= \Big((A_{11} \cdot A_{11}^\star) \cup \{\varepsilon\}\Big) \cdot B_1 && \text{(by definition of $\star$-closure)} \\
&= \Big((A_{11} \cdot A_{11}^\star) \cdot B_1\Big) \cup (\{\varepsilon\} \cdot B_1) && \text{(concatenation distributes over union)} \\
&= \Big(A_{11} \cdot (A_{11}^\star \cdot B_1)\Big) \cup B_1 && \text{(concatenation is associative;} \\
& && \text{ε is an identity for concatenation).}
\end{aligned}
$$

It follows that $A_{11}^\star \cdot B_1$ is a subset of every solution of equation (6.5). We show via contradiction that $A_{11}^\star \cdot B_1$ is, in fact, the *unique* solution to (6.5). Were this not the case, there would be another solution

$$
X_1 = A_{11}^\star \cdot B_1 \cup C,
$$

for some $C \subset \Sigma^\star$ that is *nonempty* and *disjoint from* $A_{11}^\star \cdot B_1$. We would then have

$$
\begin{aligned}
A_{11}^\star \cdot B_1 \cup C &= A_{11} \cdot (A_{11}^\star \cdot B_1 \cup C) \cup B_1 && (A_{11}^\star \cdot B_1 \cup C \text{ is a solution}) \\
&= (A_{11} \cdot A_{11}^\star \cdot B_1) \cup (A_{11} \cdot C) \cup B_1 && \text{(concatenation distributes} \\
& && \text{over union)} \\
&= \Big(A_{11} \cdot (A_{11}^\star \cdot B_1) \cup B_1\Big) \cup (A_{11} \cdot C) && \text{(union is commutative} \\
& && \text{and associative)} \\
&= (A_{11}^\star \cdot B_1) \cup (A_{11} \cdot C) && (A_{11}^\star \cdot B_1 \text{ is a solution).}
\end{aligned}
$$

If we intersect the first and last expressions in the preceding chain by C, then we discover—because C is disjoint from $A_{11}^\star \cdot B_1$—that

$$
C = C \cap (A_{11} \cdot C).
$$

This equation means—cf. Section 2.1—that C is a subset of $A_{11} \cdot C$. But this is absurd: The fact that A_{11} is ε-free implies that the shortest string in language C is *strictly shorter* than the shortest string in language $A_{11} \cdot C$. (Aha!! We finally see the impact of ε-freeness!) We conclude that the language C does not exist—so that $A_{11}^\star \cdot B_1$ is the *unique* solution to equation (6.5), as was claimed.   □

For completeness in covering the case $n = 1$, we note that if A_{11} is *not* ε-free. i.e., if $\varepsilon \in A_{11}$, then for every language $D \subseteq \Sigma^\star$, the language

$$
X_1 = A_{11}^\star \cdot (B_1 \cup D)
$$

is a solution of equation (6.5).

To see this, note that in this case,

$$
A_{11} = A_{11} \cup \{\varepsilon\}.
$$

Therefore, using elementary reasoning about the relevant set operations, we obtain

$$A_{11} \cdot (A_{11}^* \cdot (B_1 \cup D)) \cup B_1 = (A_{11} \cup \{\varepsilon\}) \cdot (A_{11}^* \cdot (B_1 \cup D)) \cup B_1$$
$$= (A_{11} \cdot A_{11}^* \cdot (B_1 \cup D)) \cup (A_{11}^* \cdot (B_1 \cup D)) \cup B_1$$
$$= A_{11}^* \cdot (B_1 \cup D).$$

Case: $n = 2$. For pedagogical reasons, we redundantly solve the case $n = 2$ explicitly, even though the case $n = 1$ provides an adequate base for our induction. Specifically, the case $n = 2$ lends intuition for our upcoming analysis of general values of n, and it illustrates some features of solutions to general systems that are too complex to see except for small specific values of n. We begin with the $(n = 2)$ version of a system of the form (6.4):

$$X_1 = A_{11} \cdot X_1 \cup A_{12} \cdot X_2 \cup B_1,$$
$$X_2 = A_{21} \cdot X_1 \cup A_{22} \cdot X_2 \cup B_2, \tag{6.6}$$

wherein each coefficient set A_{ij} is ε-free. We solve this system by the well-known strategy of elimination of unknowns. Starting arbitrarily with the first equation in (6.6)—we could just as easily, and correctly, start with the second—we invoke Lemma 6.4 to derive the following solution for language X_1, which is unique because A_{11} is ε-free:

$$X_1 = A_{11}^* \cdot ((A_{12} \cdot X_2) \cup B_1) = (A_{11}^* \cdot A_{12}) \cdot X_2 \cup (A_{11}^* \cdot B_1). \tag{6.7}$$

Substituting the value (6.7) for X_1 in the second equation of (6.6) produces the following equation for X_2:

$$X_2 = A_{21} \cdot ((A_{11}^* \cdot A_{12}) \cdot X_2 \cup (A_{11}^* \cdot B_1)) \cup A_{22} \cdot X_2 \cup B_2$$
$$= \left((A_{21} \cdot A_{11}^* \cdot A_{12}) \cup A_{22}\right) \cdot X_2 \cup \left((A_{21} \cdot A_{11}^* \cdot B_1) \cup B_2\right).$$

Because A_{12}, A_{21}, and A_{22} are all ε-free, so also is the set $(A_{21} \cdot A_{11}^* \cdot A_{12}) \cup A_{22}$. We can therefore invoke Lemma 6.4 again, to derive the following unique solution for X_2:

$$X_2 = \left((A_{21} \cdot A_{11}^* \cdot A_{12}) \cup A_{22}\right)^* \cdot \left((A_{21} \cdot A_{11}^* \cdot B_1) \cup B_2\right).$$

We now have a unique *complete* solution for X_2, i.e., a solution that expresses X_2 as a subset of Σ^*, independently of X_1. We can now *back-substitute* this solution into (6.7), in order to get a unique complete solution for X_1. We thereby arrive at the following unique solution-pair for both unknowns in system (6.6):

$$X_1 = (A_{11}^* \cdot A_{12}) \cdot \left((A_{21} \cdot A_{11}^* \cdot A_{12}) \cup A_{22}\right)^* \cdot \left((A_{21} \cdot A_{11}^* \cdot B_1) \cup B_2\right) \cup \left(A_{11}^* \cdot B_1\right),$$
$$X_2 = \left((A_{21} \cdot A_{11}^* \cdot A_{12}) \cup A_{22}\right)^* \cdot \left((A_{21} \cdot A_{11}^* \cdot B_1) \cup B_2\right).$$

As an exercise, the reader should now solve the system (6.6) by solving for the unknown X_2 first. The resulting solution-pair will describe the same two solution languages, X_1 and X_2—because Lemma 6.4 guarantees the uniqueness of this pair! But the new expressions that denote these languages will be unrecognizably different from the ones that we have just derived! We shall note in Chapter 11, specifically in Section 11.2, that this naming problem—finding denotations for regular languages— plagues almost all systems for representing regular languages. The one major exception is representation by minimal-state FAs, which, as we have seen in Section 5.2, yields "expressions" (i.e., the minimum-state FAs) that are unique up to the renaming of states. The problem, as we shall also see in Chapter 11, specifically in Section 11.1, is that the smallest description of a regular language L via a minimum-state FA (an FA M such that $L = L(M)$) can be *exponentially larger* than the smallest description of L via a regular expression or, equivalently, via a solution to a system such as (6.4). We just have to live with this state of affairs.

Case: General $n > 1$. The case of general n—i.e., of solving a system of the form (6.4)—is an extension of the case $n = 2$ that is quite straightforward conceptually but quite onerous computationally. You definitely want to write a program to generate the solutions! But here is how the basic strategy works.

We choose an arbitrary order in which to eliminate the unknowns in system (6.4). Regrettably, one can generally not predict which order of elimination of unknowns will yield the simplest expressions for the solution languages. Therefore, to simplify the exposition, we shall eliminate the unknowns in the order of their appearance in (6.4); i.e., we first eliminate X_1, then we eliminate X_2, then we eliminate X_3, and so on.

One can show by induction—building in a very direct way upon our analysis of the case $n = 2$—that at the step in which we eliminate unknown X_i, the coefficient of X_i is ε-free. We have seen this already in our analyses of the cases $n = 1$ and $n = 2$. An inductive argument shows that the coefficient of X_i, when it is about to be eliminated, is the union of sets that are "preconcatenated"—i.e., multiplied on the left—by the sets $A_{i1}, A_{i2}, \ldots, A_{ii}$; hence this coefficient set is ε-free, because all of the A_{ij} are. We can therefore eliminate each successive unknown X_i by invoking Lemma 6.4 to derive a *unique* expression for X_i in terms of all of the coefficient sets and all of the X_j with $j > i$ (which follow X_i in the order of elimination). Let us call this expression a *quasisolution* for unknown X_i, since it involves other unknowns, rather than just subsets of $\Sigma^\star$.

After the nth unknown-elimination via invocation of Lemma 6.4, we finally have a unique *complete* solution for unknown X_n, i.e., a solution *in terms of all of the coefficient sets*, as a subset of $\Sigma^\star$. We can now back-substitute this complete solution for X_n into the quasisolution for X_{n-1}. Because X_n is the only unknown that appears in the quasi-solution for X_{n-1}, this back-substitution gives us a *complete* solution for X_{n-1}. By continuing the process of back-substitution—next back-substituting the complete solutions for X_n and X_{n-1} into the quasisolution for X_{n-2}, then back-substituting the complete solutions for X_n, X_{n-1}, and X_{n-2} into the quasisolution for X_{n-3}, and so on—we eventually obtain complete solutions for all unknowns, in the reverse of the order in which we obtained quasisolutions by eliminating unknowns.

Thus, after n eliminations, followed by n back-substitutions, we finally have complete solutions for all n unknowns. We emphasize that Lemma 6.4 guarantees that all complete solutions are unique—even though, we repeat, the expressions that we obtain for these solutions depend on the order in which we eliminate the unknowns and perform back-substitutions, and on any possible simplifications that we perform in the course of these "big" operations.

Since all of the set operations that we perform involve the three operations of union, concatenation, and $\star$, we shall see in Section 11.2 that all solution languages are regular whenever all the coefficient languages are.   □

"Sometimes a cigar is just a cigar."
(Sigmund Freud [attributed])

While the good Dr. Freud's observation may be correct in reference to vehicles for smoking tobacco, it is totally off base regarding computation—as an exemplar of the large array of disparate subject areas that can be viewed as studying infinite sets via some system of string-based representations of their elements. We shall see in our treatment of the "pillar" topic ENCODING that in a computational setting, a positive integer may (to paraphrase Freud) be just a positive integer, but it may simultaneously be an encoding of any of myriad finite objects as complex and diverse as:

- a complex data structure D,
- a program P,
- program P operating on data structure D,
- the sequence of operations that program P performs when operating on data structure D (*providing that that sequence is finite*).

> The preceding list focuses only on phenomena and issues that revolve around automatic digital computation and its mathematical underpinnings because of this book's focus. As noted above, we could compile analogous lists for many other subject areas, including almost any area that shares computation theory's (indeed, computer science's) focus on string-based representations of its items of interest. We have already referred numerous times to mathematical logic (via the work of Gödel and others) as such a representation-focused area. Another obvious such area is the field of formal linguistics (via the work of Chomsky and others). Less obvious such areas include the representation-oriented subfields of algebra such as the theories of groups and semigroups (and thereby fields and rings). The reader who would like to learn more about the algebraic implications of the theory we will be developing under the "pillar" topic ENCODING might want to study sources such as [7, 66, 78] in parallel with Chapter 9.

As the preceding discussion suggests, the topic of encoding must play a very significant role in our development of computation theory. Importantly for that development, the mathematics needed to achieve the encodings that we need for our introductory study of computation theory is minimal. And do not worry that we are going to be taking only "baby steps" into the theory: We shall be studying each of the four encodings alluded to in the first paragraph. In a sense that will become very clear as soon as we begin to probe the technical portion of this "pillar" topic, a bare-bones kit bag of mathematical tools will suffice. Specifically, once we have (intellectual) access to the nonnegative integers (the set $\mathbb{N}$), we need only be able to add and multiply integers, and to test the equality of integer expressions formed using these operations.

From both historical and intellectual perspectives, it is interesting that our development of encodings and their underlying mathematics builds on a seminal study that does not obviously relate to encodings at all! Indeed, this foundational material, which appears in Chapter 7, was developed in order to answer the following question—which, note, never refers to representations. "Can one create a one-to-one association between the integer-coordinate points in the (two-dimensional) plane and the integer points on the line?" (We shall see that it is easy to rephrase this question in terms of the *rational* numbers, instead of integer pairs.) Not obviously, the answer is yes. All the other encodings that we shall need will be readily accessible from the one that leads to this affirmative answer. It turns out that this encoding power of positive integers has (literally!) foundation-shaking implications, especially when paired with the companion question, "Can one create a one-to-one association between the real numbers in the interval $\{x \mid 0 \leq x \leq 1\}$ and the integer points on the line?" Certainly not obvious, the answer to this second question is no. Among these foundation-shaking implications is the following. There exist very well defined specifications for programs and for digital computers that can *never* be realized! Specifically, no matter how technology advances, one will never be able to craft either programs or computers that meet these specifications!

The integer-related questions of the preceding paragraph—the one with the affirmative answer and the one with the negative answer—were both the brainchild of Georg Cantor. The implications of these questions for encodings of discrete structures by integers are generally attributed to Kurt Gödel in the case of formal logic and to Alan Turing in the case of digital computation.

Let's begin to flesh out the promises of the preceding paragraphs by developing the machinery for encodings and proofs of nonencodability and by developing the implications of these results.

Chapter 7
Countability and Uncountability: The Precursors of "Encoding"

The theory of countability (and uncountability) began with Georg Cantor's deliberations on the nature of infinity [10]. Cantor concentrated on questions that can be framed intuitively as follows: Are there "more" rational numbers than integers? Are there "more" real numbers than integers? (Note that we need to put "more" in quotes, because all three sets, the integers, the rational numbers, and the real numbers, are infinite—so what does "more" mean?)

> Let us digress a bit so that we can appreciate these questions.
>
> From one perspective, one might intuit that there are "more" rationals than integers and "more" reals than rationals. Here is the evidence for this position. There are only finitely many integers between every two integers; there are infinitely many rational numbers between every two integers; there are infinitely many real numbers between every two rational numbers.
>
> There is some counter-evidence, though. First, consider the integers and the rational numbers. Since every rational number is a quotient of two integers, there do not seem to be "so many" more rational numbers than integers; indeed, there are certainly no more rational numbers than ordered pairs of integers! With this insight, one can start "playing around" with the integer-pair coordinates that name the so-called *integer lattice points* in the two-dimensional plane (which, we have just noted, subsume the rational numbers). By such "playing around," we find linear listings of the integer lattice points that make it plausible that there are "equally many" integers and rational numbers. One linear listing that is particularly pleasing aesthetically is given by the function (7.3) that we shall study a bit in Chapter 8. Second, consider the rational numbers and the real numbers. Our earlier argument about infinitely many real numbers between every two rational numbers is clearly flawed—for there also are infinitely many *rational* numbers between every two real numbers. Hence, this "density" argument clearly tells us nothing definitive about the rationals vs. the reals. Indeed, we shall see that the argument tells us nothing useful even about the rationals and the integers!

Since we are concerned with the set of computable functions rather than any set of numbers, we actually develop technically simpler analogues of these questions. But the tools that we develop here—which are, essentially, the ones that Cantor used to answer his questions—can be adapted in very simple ways to answer Cantor's questions directly.

A.L. Rosenberg, *The Pillars of Computation Theory*, Universitext, DOI 10.1007/978-0-387-09639-1_7, © Springer Science+Business Media, LLC 2010

In order to start thinking about Cantor's questions, we must find a formal, precise way to talk about one infinite set's having "more" elements than another. We would like this way to be an *extension* of how we make this comparison with finite sets. In other words, if we apply this mechanism to a pair of sets, one infinite and the other finite, then we demand that the mechanism tell us that the infinite set has "more" elements than the finite set; and if we apply this mechanism to a pair of finite sets, then we demand that it tell us that the bigger set has "more" elements than the smaller. Here is the simple mechanism that Cantor devised and that we use.

Let A and B be (possibly infinite) sets. We write

$$|A| \leq |B| \qquad (7.1)$$

just when there is an *injection*

$$f : A \xrightarrow{\text{1-1}} B$$

that maps A *one-to-one* into B.

When A is a finite set, expression (7.1) as we have defined it means: *set A has no more elements than does set B.* It is important *not* to read the assertion that way when A is infinite, since "more" is not defined in that case. In fact, since we nowhere define the "cardinality" of an infinite set A—we don't need such a definition for our purposes in this chapter—one should not read "$|A|$" as "the cardinality of A" (even though that is how it can be read when A is finite), nor should one read (7.1) as asserting something like "the cardinality of A does not exceed the cardinality of B" (again, even though a reading of that sort is fine when A is finite). Because of the danger of thinking about infinite sets in terms that really apply only to finite sets, and thereby generating fallacies, I always strongly recommend that my students view the string of symbols (7.1) as a sentence that should be read silently and nonverbally, but that should not be verbalized, particularly not using the finitistic words we are all accustomed to. (You can always create your own name for $|A|$ and then read (7.1) as loud as you want—but you'd better give your audience access to your expanded lexicon.)

Is the preceding formal definition an extension of the familiar relation $\leq$ on the cardinalities of finite sets? The hallmark of an injection such as f is that given any $b \in B$, there is *at most one* $a \in A$ such that $f(a) = b$. This means, whenever set A is *finite*, that set B has at least as many elements as A does. Thus, we do, indeed, have an extension of the finite situation.

Of course, the preceding formal definition, being precise, would make sense even if it did not provide the desired extension of the finite situation. Therefore, we *could* just take the definition as written and start to study it mathematically: the resulting theorems would, indeed, be theorems. They would just not be as impressive an accomplishment as Cantor in fact achieved—for his definition in fact put infinite sets on an equal footing with finite sets, at least with regard to questions about the relative "sizes" of sets.

When

$$|A| \leq |B| \quad \textbf{and} \quad |B| \leq |A|,$$

we write

$$|A| = |B|. \qquad (7.2)$$

Since we have still not defined the notation $|A|$ for arbitrary sets, the assertion in (7.2) is another one that should be read nonverbally.

By arguing about composition of injections, the reader should prove the following fundamental results about the relations in (7.1) and (7.2). The importance of Lemma 7.1 is that it shows that Cantor's extensions of the relations $\leq$ and $=$ from the domain of (the cardinalities of) finite sets to that of infinite sets obey the rules that allow one to infer the existence of certain important relations from the existence of others. As one important example (you can supply others), if for given sets A, B, and C, we have both $|A| \leq |B|$ and $|B| \leq |C|$, then we know by Lemma 7.1(a) that $|A| \leq |C|$.

Lemma 7.1. (a) *The relation "$\leq$" of (7.1) is reflexive and transitive.*
(b) *The relation "$=$" of (7.2) is an equivalence relation.*

We single out the important case $B = \mathbb{N}$. When

$$|A| \leq |\mathbb{N}|$$

we say that the set A is *countable*. When

$$|A| = |\mathbb{N}|$$

we say that the set A is *countably infinite*.

The following important result of Ernst Schröder [95, 96] and Felix Bernstein [4] plays a crucial role in the study of countability and related topics (such as encodings).

Theorem 7.1. (The Schröder–Bernstein theorem) *If $|A| = |B|$, then there is a* bijection, *i.e., one-to-one, onto function,*

$$f : A \xrightarrow{1\text{-}1,\text{onto}} B.$$

The Schröder–Bernstein theorem is quite easy to prove for finite sets but is decidedly nontrivial for infinite ones. (See an algebra text such as [5] for the theorem's proof, which is beyond the scope of this book.)

In deference to our intended use of the bijections promised by Theorem 7.1, we henceforth view these functions as *encoding* (elements of) set A as (elements of) set B, rather than just mapping (elements of) set A to (elements of) set B. While this is more a change of viewpoint than of substance, it may help you to start thinking in a computation-theoretic way. In order to emphasize the proposed change of viewpoint, we henceforth call such functions *encoding functions*.

7.1 Encoding Functions and Proofs of Countability

This section is devoted to establishing the countability of a variety of infinite sets that we shall use repeatedly as we develop the material related to the "pillar" topic ENCODING.

Theorem 7.2. *The following sets are countable:*

1. $\Sigma^\star$, *for any finite set* Σ.
2. *The set of all* finite *subsets of* $\mathbb{N}$.
3. $\mathbb{N}^\star$. *Note that this set includes all sets* $\mathbb{N} \times \mathbb{N} \times \mathbb{N} \times \cdots \times \mathbb{N}$, *where the product is performed any finite number of times.*

We focus on each of the theorem's three sets in the next three lemmas.

Lemma 7.2. *The set* $\Sigma^\star$ *is countable for any finite set* Σ.

Proof. For our purposes, $\Sigma^\star$ is the most important of the theorem's three sets to prove countable, because *every program in any programming language is a finite string over some finite alphabet*. Because a function must be programmable in order to be computable, we shall, therefore, have the following important corollary to our proof.

Corollary 7.1 *The set of computable functions is countable.*

Focus on a finite set $\Sigma = \{\sigma_1, \sigma_2, \ldots, \sigma_n\}$. The easiest way to prove the countability of $\Sigma^\star$ is to interpret every string over Σ as a base-$(n+1)$ numeral, where $n = |\Sigma|$.

We use $n + 1$, rather than n, as the base in order to avoid the vexatious problem of leading 0's. We would like every string, viewed as a numeral, to represent a distinct integer. Leading 0's prevent this, as one can see from the fact that the distinct numerals 1, 01, 001, and so on, all denote the integer "one." So, we sneak around this annoying but insubstantial problem by having our encoding function avoid leading 0's.

To implement our proof strategy, consider the function

$$f_\Sigma : \Sigma^\star \longrightarrow \mathbb{N}$$

that is defined as follows. Order the elements of Σ in any way, so that we can refer unambiguously to the "kth" element of Σ. Associate each $\sigma \in \Sigma$ with the integer assigned via this ordering; denote this integer by $|\sigma|$. For instance, if σ is the "kth" element of Σ, then we denote this fact by the notation $|\sigma| = k$. Then define the value of f_Σ on each string $\sigma_{i_m} \sigma_{i_{m-1}} \cdots \sigma_{i_1} \in \Sigma^\star$ as follows:

$$f_\Sigma(\sigma_{i_m} \sigma_{i_{m-1}} \cdots \sigma_{i_1}) \stackrel{\text{def}}{=} \sum_{j=1}^{m} |\sigma_{i_j}| (n+1)^{j-1}.$$

Because every numeral that contains no 0's specifies a unique integer in any "–ary" positional number system (binary, ternary, octal, decimal, etc.), the function f_Σ is one-to-one; hence f_Σ witnesses the countability of Σ^*. $\square$

An aside. Although we all "grew up" using "–ary" positional number systems, there are alternative systems that are interesting in various contexts. One example is the "–adic" family of positional number systems. Just as each "–ary" system has a name derived from Latin, each "–adic" system has a name derived from Greek ("dyadic" instead of "binary" for base-2, "triadic" instead of "ternary" for base-3, and so on.) Focus on base-2 for definiteness; higher bases behave analogously. The base-2 numeral

$$\beta \overset{\text{def}}{=} \beta_n \beta_{n-1} \cdots \beta_1 \beta_0$$

(each $\beta_i \in \{0,1\}$) represents the integer

$$f^{(\text{binary})}(\beta) = \sum_{i=0}^{n} \beta_i \cdot 2^i$$

in the binary system, and it represents the integer

$$f^{(\text{dyadic})}(\beta) = \sum_{i=0}^{n} (\beta_i + 1) \cdot 2^i$$

in the dyadic system. A nice feature of the dyadic system is that every numeral represents a distinct number. This yields an even simpler proof of the countability of Σ^* for any finite Σ. An inconvenience of the dyadic system is that it has no numeral for the number 0.

Lemma 7.3. *The set of all* finite *subsets of* $\mathbb{N}$ *is countable.*

Proof. We build on the just-established countability of all sets Σ^* to prove that the set of all *finite* subsets of $\mathbb{N}$ is countable. Technically, we use a mapping-based technique called *reducing one problem to another*, which is exceedingly important in all of mathematics, and is particularly central to the theories of both computability and computational complexity. Specifically, we now reduce the problem of establishing the countability of the set of finite subsets of $\mathbb{N}$ to the problem of establishing the countability of Σ^*. We begin by recalling the following notion from Chapter 2.

The *characteristic vector*, $\beta(S)$, of a finite set $S \subset \mathbb{N}$ is the following binary string, whose length is 1 greater than the maximum integer in S, call this integer $\max(S)$:

$$\beta(S) \overset{\text{def}}{=} \delta_0 \delta_1 \cdots \delta_{\max(S)},$$

where for each $i \in [0, \max(S)]$,

$$\delta_i = \begin{cases} 1 \text{ if } i \in S, \\ 0 \text{ if } i \notin S. \end{cases}$$

We thus see that characteristic vectors afford us an injection

$$g : [\text{Finite subsets of } \mathbb{N}] \xrightarrow{\text{1-1}} \{0,1\}^*.$$

It follows by the definition of the unpronounceable sentence (7.1) that

$$\text{|Finite subsets of } \mathbb{N}| \ \leq \ |\{0,1\}^*|.$$

Because we already know that the set $\{0,1\}^*$ is countable, and because the relation "$\leq$" is transitive (Lemma 7.1), we conclude that the set of finite subsets of $\mathbb{N}$ is countable. $\square$

Lemma 7.4. *The set $\mathbb{N}^*$ of all finite integer sequences is countable.*

Proof. The ploy of encoding the objects of interest as numerals will not work here, because there is no natural candidate for the base of the number systems; no finite base will work in a straightforward way. Therefore, we call in two pieces of heavier mathematical machinery to accomplish our task.[1] The first piece of machinery is the following theorem, which is traditionally attributed to the Greek mathematician Euclid.

Theorem 7.3. (Euclid's theorem on primes) *There are infinitely many primes.*

Euclid's theorem on primes can be proved by showing that any finite set of primes $\{p_1, p_2, \ldots, p_n\}$ is inadequate to provide a prime factorization for the positive integer $1 + \prod_{i=1}^{n} p_i$. We leave the completion of the proof to the reader.

The second piece of mathematical machinery is the *fundamental theorem of arithmetic*, which is sometimes called the *prime-factorization theorem*.

Theorem 7.4. (The fundamental theorem of arithmetic) *Every integer $n > 1$ can be represented as a product of primes in one and only one way, up to the order of the primes in the product.*

We use the preceding two theorems to establish the injectivity of a specific function $h : \mathbb{N}^* \rightarrow \mathbb{N}$; this injectivity verifies the countability of $\mathbb{N}^*$. We define the function h as follows. For any finite sequence $m_1, m_2, \ldots, m_k$, of nonnegative integers,

$$h(m_1, m_2, \ldots, m_k) \ = \ \prod_{i=1}^{k} p_i^{m_i},$$

where for each $i \in [1,k]$, p_i is the ith smallest prime. By the fundamental theorem of arithmetic, the function h assigns a unique integer to each sequence of integers $m_1, m_2, \ldots, m_k$; hence h is an injection of $\mathbb{N}^*$ into $\mathbb{N}$, whence the former set is countable. $\square$

In addition to meeting our primary goal, namely, establishing the countability of $\mathbb{N}^*$, the function h can clearly be used to establish the countability of each *finite* cross product $\mathbb{N} \times \mathbb{N} \times \cdots \times \mathbb{N}$. For instance, when we restrict h to the set of ordered pairs $\mathbb{N} \times \mathbb{N}$, we obtain the injection

---

[1] As with other mathematical tools, we do not prove these results here; see, e.g., [5].

$$h(m_1, m_2) = 2^{m_1} \cdot 3^{m_2}$$

that maps $\mathbb{N} \times \mathbb{N}$ one-to-one into $\mathbb{N}$; and when we restrict h to the set of ordered triples $\mathbb{N} \times \mathbb{N} \times \mathbb{N}$, we obtain the injection

$$h(m_1, m_2, m_3) = 2^{m_1} \cdot 3^{m_2} \cdot 5^{m_3}$$

that maps $\mathbb{N} \times \mathbb{N} \times \mathbb{N}$ one-to-one into $\mathbb{N}$. However, there is something unaesthetic about using the function h for these finite cross products, namely, the *sparseness* of the set of integers that occur as images of h's argument vectors of integers. When we use h to encode ordered pairs of integers as integers, for instance, we use just 2 and 3, among the infinitude of available primes, to build the image integers—thereby "wasting" all the rest of the primes. There must be better encoding functions for these finite cross products! And indeed there are.

Of course, all of the finite cross products $\mathbb{N} \times \mathbb{N} \times \cdots \times \mathbb{N}$ are infinite sets. By the Schröder–Bernstein theorem, therefore, there exist *bijections*

$$f_2 : \mathbb{N} \times \mathbb{N} \xrightarrow{\text{1-1,onto}} \mathbb{N}$$
$$f_3 : \mathbb{N} \times \mathbb{N} \times \mathbb{N} \xrightarrow{\text{1-1,onto}} \mathbb{N}$$
$$\vdots$$

Of course, being *bijections*, each of these functions employs *all* of the integers as images. In fact, in the case of these special sets, the bijections advertised by the Schröder–Bernstein theorem actually have simple forms. We now present one such bijection for $\mathbb{N} \times \mathbb{N}$, just to indicate its charming form. One can easily build the following "pairing" function $\mathscr{D}$ (in several ways) into a "tripling" function and a "quadrupling" function, and so on.

$$\mathscr{D}(x,y) = \binom{x+y+1}{2} + y = \frac{1}{2}(x+y)(x+y+1) + y. \tag{7.3}$$

(Clearly, there is a twin of $\mathscr{D}$ that interchanges x and y.)

The bijection $\mathscr{D}$ is usually attributed to Cantor in work from the last quarter of the nineteenth century [10], but there is evidence that it was known already to Augustin Cauchy in the first quarter of that century [11].

We have dubbed the "pairing" function $\mathscr{D}$ of (7.3) the *diagonal* pairing function (whence its name "$\mathscr{D}$") for the following reason. If we illustrate $\mathbb{N} \times \mathbb{N}$ as follows,

$$
\begin{array}{ccccc}
(0,0) & (0,1) & (0,2) & (0,3) & (0,4) \cdots \\
(1,0) & (1,1) & (1,2) & (1,3) & (1,4) \cdots \\
(2,0) & (2,1) & (2,2) & (2,3) & (2,4) \cdots \\
(3,0) & (3,1) & (3,2) & (3,3) & (3,4) \cdots \\
(4,0) & (4,1) & (4,2) & (4,3) & (4,4) \cdots \\
\vdots & \vdots & \vdots & \vdots & \vdots & \ddots
\end{array}
$$

then the action of the bijection $\mathscr{D}$ can be seen in the following illustration, where each position (x, y) contains $\mathscr{D}(x, y)$, and where the "diagonal" $x + y = 4$ is highlighted:

$$
\begin{array}{ccccccccc}
0 & 2 & 5 & 9 & \boxed{14} & 20 & 27 & 35 & \cdots \\
1 & 4 & 8 & \boxed{13} & 19 & 26 & 34 & 43 & \cdots \\
3 & 7 & \boxed{12} & 18 & 25 & 33 & 42 & 52 & \cdots \\
6 & \boxed{11} & 17 & 25 & 33 & 42 & 52 & 63 & \cdots \\
\boxed{10} & 16 & 23 & 31 & 40 & 50 & 61 & 73 & \cdots \\
15 & 22 & 30 & 39 & 49 & 60 & 72 & 85 & \cdots \\
21 & 29 & 38 & 48 & 59 & 71 & 84 & 98 & \cdots \\
28 & 37 & 47 & 58 & 70 & 83 & 97 & 112 & \cdots \\
\vdots & \vdots & \vdots & \vdots & \vdots & \vdots & \vdots & \vdots & \ddots
\end{array}
$$

The preceding illustration lends us the intuition to prove that d is, indeed, a bijection. To wit:

- Along each "diagonal" of $\mathbb{N} \times \mathbb{N}$—which corresponds to a fixed value of $x + y$—the value of $\mathscr{D}(x, y)$ increases by 1 as x decreases (by 1) and y increases (by 1).
- As we leave diagonal $x + y$—at node $(0, x + y)$—and enter diagonal $x + y + 1$—at node $(x + y + 1, 0)$—the value of $\mathscr{D}(x, y)$ increases by 1, because

$$
\begin{aligned}
\mathscr{D}(x+y+1, 0) &= \binom{x+y+2}{2} \\
&= \frac{(x+y+1)(x+y+2)}{2} \\
&= \frac{(x+y)(x+y+1) + 2(x+y+1)}{2} \\
&= \frac{(x+y)(x+y+1)}{2} + (x+y) + 1 \\
&= \binom{x+y+1}{2} + (x+y) + 1 \\
&= \mathscr{D}(0, x+y) + 1.
\end{aligned}
$$

Note that the various injections that we have used to prove the countability of sets—numeral evaluation in an "–ary" number system, encodings via powers of primes, the Cauchy–Cantor polynomial—are eminently computable. The importance of this fact is that we can use the functions as *encoding mechanisms*. In other words:

We can now formally and rigorously encode "everything"—programs, data structures, data— as strings (say, for definiteness, binary strings) or as integers.

On the one hand, this gives us tremendous power, by "flattening" out our universe of discourse; henceforth, we can discuss *only* functions from $\mathbb{N}$ to $\mathbb{N}$, without losing any generality. On the other hand, we must be very careful from now on, because as we apparently discuss and manipulate integers, we are also—via the appropriate encodings—discussing and manipulating programs and computations, etc. We shall see before long the far-reaching implications of this new power.

A brief survey of surprising uses of encoding functions (using their old name, "pairing functions") can be found in Chapter 8.

7.2 Diagonalization: Proofs of Uncountability

Just as the major ideas underlying proofs of countability/encodability had their seeds in the work of G. Cantor, so also do the major ideas underlying proofs of uncountability/unencodability. This section is devoted to developing a proof technique called *diagonalization* (or more fully, *Cantor's diagonalization argument*), that is the primary tool for almost all of the negative proofs regarding countability and encodability in (at least the introductory parts of) the theories of computability and computational complexity. Cantor developed his diagonalization argument to prove that certain sets—notably, the real numbers—are not countable. Our interest in the argument stems from its usefulness in establishing the noncomputability of certain functions and/or the existence of functions whose computational complexity exceeds certain limits. (A function's complexity can be measured, e.g., in terms of the time requirements of any program that computes the function or of the memory requirements of any such program.) The latter, complexity-theoretic, role of diagonalization is hard to talk about until we establish a framework for studying the complexity of computation; in contrast, we shall have our first negative computability-theoretic result by the end of this section!

Theorem 7.5. *The following sets are not countable (or, equivalently, are uncountable):*

1. the set of functions $\{f : \mathbb{N} \longrightarrow \{0,1\}\}$;
2. the set of all subsets of $\mathbb{N}$;
3. the set of (countably) infinite binary strings.
4. the set of functions $\{f : \mathbb{N} \longrightarrow \mathbb{N}\}$;

Proof. Once we establish the uncountability of the set of *binary-valued* functions—i.e., the set $\{f : \mathbb{N} \longrightarrow \{0,1\}\}$—we can immediately infer the uncountability of the set of *integer-valued* functions—i.e., the set $\{f : \mathbb{N} \longrightarrow \mathbb{N}\}$—for the former set can be mapped into the latter via the identity function, which is clearly an injection. (You should carefully verify this consequence of the transitivity of "$\leq$" [as exposed in Lemma 7.1].) We therefore leave assertion 4 of the theorem to the reader.

Somewhat surprisingly, we can attack the other three parts of the theorem, namely, assertions 1, 2, and 3, in tandem. This is because we can encode/represent any of the three sets in question as any of the others! *In some sense, the three sets are just different interpretations of a single set.* Let's consider in detail why the highlighted sentence is true, because the ability to represent/view one type of object as an ostensibly quite distinct other type is central to the "pillar" topic ENCODING.

The basic observation underlying our identification of the sets in assertions 1–3 of Theorem 7.5 is that one can represent either of the following objects,

- any subset of $\mathbb{N}$,
- any function $f : \mathbb{N} \longrightarrow \{0, 1\}$,

as a unique (countably) infinite binary string.

To establish the preceding claim, focus first on a fixed, but arbitrary, set $A \subseteq \mathbb{N}$. (A can be as small as the empty set $\emptyset$ or as big as $\mathbb{N}$; it can be finite or infinite.) We can represent the set A by its characteristic vector (as defined in Chapter 2). Because each subset of $\mathbb{N}$ is represented thereby by a unique countably infinite binary string, and each countably infinite binary string represents a unique subset of $\mathbb{N}$, we thus have

$$|\{\text{subsets of } \mathbb{N}\}| \;=\; |\{\text{countably infinite binary strings}\}|. \qquad (7.4)$$

Focus next on a fixed, but arbitrary, function $f : \mathbb{N} \longrightarrow \{0, 1\}$. One can represent the function f by the unique countably infinite binary string comprising the sequence of values of f on successive integers, namely, the string

$$f(0) \, f(1) \, f(2) \, \cdots$$

whose kth bit-value is $f(k)$. Because each function $f : \mathbb{N} \longrightarrow \{0, 1\}$ is represented thereby by a unique countably infinite binary string, and each countably infinite binary string represents a unique binary-valued function, we thus have

$$|\{f : \mathbb{N} \longrightarrow \{0, 1\}\}| \;=\; |\{\text{countably infinite binary strings}\}|. \qquad (7.5)$$

In the presence of (7.4) and (7.5), Lemma 7.1 assures us that the uncountability of any of the sets in assertions 1–3 of Theorem 7.5 implies the uncountability of all three sets. So, let us concentrate on the set $\mathscr{B}$ of (countably) infinite binary strings and prove that $\mathscr{B}$ is uncountable. The proof is by contradiction.

Assume, for contradiction, that the set $\mathscr{B}$ is countable, so that $|\mathscr{B}| \leq |\mathbb{N}|$.

We note first that $|\mathbb{N}| \leq |\mathscr{B}|$, because $\mathscr{B}$ is an infinite set. To see this, just focus on the subset of $\mathscr{B}$ that comprises the infinite characteristic vectors of the *singleton* sets $\big\{\{k\} \mid k \in \mathbb{N}\big\}$. The infinite characteristic vector of each such set, $\{m\}$, is the infinite binary string that consists of all 0's, except for precisely one 1, in bit-position m.

Combining the preceding proof that $|\mathbb{N}| \leq |\mathscr{B}|$ with the assumed fact that $|\mathscr{B}| \leq |\mathbb{N}|$, we have $|\mathscr{B}| = |\mathbb{N}|$.

Worded in a way that might be clearer to some readers: Given our proof that $|\mathbb{N}| \leq |\mathscr{B}|$, if it were true that $|\mathscr{B}| \leq |\mathbb{N}|$, then we would have $|\mathscr{B}| = |\mathbb{N}|$. Therefore, assuming that $|\mathscr{B}| \leq |\mathbb{N}|$ is equivalent to assuming that $|\mathscr{B}| = |\mathbb{N}|$.

By the Schröder–Bernstein theorem (Theorem 7.1), there therefore exists a *bijection*

$$h : \mathscr{B} \xrightarrow{\text{1-1,onto}} \mathbb{N}.$$

It is not hard to view the bijection h as producing an "infinite-by-infinite" binary matrix Δ, whose kth row is the infinite binary string $h^{-1}(k)$. Let's visualize Δ:

$$\Delta = \begin{matrix} \delta_{0,0} & \delta_{0,1} & \delta_{0,2} & \delta_{0,3} & \delta_{0,4} & \cdots \\ \delta_{1,0} & \delta_{1,1} & \delta_{1,2} & \delta_{1,3} & \delta_{1,4} & \cdots \\ \delta_{2,0} & \delta_{2,1} & \delta_{2,2} & \delta_{2,3} & \delta_{2,4} & \cdots \\ \delta_{3,0} & \delta_{3,1} & \delta_{3,2} & \delta_{3,3} & \delta_{3,4} & \cdots \\ \delta_{4,0} & \delta_{4,1} & \delta_{4,2} & \delta_{4,3} & \delta_{4,4} & \cdots \\ \vdots & \vdots & \vdots & \vdots & \vdots & \ddots \end{matrix}$$

Now let us construct the following infinite binary string:

$$\Psi = \psi_0\ \psi_1\ \psi_2\ \psi_3\ \psi_4 \cdots,$$

by setting, for each index i,

$$\psi_i = \overline{\delta}_{i,i} = 1 - \delta_{i,i}.$$

The term "*diagonal argument*" for this proof stems from the fact that we create the new string Ψ by making changes to the *diagonal* elements of the matrix Δ.

Note that the new string, Ψ, clearly does not occur as a row of matrix Δ. This is because Ψ differs from each row of Δ in at least one position. Specifically, for each index i, Ψ differs from row i of Δ at least in position i: $\psi_i \neq \delta_{i,i}$. But if the binary string Ψ does not occur as a row of matrix Δ, then Δ *does not* contain *every* infinite binary string as one of its rows—which contradicts Δ's assumed defining characteristic!

Where could we have gone wrong? Every step of our argument, save one, is backed up by a theorem—so the one step that is not so bolstered must be the link that has broken. This one unsubstantiated step is our assumption that the set $\mathscr{B}$ is countable. Since this assumption has led us to a contradiction, we must conclude that the set $\mathscr{B}$ is *not* countable!

We thus have the uncountability of the four sets enumerated in Theorem 7.5, which completes the proof of the theorem. $\quad\square$

7.3 Where Has (Un)countability Led Us?

For the purposes of our developing the highlights of computation theory, the most important consequence of our work in this section is the following.

Corollary 7.2 *Because the set of (0-1 valued) integer functions is uncountable, while the set of programs is countable, there must exist noncomputable (0-1 valued) integer functions.*

As we promised in Section 5.3, we now urge the reader to return to Lemma 5.3 in that section. You now have ample background to understand that result's proof!

Chapter 8
Enrichment Topic: "Efficient" Pairing Functions, with Applications

—Entia non sunt multiplicanda praeter necessitatem.

(Occam's razor, William of Occam)

We have already seen the preceding admonition by William of Occam (fourteenthth century), in Section 6.1.2. The principle that underlies the admonition—always to strive for simplicity—is particularly worth heeding when one seeks mathematical models of computational phenomena, for it is always tempting to embellish one's models with "real" features of the phenomenon or structure being modeled.

> The (mathematical) success of models such as the Turing machine testifies eloquently to how far a truly bare-bones model can take you, even when studying sophisticated notions such as computation; cf. [104] and Section 3.3.

Within the spirit of Occam's razor, this chapter provides a (very) short guided tour through the world of *pairing functions*—bijections between $\mathbb{N}^+ \times \mathbb{N}^+$ and $\mathbb{N}^+$—as tools for reducing the representational complexity of complex computational "situations." In a word, pairing functions allow one to represent families of structures that seem inherently to have multidimensional structure as sets of integers. We illustrate the benefits that can accrue from such simplification via two examples:

1. It is not clear how to devise *efficient* mappings into computer memory (or computer storage) for multidimensional arrays/tables that can change their shapes dynamically, i.e., at run time. Focus, for instance, on an *extendible* $2^n \times n$ table (whch could represent a relation in a database). Say that one wanted to add a row to the table in the course of a calculation. (Many programming languages allow such dynamic changes to the dimensions of multidimensional arrays/tables.) How should this change in the "logical" table be accommodated in its "physical" storage layout? A naive approach would relocate/remap the entire new table, but this option—*which is the one adopted by many programming-language implementations*—would force one to remap roughly $n2^n$ table entries in order to accommodate a change to only n entries! What is the alternative? Section 8.3 is devoted to a sophisticated approach to storage mappings that is based on the

A.L. Rosenberg, *The Pillars of Computation Theory*, Universitext,
DOI 10.1007/978-0-387-09639-1_8, © Springer Science+Business Media, LLC 2010

use of pairing functions and that obviates reallocating any already-stored table entries.

2. The Internet has given rise to new modalities of computing wherein the owners of computers "volunteer" their computing resources to others, for reasons ranging from curiosity to charity to the hope of compensatory computing support; examples appear in [8, 16, 54], among other sources. One hallmark of many of these Internet-based computing projects is that the participants are unknown to one another, hence untrusted. Indeed, David Anderson, the director of the well-known SETI@home project [54] has been reported as saying:

> "Fifty percent of the project's resources have been spent dealing with security problems ... the really hard part has to do with verifying computational results." The report said that Anderson went on to elaborate: "Seti@home software had been hacked – some were malicious, others not – to make it run faster, to spoof positive results and to make it look [as if] more work had been performed to improve leader board rankings."
>   http://www.wired.com/news/technology

Given a project with such experiences, one could imagine a desire to keep track of which "volunteers" produced which results, so that one could ban repeat offenders from subsequent participation in the project. Putting aside thorny issues such as how to reliably identify "volunteers" (IP addresses can easily be spoofed), one is left with the challenge of efficiently associating "volunteers" with the results they have produced. In Section 8.4, we somewhat simplify this *accountability* problem, by using pairing functions to translate the (volunteer, result-index) ordered pairs to single indices.

Of course, we are not going to abandon the reader with just a couple of proposed uses for pairing functions in modern computing settings. We shall, in fact, embellish these proposals with discussions of how to craft pairing functions that are particularly appropriate—mostly in terms of some notion of efficiency—for the proposed application area. We shall further augment these "applied" enrichment topics with one "pure" one: a short discussion of the computationally simplest pairing functions, the Cauchy–Cantor "diagonal" polynomials that we introduced in Section 7.1, via the specification (7.3).

8.1 Background

Because we shall be using the phrase "pairing function" so often in this chapter, we shall henceforth abbreviate the phrase—just for this chapter—by "PF."

As we have described earlier, PFs have played a major role in a variety of studies that are "classical" within the context of computation theory. They played a pivotal role in Cantor's seminal study of infinities [10], supplying a rigorous formal basis for asserting the counterintuitive "equinumerousness" of the integers and the rationals. It took revolutionary thinkers such as Gödel and Turing to recognize that the correspondences embodied by PFs can be viewed as *encodings*, or *translations*, of ordered pairs (and thence of arbitrary finite tuples or strings) as integers. This insight allowed Gödel and Turing to build on the existence of studies of, respectively, logical systems [30] and eminently computable—indeed, easily computed—PFs in their famous algorithmic systems [104]. The uses we propose for PFs in this chapter, while certainly less profound than those of Gödel and Turing, also build on the insight that PFs can be used as encoding mechanisms, specifically allowing one to slip gracefully, yet formally, among the worlds of strings, integers, and tuples of integers.

Throughout the chapter, we illustrate selected values from selected PFs using the following convention. We illustrate a PF $\mathscr{F} : \mathbb{N}^+ \times \mathbb{N}^+ \leftrightarrow \mathbb{N}^+$ via a two-dimensional array whose entries are the values of $\mathscr{F}$ as described in Figure 8.1.

$$
\begin{array}{|c|c|c|c|c|c}
\mathscr{F}(1,1) & \mathscr{F}(1,2) & \mathscr{F}(1,3) & \mathscr{F}(1,4) & \mathscr{F}(1,5) & \cdots \\
\mathscr{F}(2,1) & \mathscr{F}(2,2) & \mathscr{F}(2,3) & \mathscr{F}(2,4) & \mathscr{F}(2,5) & \cdots \\
\mathscr{F}(3,1) & \mathscr{F}(3,2) & \mathscr{F}(3,3) & \mathscr{F}(3,4) & \mathscr{F}(3,5) & \cdots \\
\mathscr{F}(4,1) & \mathscr{F}(4,2) & \mathscr{F}(4,3) & \mathscr{F}(4,4) & \mathscr{F}(4,5) & \cdots \\
\mathscr{F}(5,1) & \mathscr{F}(5,2) & \mathscr{F}(5,3) & \mathscr{F}(5,4) & \mathscr{F}(5,5) & \cdots \\
\vdots & \vdots & \vdots & \vdots & \vdots & \ddots
\end{array}
$$

Fig. 8.1 *Our generic template for sampling from a PF.*

8.2 The Prettiest Pairing Function(s)

8.2.1 The Diagonal PF $\mathscr{D}(x,y)$

As we noted early in Chapter 7, it is not a priori obvious that there exist bijections between $\mathbb{N}^+ \times \mathbb{N}^+$ and $\mathbb{N}^+$. A reading of Cauchy's major tome [11] makes it likely that such bijections have been known to exist for (at least) close to two centuries.

> The uncertainty results from the fact that Cauchy describes a bijection—indeed, the one specified in (8.1)—only pictorially, without an accompanying analysis of bijectivity.

Indeed, it has been known for at least 125 years that there exist such bijections that are *polynomials!* Specifically, in [10], Cantor shows that the following function from $\mathbb{N}^+ \times \mathbb{N}^+$ to $\mathbb{N}^+$ is a bijection.

$$
\mathscr{D}(x,y) = \binom{x+y-1}{2} + y. \tag{8.1}
$$

(Of course, the bijection $\mathscr{D}$ has a bijective twin that is obtained by exchanging x and y in (8.1).) If one observes how the function $\mathscr{D}$ "labels" the two-dimensional integer lattice points with integers—as we begin to do in Figure 8.2—then one sees why we call $\mathscr{D}$ the *diagonal-shell* PF.

One proves that the function $\mathscr{D}$ is really a bijection between $\mathbb{N}^+ \times \mathbb{N}^+$ and $\mathbb{N}^+$ via a simple double induction that specifies in detail how $\mathscr{D}$ labels the integer lattice

1	3	6	10	15	21	28	36	⋯
2	5	9	14	20	27	35	44	⋯
4	8	13	19	26	34	43	53	⋯
7	12	18	25	33	42	52	63	⋯
11	17	24	32	41	51	62	74	⋯
16	23	31	40	50	61	73	86	⋯
22	30	39	49	60	72	85	99	⋯
29	38	48	59	71	84	98	113	⋯
⋮	⋮	⋮	⋮	⋮	⋮	⋮	⋮	⋱

Fig. 8.2 *The diagonal-shell PF $\mathscr{D}$. The shell $x+y=6$ is highlighted.*

points. (We provided the calculations for this argument in Section 7.1.) Specifically, $\mathscr{D}$ assigns integers to the lattice points in an "upward direction" along the successive "diagonal shells," $x+y=2$, $x+y=3$, $x+y=4$, $\ldots$. One induction establishes that all of the integer labels assigned to the lattice points of the diagonal shell $x+y=c$ are smaller than the labels assigned to the lattice points of every diagonal shell $x+y<c$. In other words,

$$\mathscr{D}(x,y) \; < \; \mathscr{D}(x',y')$$

whenever $x+y<x'+y'$. The second induction establishes that the label assigned to the topmost lattice point in the diagonal shell $x+y=c$ is precisely 1 less than the bottommost lattice point in the diagonal shell $x+y=c+1$; i.e.,

$$\mathscr{D}(x+y+1,1) \; = \; 1+\mathscr{D}(1,x+y).$$

The preceding two assertions are clearly adequate to prove that $\mathscr{D}$ is, indeed, a bijection.

One finds in [23] a proof of $\mathscr{D}$'s bijectivity that is computationally more detailed than ours, in that it develops an explicit recipe for computing $\mathscr{D}$'s inverse functions.

Note the plural form of the phrase "inverse function" here. $\mathscr{D}$ maps a *pair* of integers, (m,n), to a single integer, p; therefore, $\mathscr{D}$ needs a "left" inverse function that maps p to m, and a "right" inverse function that maps p to n.

For the interested reader, we present the explicit recipe from [23] for computing $\mathscr{D}$'s inverse functions. Let us denote by $\ominus$ the operation of *positive subtraction*, which is defined as follows. For all real numbers x and y,

$$x \ominus y \; = \; \begin{cases} x-y & \text{if } x \geq y, \\ 0 & \text{if } x < y. \end{cases}$$

To obtain $\mathscr{D}$'s inverse functions, define, for any integer $n \in \mathbb{N}^+$,

$$\Phi(n) = \left\lfloor \tfrac{1}{2}(\lfloor \sqrt{8n+1} \rfloor) \right\rfloor \ominus 1,$$

$$\Psi(n) = 2n \ominus (\Phi(n))^2.$$

It is shown in [23] that for all $x, y \in \mathbb{N}^+$,

$$x = \lfloor \tfrac{1}{2}(\Psi(\mathscr{D}(x,y)) \ominus \Phi(\mathscr{D}(x,y)))\rfloor,$$

$$y = \Phi(\mathscr{D}(x,y)) \ominus \lfloor \tfrac{1}{2}(\Psi(\mathscr{D}(x,y)) \ominus \Phi(\mathscr{D}(x,y)))\rfloor.$$

We do not present the proof from [23] because—in contrast to our less-detailed proof, whose skeleton we shall use to validate several alleged PFs in coming sections—the proof [23] is tightly coupled to the function $\mathscr{D}$, hence provides little insight for other PFs.

8.2.2 Is $\mathscr{D}(x,y)$ the Only Polynomial PF?

The diagonal-shell PF $\mathscr{D}$ attests to the existence of PFs that are conceptually and computationally simple. This is very satisfying to us as computation theorists. But $\mathscr{D}$ attests also to the existence of PFs that are *low-degree polynomials*—arguably as computationally simple and well structured as a function can get! This is enough to increase a mathematician's heart rate! Invariably, encountering such a pleasant surprise—a *low-degree polynomial PF*—leads a mathematically inclined person to question $\mathscr{D}$'s *uniqueness,* or lack thereof—as a *polynomial* PF: Are there others? In order to to avoid trivial answers to this question (in either direction), we must eliminate $\mathscr{D}$'s twin from contention. We therefore refine the question of $\mathscr{D}$'s uniqueness by considering as identical any two PFs that differ only in the interchange of x and y.

Its apparent simplicity notwithstanding, the question of $\mathscr{D}$'s uniqueness as a *polynomial* PF remains largely open. There are a few nontrivial beginnings to an answer, which we enumerate now.

1. There is no quadratic polynomial PF other than $\mathscr{D}$ (and its twin) [27].
2. The preceding assertion remains true if the "onto" condition for bijections is replaced by a "unit density" condition [57].
3. No cubic or quartic polynomial is a PF [58].
4. The development in [58] excludes large classes of higher-degree polynomials from being PFs. One simple example: a superquadratic polynomial whose coefficients are all positive cannot be a PF.

The mathematics underlying these results is beyond the scope of this book, combining elements of geometric number theory with nonstandard Diophantine analysis.

Diophantine analysis—named in honor of the Greek number theorist Diophantus who lived roughly 1700 years ago—focuses on a variety of questions concerning integer solutions to polynomial equations, typically of the form

- Do polynomial equations of such and such a type admit *any* integer solution?
- Do polynomial equations of such and such a type admit *infinitely many* integer solutions?

The questions needed to address the uniqueness question for the function $\mathscr{D}$ have the form

- Do polynomial equations of such and such a type admit *a unique* integer solution?

"Standard" Diophantine analysis does not address questions of uniqueness.

The proofs of the cited results focus on two types of situations: those in which a function $\mathscr{F}(x, y)$ in a given class (*a*) *grows too slowly* with its arguments, hence cannot be *injective* (one-to-one), or (*b*) *grows too fast* with its arguments, hence cannot be *surjective* (onto). For instance, one simple result from [58] shows that a superquadratic polynomial $\mathscr{F} : \mathbb{N}^+ \times \mathbb{N}^+ \to \mathbb{N}^+$ whose coefficients are all positive cannot be surjective. To wit, the number of two-dimensional integer lattice points within distance d of the origin grows quadratically with d (because of the *two*-dimensionality). Consequently, $\mathscr{F}$'s superquadratic growth must leave large gaps in its range, so it cannot be onto. The complication in extending this result to broader classes of superquadratic polynomials is that one must show that $\mathscr{F}$'s *lower-degree negative* terms do not lead to large "troughs" that capture all of the integers that its lead terms "jump over." The cases that must be eliminated grow quickly in number with the degree of $\mathscr{F}$— cf. [58]—hence yield a barrier to the ultimate resolution of whether there are any superquadratic PFs. In principle, this barrier is not impenetrable, but new ideas are likely needed in order to make further progress on the question.

8.3 Pairing Functions and the Storage of Extendible Arrays/Tables

It has been recognized since the 1950s (cf. [17, 29, 45]) that many classes of computations, arising in applications as diverse as linear-algebraic scientific computations (e.g., linear system solvers [29]) and relational databases [17], can be expressed most naturally with data organized in multidimensional arrays and tables. Moreover, many of these computations benefit—in ease of specification and execution—from the ability to *reshape* the arrays and tables dynamically. Relational databases, for instance, perform complex transactions by combining smaller tables to create bigger ones and by extracting smaller tables from bigger ones [17]. Scientific packages sometimes solve big complex computations on multidimensional arrays by building up the actual arrays of interest from smaller simpler ones [29]. To simplify the exposition, let us talk henceforth only about arrays, keeping in mind that we can easily adapt whatever we say to tables, because the only feature of arrays and tables that is germane to our discussion is that their positions can be labeled, or indexed, by a contiguous collection of integer lattice points.

Several programming languages afford a user easy mechanisms for specifying at least some types of reshapings of arrays—say, the addition and/or deletion of rows and/or columns in two dimensions. However, the mechanisms that most language processors use to implement even such simple reshapings are really quite naive: the processors completely remap an array each time that it is reshaped. This is, of course, very wasteful of time, since one does $\Omega(n^2)$ work to accommodate $O(n)$ changes to the array. Can one avoid this apparently wasteful remapping? Yes—provided that one maps arrays to storage using sophisticated mapping mechanisms, instead of the *dimension-order* mappings that have been the standard since the 1950s.

Dimension-order mappings operate as follows. Say that one wants to map an $m \times n$ array A onto the linear address space employed by virtually all computers. Since the days of FOR-TRAN 1, in the 1950s, one employed the *row-major* mapping function

$$\mathscr{F}(i, j) = A + (i - 1) + n \cdot (j - 1),$$

which stores each array position $A(i, j)$ via an offset from the base address A of the array. The mapping $\mathscr{F}$ works fine as long as the array retains its $m \times n$ "shape," but if one appends a column to array A, so that it adopts the new shape $m \times (n + 1)$, then the storage map $\mathscr{F}$ no longer assigns a unique storage slot to each array position; i.e., $\mathscr{F}$ is no longer injective on the set of positions of array A. To wit, we now have $\mathscr{F}(n + 1, 1) = n = \mathscr{F}(1, 2)$. In order to accommodate the new column while retaining the row-major regimen, one must replace the mapping $\mathscr{F}$ by the mapping

$$\mathscr{F}'(i, j) = A + (i - 1) + (n + 1) \cdot (j - 1),$$

which requires one to remap virtually the entire array A.

Note that the need to remap A does not occur when we append a new *row* to A; the mapping $\mathscr{F}$ continues to work in that case. Of course, a *column-major* mapping function would behave complementarily to a row-major function: it would gracefully accommodate new columns but not new rows. Thus, the challenge we focus on in this section arises only when one seeks storage mappings that accommodate *both* new rows *and* new columns.

Note that we are not claiming that the ability to append new rows and columns comes at no cost! Each evaluation of the mapping function—i.e., each function evaluation $\mathscr{F}(i, j)$—will be more complex using the strategy we are about to describe. However, in applications that involve much reshaping and relatively few probes of individual positions—such as relational databases—our storage mappings could be a good alternative to dimension-order mappings.

Where should one look for the kind of sophisticated storage mechanism that will avoid expensive remappings? It turns out that PFs can often serve as efficient storage-mapping functions for two-dimensional rectangular arrays, providing mappings that allow one to add and/or delete rows and/or columns dynamically, without ever remapping array/table positions that are unaffected by the reshaping. (We restrict our discussion to two-dimensional arrays to simplify the exposition. Simple techniques will extend this work to higher fixed dimensionalities.) This section surveys some of the results from [82, 83] on the use of PFs as storage mappings for extendible arrays.[1]

8.3.1 Array-Storage Mappings via Pairing Functions

Storing all square arrays extendibly. We begin our study of this topic with an example. Say that one is performing a large number of computations that all operate on *square* matrices (or tables)—i.e., matrices of dimensions $m \times m$ for some positive

---

[1] Our use of the word "extendible" follows [82, 83]; many sources prefer the term "extensible."

integer m. Say, moreover, that the computations build successive matrices by appending "shells" of new positions that expand an $m \times m$ matrix to an $(m+1) \times (m+1)$ matrix, thence to an $(m+2) \times (m+2)$ matrix, and so on. It turns out that there is a computationally simple PF—call it $\mathscr{A}_{1,1}$, for reasons that will become clear eventually—that can act as a storage mapping for *all* of the (square) matrices that one will ever use, and that morphs from being a storage mapping for an $m \times m$ matrix to a storage mapping for an $(m+1) \times (m+1)$ matrix without requiring one to remap any positions of the $m \times m$ matrix. This *square-shell* PF, $\mathscr{A}_{1,1}$ is specified by the following explicit expression.

$$\mathscr{A}_{1,1}(x,y) = m^2 + m + y - x + 1,$$
$$\text{where} \qquad m \overset{\text{def}}{=} \max(x-1, y-1). \tag{8.2}$$

As Figure 8.3 indicates, $\mathscr{A}_{1,1}$ maps integers in a counterclockwise direction along

1	4	9	16	25	36	49	64	$\cdots$
2	3	8	15	24	35	48	63	$\cdots$
5	6	7	14	23	34	47	62	$\cdots$
10	11	12	13	22	33	46	61	$\cdots$
17	18	19	20	21	32	45	60	$\cdots$
26	27	28	29	30	31	44	59	$\cdots$
37	38	39	40	41	42	43	58	$\cdots$
50	51	52	53	54	55	56	57	$\cdots$
$\vdots$	$\vdots$	$\vdots$	$\vdots$	$\vdots$	$\vdots$	$\vdots$	$\vdots$	$\ddots$

Fig. 8.3 *The square-shell PF $\mathscr{A}_{1,1}$. The shell $\max(x,y) = 5$ is highlighted.*

the *square shells* $m = \max(x-1, y-1) = 0$, $m = \max(x-1, y-1) = 1$, ... Having noticed this pattern, one verifies $\mathscr{A}_{1,1}$'s bijectivity on $\mathbb{N}^+ \times \mathbb{N}^+$ via a double induction that mirrors the one we used to validate the bijectivity of the diagonal-shell PF $\mathscr{D}$ of (8.1) and Figure 8.2. (Of course, $\mathscr{A}_{1,1}$ has a twin that proceeds in a *clockwise* direction along the square shells.)

Extendibly storing arrays of *any* single shape. It turns out that there is nothing special about diagonal shells or square shells with respect to extendibility: one can replicate, for shells of *any* fixed shape, the desirable support for extendibility that is illustrated in Figures 8.2 and 8.3—though usually not the simplicity of computation. We now explain and verify the preceding assertion, by providing a systematic process for constructing PFs that "favor" any given shell structure. We excerpt from [82].

Procedure PF-Constructor($\mathscr{A}$)
/*Construct a PF $\mathscr{A}$ to favor any fixed set of shells*/

1. Partition $\mathbb{N}^+ \times \mathbb{N}^+$, the set of potential array positions, into finite sets called *shells*. Order the shells linearly in some way. (Many natural shell-partitions carry a natural order, as the square and diagonal shells suggest.)

Samples. For each relevant integer c, shell c comprises all pairs $\langle x, y \rangle$ such that:

$x + y = c$	the *diagonal* shells that define the PF $\mathscr{D}$ of (8.1) and Figure 8.2,
$\max(x, y) = c$	the *square* shells that define the PF $\mathscr{A}_{11}$ of (8.2) and Figure 8.3,
$xy = c$	the *hyperbolic* shells of (8.5) and Figure 8.6, which play an important role later in the section.

2. Construct a PF from the shells as follows.

   a. Enumerate the array positions shell by shell, honoring the ordering of the shells.

   b. Enumerate each shell in some systematic way, say "by columns." This means enumerating the pairs $\langle x, y \rangle$ in the shell in increasing order of y and, for pairs having equal y values, in, say, decreasing order of x. (Increasing order of x works as well, of course.)

Theorem 8.1 ([82]). *Any function* $\mathscr{A} : \mathbb{N}^+ \times \mathbb{N}^+ \leftrightarrow \mathbb{N}^+$ *that is designed via Procedure* PF-Constructor *is a valid PF.*

Proof. Step 1 of Procedure PF-Constructor constructs a partial order on $\mathbb{N}^+ \times \mathbb{N}^+$, in which (*a*) each set of incomparable elements—called a shell—is finite; (*b*) there is a linear order on the shells. Step 2 extends the partial order of Step 1 to a linear order, by honoring the linear order on the shells and imposing a linear order within each shell. The function $\mathscr{A}$ constructed by the procedure can thus be viewed as an enumeration of $\mathbb{N}^+ \times \mathbb{N}^+$—which means that $\mathscr{A}$ is a PF. $\square$

Of course, Procedure PF-Constructor does not address the question of how to guarantee that the PF $\mathscr{A}$ is "efficient." This question will be our major focus for the remainder of this section, using two important interpretations of "efficient." Most of our effort will be devoted to crafting PFs that are *compact*, in the sense made precise in Section 8.3.2; the compactness of a PF $\mathscr{A}$ is related to the ease of "managing storage" while using $\mathscr{A}$ as a storage mapping. We shall devote some attention also to the computational complexity of a PF $\mathscr{A}$: how easy it is to compute $\mathscr{A}(x, y)$. A version of this latter question will recur in Section 8.4.1.

8.3.2 Pursuing Compact Pairing Functions

When one considers using a PF for mapping arrays/tables into storage, one notes immediately the poor resulting management of storage. For instance, the diagonal-shell PF $\mathscr{D}$ spreads the n^2-position $n \times n$ array/table over $2n^2$ addresses: $\mathscr{D}(1, 1) = 1$ and $\mathscr{D}(n, n) = 2n^2$; even worse (percentagewise), $\mathscr{D}$ spreads the n-position $1 \times n$ array/table over $> \frac{1}{2}n^2$ addresses: $\mathscr{D}(1, 1) = 1$ and $\mathscr{D}(1, n) = \frac{1}{2}(n^2 + n)$. This loss of "compactness" is a more serious deficiency than is the loss of the bidirectional arithmetic progressions enjoyed by the standard row- or column-major indexings used by most compilers, since the waste of storage plagues one no matter how one intends to access the array. It turns out that one can do a lot better than the PF $\mathscr{D}$ in controlling the "spread" of a PF, a measure of how poorly the PF utilizes storage. Following [82, 83], we seek PFs $\mathscr{A}$ that are *compact*, as measured by small growth rates of their *spread functions*

$$\mathbf{S}_{\mathscr{A}}(n) \overset{\text{def}}{=} \max\{\mathscr{A}(x,y) \mid xy \leq n\}. \tag{8.3}$$

In other words, $\mathbf{S}_{\mathscr{A}}(n)$ is the largest address that the PF $\mathscr{A}$ assigns to any position of an array that has n or fewer positions.

The next three paragraphs summarize the relevant results from [82, 83].

A. PFs that favor one fixed aspect ratio. Say that one moderates one's demands on the PF $\mathscr{A}$ by focusing only on its compactness when storing arrays of a single fixed aspect ratio $\langle a, b \rangle$, i.e., arrays whose dimensions have the form $ak \times bk$ for some integer k. In this highly constrained case, one can manage storage perfectly, in the sense that there exists a PF $\mathscr{A}_{a,b}$ such that

$$\mathbf{S}_{\mathscr{A}_{a,b}}(n) \overset{\text{def}}{=} \max\{\mathscr{A}_{a,b}(x,y) \mid [x \leq ak] \wedge [y \leq bk] \wedge [abk^2 \leq n]\} = n. \tag{8.4}$$

In other words, $\mathscr{A}_{a,b}$ maps every position (x,y) of an $ak \times bk$ array that has n or fewer positions to an address $\leq n$.

Note that arrays with aspect ratio $\langle 1, 1 \rangle$ are *square*—which is why we assigned the name $\mathscr{A}_{11}$ to the square-shell PF of (8.2) and Figure 8.3.

It is easy to construct $\mathscr{A}_{a,b}$ via the shells specified as follows, with each shell linearized by columns, i.e., enumerated in column-major order.

1. Shell 1 comprises the positions of the $a \times b$ array, i.e., the set

$$\{\langle x, y \rangle \mid [x \leq a] \wedge [y \leq b]\}.$$

2. Inductively, Shell $k+1$ comprises the positions of the $a(k+1) \times b(k+1)$ array that are not elements of the $ak \times bk$ array, i.e., the set

$$\{\langle x, y \rangle \mid [ak < x \leq a(k+1)] \vee [bk < y \leq b(k+1)]\}.$$

Figure 8.4 depicts an $ak \times bk$ array (A in the figure) and Shell $k+1$ (the union of B and C in the figure), which extends A to an $a(k+1) \times b(k+1)$ array. We leave as an exercise the verification of each $\mathscr{A}_{a,b}$'s optimal spread on every array of shape $ak \times bk$.

Although the preceding specification of each $\mathscr{A}_{a,b}$'s layout strategy ignores the question of how easy it is to compute each address $\mathscr{A}_{a,b}(x,y)$, one sees as follows that this computation is not onerous.

1. A simple computation determines the shell number $k+1$ of array-position $\langle x, y \rangle$ as the maximum of:

   - the smallest k_1 such that $ak_1 < x \leq a(k_1+1)$,

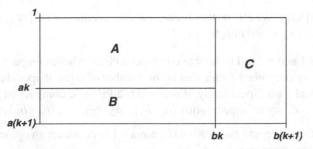

Fig. 8.4 *An $ak \times bk$ array, A, and its extension to an $a(k+1) \times b(k+1)$ array via the shell $B \cup C$.*

- the smallest k_2 such that $bk_2 < y \le b(k_2 + 1)$.

2. One then notes that Shell k is composed of two disjoint parts, one having dimensions $a \times bk$ (this is the subarray B in Figure 8.4), the other having dimensions $a(k+1) \times b$ (this is the subarray C in Figure 8.4). $\mathscr{A}_{a,b}$ stores each of these subarrays in column-major order.

We can summarize the results of this paragraph as follows.

Lemma 8.1 ([83]). *Each PF $\mathscr{A}_{a,b}$ has optimal spread, $\mathbf{S}_{\mathscr{A}_{a,b}}(n) = n$, on arrays having aspect ratio $\langle a, b \rangle$.*

B. PFs that favor finite sets of aspect ratios. It is not known how to store extendible arrays whose aspect ratios are not fixed with optimal compactness, but one can retain "good" compactness, as long as the arrays cannot assume too many potential shapes. The following procedure from [83] accomplishes this.

Say that one is given any set $\{\mathscr{A}_1, \mathscr{A}_2, \ldots, \mathscr{A}_m\}$ of m PFs. The following procedure—which is called *dovetailing* in [83]—crafts a PF $\mathscr{A}$ whose compactness is no worse than m times that of the most compact of the PFs $\mathscr{A}_i$; i.e., for all n,

$$\mathbf{S}_{\mathscr{A}}(n) \le m \cdot \min_i \mathbf{S}_{\mathscr{A}_i}(n).$$

Dovetailing is performed in two steps.

1. Alter each $\mathscr{A}_k$ to be a bijection $\mathscr{A}_k^{(m)}$ between $\mathbb{N}^+ \times \mathbb{N}^+$ and the congruence class $(k-1) \bmod m$, i.e., the set of integers of the form $mx + k - 1$. Specifically, define $\mathscr{A}_k^{(m)}$ as follows:

$$\mathscr{A}_k^{(m)}(x,y) = m \cdot \mathscr{A}_k(x,y) + k - 1.$$

2. Define the PF $\mathscr{A}$ as follows: for all integer pairs $x, y \in \mathbb{N}^+$,

$$\mathscr{A}(x,y) = \min_k \{\mathscr{A}_k^{(m)}(x,y)\}.$$

Direct calculation now yields the following summarizing result.

Lemma 8.2 ([83]). *The PF $\mathscr{A}$ that is created by dovetailing PFs $\{\mathscr{A}_1, \mathscr{A}_2, \ldots, \mathscr{A}_m\}$ has spread $\mathbf{S}_{\mathscr{A}}(n) \leq m \cdot \min_i \mathbf{S}_{\mathscr{A}_i}(n)$.*

Lemmas 8.1 and 8.2 tell us how to construct a PF $\mathscr{A}$ whose compactness degrades from optimality only via a factor that is the number of array shapes that $\mathscr{A}$ must be prepared to deal with. Specifically, if one wants a PF to be compact on arrays of any fixed finite set of, say m, aspect ratios $\langle a_1, b_1 \rangle, \langle a_2, b_2 \rangle, \ldots, \langle a_m, b_m \rangle$, then:

1. One uses the procedure that leads to Lemma 8.1 to construct an optimally compact PF $\mathscr{A}_i$ for each aspect ratio $\langle a_i, b_i \rangle$.
2. One uses the procedure that leads to Lemma 8.2 to dovetail the m PFs $\mathscr{A}_1, \mathscr{A}_2, \ldots, \mathscr{A}_m$ into a PF $\mathscr{A}_{a_1,b_1;a_2,b_2;\ldots;a_m,b_m}$ that maps every position (x,y) of an array that has one of the m fixed aspect ratios and that has n or fewer positions to an address $\leq mn$.

C. A PF that minimizes worst-case spread. The shape-based guarantees of paragraphs A and B do little to minimize spread when used with applications such as relational databases, wherein one cannot limit a priori the potential shapes of one's tables (or arrays). How far can one go toward minimizing spread—i.e., how compact can a PF be—when storing arrays of arbitrary shapes? Recalling that both the diagonal-shell PF $\mathscr{D}$ of (8.1) and Figure 8.2 and the square-shell PF $\mathscr{A}_{11}$ of (8.2) and Figure 8.3 have spread $\Theta(n^2)$ establishes the benchmark that we have to beat. And beat it we can! One finds in [83] the specification and analysis of a PF $\mathscr{H}$ that is within a constant factor of optimal in its worst-case spread. In detail:

Theorem 8.2 ([83]). (a) *There exists a PF $\mathscr{H}$ whose spread satisfies*

$$\mathbf{S}_{\mathscr{H}}(n) = O(n \log n).$$

(b) *No PF can beat $\mathscr{H}$'s level of compactness (in the worst case) by more than a constant factor.*

Proof. In order to develop the intuition necessary to understand where the PF $\mathscr{H}$ "comes from," we prove part (b) of the theorem before part (a).

(b) We establish this part of the theorem by answering—to within constant factors—the question, *How fast must the spread of a PF $\mathscr{A}$ grow, as a function of n?*

We answer this question in stages. We note first that if integer lattice points $\langle x,y \rangle$ and $\langle x',y' \rangle$ both reside in the same array that has $\leq n$ elements, then—because $\mathscr{A}$ maps $\mathbb{N}^+ \times \mathbb{N}^+$ *one-to-one* onto $\mathbb{N}^+$—we must have

$$\mathbf{S}_{\mathscr{A}}(n) \geq \max(\mathscr{A}(x,y), \mathscr{A}(x',y')).$$

Consequently, if we let S_n denote the set that comprises all integer lattice points that reside in *some* array that has $\leq n$ elements, then no matter how cleverly we craft the PF $\mathscr{A}$, we must have

$$\mathbf{S}_{\mathscr{A}}(n) = \max\{\mathscr{A}(x,y) \mid \text{point } \langle x,y \rangle \text{ resides in an array that has} \le n \text{ elements}\}$$
$$\ge |S_n|.$$

Our original question has now been reduced to the question, *How many elements does the set S_n have, as a function of n?*

To answer this question, we must consider when an integer lattice point $\langle x,y \rangle$ resides in some array that has $\le n$ elements. We sneak up on this question by focusing on the $a \times b$ array, where $ab \le n$. On the one hand, this array has ab elements; on the other hand, every element of this array is an integer lattice point $\langle x,y \rangle$ for which $x \le a$ and $y \le b$. Thus, every element of the array has $xy \le ab \le n$. Conversely, every integer lattice point $\langle x,y \rangle$ for which $xy \le n$ resides in the $x \times y$ array—which has $xy \le n$ elements. In summation: *An integer lattice point $\langle x,y \rangle$ resides in some array that has $\le n$ elements if and only if $xy \le n$.*

So, our original question has now been reduced to the question, *How many integer lattice points $\langle x,y \rangle$ have $xy \le n$?*

This question is easy to answer if one looks at it in the right way. Figure 8.5 (generalized to arbitrary n) should help. One sees from the figure that the union of

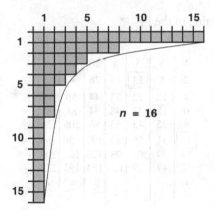

Fig. 8.5 *The aggregate set of positions of arrays having ≤ 16 elements.*

the elements of all arrays that have $\le n$ elements—which is the set S_n—is the set of integer lattice points that lie under the hyperbola $xy = n$. This means that

$$|S_n| = \sum_{i=1}^{n} \left\lfloor \frac{n}{i} \right\rfloor = \Theta(n \log n).$$

The preceding summation follows from our discussion of which lattice points appear in arrays that have $\le n$ elements; the size of the sum follows most easily via estimating the sum by the integral $n \int (1/x)\mathrm{d}x$, with appropriate limits; cf. [20].

(a) The preceding analysis gives us strong hints for constructing the PF $\mathscr{H}$: one strives, for each integer n, to conform as closely as possible to the hyperbola $xy = n$, which is the curve depicted in Figure 8.5. The figure suggests how to accomplish this using a collection of superimposed rectangles; each rectangle is one of the "maximal" arrays that has $\leq n$ elements, where "maximal" means that none of these arrays is a subset of any other array that has $\leq n$ elements. (Hence, each "maximal" array contributes at least one new integer lattice point to S_n.) In detail, we construct the PF $\mathcal{H}$ using the shell-based strategy of Procedure PF-Constructor, with shells that approximate the hyperbolic shape of the curve in Figure 8.5. The resulting *hyperbolic shells* are defined as follows: Shell 1 consists of the unique integer lattice point $\langle x, y \rangle$ with $xy = 1$, namely, $\langle 1, 1 \rangle$; Shell 2 comprises the two integer lattice points $\langle x, y \rangle$ with $xy = 2$, namely, $\langle 1, 2 \rangle$ and $\langle 2, 1 \rangle$; ...; Shell 8 comprises the four integer lattice points $\langle x, y \rangle$ with $xy = 8$, namely, $\langle 1, 8 \rangle$ and $\langle 2, 4 \rangle$; $\langle 4, 2 \rangle$ and $\langle 8, 1 \rangle$; and so on.

Using Procedure PF-Constructor with these shells, we end up with the following specification for the *hyperbolic-shell PF* $\mathcal{H}$. We can easily give a recipe for computing $\mathcal{H}$, with the help of the function $\delta(n)$ that specifies the number of distinct divisors of the integer $n \in \mathbb{N}^+$. We find that[2]

$$\mathcal{H}(x, y) = \sum_{k=1}^{xy-1} \delta(k) \; + \text{ the position of } \langle x, y \rangle \text{ among}$$

(8.5)

$$\text{2-part factorizations of } xy,$$
$$\text{in reverse lexicographic order.}$$

Figure 8.6 depicts a portion of the PF $\mathcal{H}$.

1	3	5	8	10	14	16	$\cdots$
2	7	13	19	26	34	40	$\cdots$
4	12	22	33	44	56	69	$\cdots$
6	18	32	48	64	81	99	$\cdots$
9	25	43	63	86	108	130	$\cdots$
11	31	55	80	107	136	165	$\cdots$
15	39	68	98	129	164	200	$\cdots$
17	47	79	116	154	193	235	$\cdots$
$\vdots$	$\vdots$	$\vdots$	$\vdots$	$\vdots$	$\vdots$	$\vdots$	$\ddots$

Fig. 8.6 *The hyperbolic-shell PF* $\mathcal{H}$. *The shell* $xy = 6$ *is highlighted.*

The validity of $\mathcal{H}$ as a PF follows from Procedure PF-Constructor. The bound on the spread of $\mathcal{H}$ follows from the analysis in part (b) of the proof.

Aside. The work described in this section aims at giving one a broad range of ways of accessing one's arrays: by position, by row/column, by block (at varying computational costs). If one is interested in accessing an extendible array only by position, then one might be well served by the *hashing schemes* studied in [88]. Those hashing schemes enjoy the following resource consumption. When one's array has at most n elements, then, no matter what the array's aspect ratio, the hashing scheme will employ *fewer than* $2n$ *memory locations* and will allow one to *access any position of the array in expected time* $O(1)$ *and worst-case time* $O(\log\log n)$.

---

[2] A "2-part factorization" of an integer n is a *pair* of integers a, b such that $ab = n$.

8.4 Pairing Functions and Volunteer Computing

Computing and communication technology have experienced revolutionary advances over the past decades. Whereas the Internet has historically—its short history has itself been revolutionary—been mainly a medium for communication and content delivery, more recent technological advances have given rise to new modalities of computing, wherein geographically dispersed computers cooperate on massive computations. One of these modalities—often called *volunteer computing*—is particularly interesting, because the (owners of the) participating computers neither know nor trust one another. (They do not necessarily *distrust* one another; they just don't *trust* one another.) A volunteer computing project—one of the earliest and best known being *SETI@home* [54]—proceeds roughly as follows.

We employ evocative anthropomorphic terminology in this description, in place of more precise computer-oriented terminology.

Volunteers—the remote computers—register with a volunteer computing website. After having registered, each volunteer visits the website from time to time to receive a task to compute. Some time after completing its present task, a volunteer returns to the volunteer computing website in order to transmit the results from that task and to receive a new task. And the cycle continues.

As discussed in the early paragraphs of this chapter, typical implementations of volunteer computing projects—which do not require prior vetting before admitting a volunteer to the "team," are vulnerable to volunteers returning "false" results, for one of several possible reasons, including malice, well-intentioned but misguided "improvement" of program code downloaded from the volunteer computing website, or incompatibilities among computers' processors and operating systems. (See the interview with David Anderson mentioned earlier and the discussion in [102].)

Although the following topic appears to stray from the scope of the book, it points out a feature of computation theory that is at variance with much of "practical" computing.

As we have discussed repeatedly, computation theory focuses on computations whose arguments and results are integers or finite strings or bit values ("YES" or "NO"). Therefore, the theory never has to deal with values that are "equal within a tolerance": values are either equal or not. In contrast, floating-point computations on real computers are sensitive to the just-mentioned incompatibilities, which lead to myriad practical problems, including the vetting of volunteers' results in volunteer computing. This is why we put quotes around the word "false" when referring to volunteers' results: the results could have been obtained quite legitimately, but on a system whose tolerances exceed those of the volunteer computing website.

Back to the use of PFs in volunteer computing. Volunteer computing is maturing (as a modality of cooperative computing) beyond being a vehicle for pure research demonstrations: it now encompasses computations that relate to sensitive matters such as security[3] [90] and clinical drug testing [44, 74]. In these practical domains, "false" results, even if well-intentioned, could have dire consequences. It seems clear that a volunteer computing website that detects recurring "false" results from

---

[3] The three volunteer computing websites cited here were active just a few years ago.

a specific volunteer would want to bar that volunteer from subsequent participation in the project.

Such detection would typically result from either spot-checking results or from a computing regimen that involves redundant allocation plus voting [54, 102].

But how does one efficiently keep track of volunteers and the results that they produce? One finds in [84] a computationally lightweight scheme that employs PFs in a fundamental way. (Note that this work addresses concerns about *accountability*, not security.)

The basic idea from [84] begins by assigning positive-integer indices to (*a*) the set of all tasks in the volunteer computing website's workload, (*b*) all volunteers who are currently enrolled in the project, (*c*) the set of tasks that are reserved for each volunteer *v*. One then uses a PF $\mathscr{T}$ (which we shall call a *task-allocation function* in this context) to link volunteers with their assigned tasks. In other words, the *t*th task that volunteer *v* receives to compute is task $\mathscr{T}(v,t)$.

Since the potential practicality of such a scheme demands that the functions $\mathscr{T}$, $\mathscr{T}^{-1}$, and $\mathscr{S}(v,t) \stackrel{\text{def}}{=} \mathscr{T}(v,t+1) - \mathscr{T}(v,t)$ all be easily computed, the primary focus in [84] is on PFs that are *additive* (APFs, for short): an APF assigns each volunteer *v* a *base task-index* B_v and a *stride* S_v; it then uses the formula

$$\mathscr{T}(v,t) = B_v + (t-1)S_v$$

to determine the workload task-index of the *t*th task assigned to volunteer *v*. From a system perspective, APFs have the benefit that a volunteer's stride need be computed only when s/he registers at the volunteer computing website and can be stored for subsequent appearances.

Given our narrow focus on APFs in this section, we ignore several practical concerns for any scheme that seeks to endow a volunteer computing project with accountability and/or security.

Probably the thorniest concern results from a *Sybil attack* [24]: a malicious volunteer's camouflaging his/her identity via the use of multiple IP addresses. Strategies for detecting and preventing Sybil attacks remain a research topic to this day.

A less challenging problem is that any task-allocation scheme that is based entirely on APFs allows new volunteers to arrive dynamically but not to depart. If volunteers depart, then their tasks will never be computed—unless new volunteers arrive to take their places and compute their tasks. Such reassignment would demand added mechanisms to retain accountability. The complete scheme described in [84] has a "front end" that allows volunteers to arrive *and depart* dynamically; it also ensures that faster volunteers are always assigned smaller indices. These details are beyond this section's focus on APFs.

Given the proposed use of APFs to assign indices to volunteers, one can argue that the management of the memory where tasks reside is simplified if one devises APFs whose strides S_v grow slowly as a function of *v*. Such APFs are "compact," in the sense of (8.3). This observation sets the agenda for [84] and for the remainder of this section. Section 8.4.1 presents a methodology for designing easily computed APFs; Section 8.4.2 presents a sequence of APFs that suggest a tradeoff between the ease of computing an APF and the rate of growth of the APF's strides.

We henceforth abstract the preceding discussion from the volunteer computing scenario by replacing "volunteer" by "row" and "base task-index" by "base row-entry." We also revert to our generic uses of x and y, instead of v and t.

8.4.1 A Methodology for Designing Additive Pairing Functions

It is easy to show that any APF must have infinitely many distinct strides; i.e., S_x, viewed as a function of x, must have infinite range. Despite this, there do exist easily computed APFs. One strategy for designing such APFs builds on the following well-known property of the set O of positive odd integers.

Lemma 8.3 ([73]). *For any positive integer c, every odd integer can be written in precisely one of the 2^{c-1} forms*

$$2^c n + 1, \ 2^c n + 3, \ 2^c n + 5, \ldots, \ 2^c n + (2^c - 1),$$

for some nonnegative integer n.

One builds on Lemma 8.3 to construct APFs as follows.

Procedure APF-Constructor($\mathscr{T}$)
/*Construct an APF $\mathscr{T}$*/

Step 1. Partition the set of row-indices into *groups* whose sizes are powers of 2 (with any desired mix of equal-size and distinct-size groups). Order the groups linearly in some (arbitrary) way.

/*One can now talk unambiguously about group 0 (whose members share *group-index* $g = 0$), group 1 (whose members share group-index $g = 1$), and so on.*/

Step 2. Assign each group a distinct copy of the set O, as well as a *copy-index* $\kappa(g)$ expressed as a function of the group-index g.

Step 3. Allocate group g's copy of O to its members via the $(c = \kappa(g))$ instance of Lemma 8.3, using the multiplier 2^g as a *signature* to distinguish group g's copy of the set O from all other groups' copies.

Procedure APF-Constructor can be viewed as specializing the quite general scheme for constructing APFs in [99]. The specialization allows us to specify the APF in a computationally friendly way.

An explicit expression for $\mathscr{T}$. If we denote the $2^{\kappa(g)}$ rows of group g by $x_{g,1}, x_{g,2}, \ldots, x_{g,2^{\kappa(g)}}$, then for all $i \in \{1, 2, \ldots, 2^{\kappa(g)}\}$,

$$\mathscr{T}(x_{g,i}, y) \overset{\text{def}}{=} 2^g \left[2^{1+\kappa(g)}(y-1) + (2x_{g,i} + 1 \bmod 2^{1+\kappa(g)}) \right]. \tag{8.6}$$

Theorem 8.3. *Any function $\mathscr{T} : \mathbb{N}^+ \times \mathbb{N}^+ \leftrightarrow \mathbb{N}^+$ that is designed via Procedure* APF-Constructor, *hence is of the form (8.6), is a valid APF whose base row-entries and strides satisfy*

$$B_x \leq S_x = \mathscr{T}(x, y+1) - \mathscr{T}(x, y) = 2^{1+g+\kappa(g)}. \tag{8.7}$$

Proof. Any function $\mathscr{T}$ as described in the theorem maps $\mathbb{N}^+ \times \mathbb{N}^+$ *onto* $\mathbb{N}^+$, because every positive integer equals *some* power of 2 times *some* odd integer. Additionally, $\mathscr{T}$ is *one-to-one* because it has a functional inverse $\mathscr{T}^{-1}$. To wit, the trailing 0's of each image integer $k = \mathscr{T}(x, y)$ identify x's group g, hence the operative instance $\kappa(g)$ of Lemma 8.3. Then:

1. We compute

$$x = \frac{1}{2} \left[(2^{-g} k \bmod 2^{1+\kappa(g)}) - 1 \right],$$

   which is an integer because the division by 2^g produces an odd number.
2. This leaves us with a linear expression of the form $ay + b$, from which we easily compute y.

Finally, we read the relations (8.7) directly from (8.6). □

In order to implement Procedure APF-Constructor completely, we must express both the group-indices g and their associated copy-indices $\kappa(g)$ as functions of x. This is accomplished by noting that all x whose indices lie in the range

$$2^{\kappa(0)} + 2^{\kappa(1)} + \cdots + 2^{\kappa(g-1)} + 1 \leq x \leq 2^{\kappa(0)} + 2^{\kappa(1)} + \cdots + 2^{\kappa(g-1)} + 2^{\kappa(g)} \tag{8.8}$$

share group-index g and copy-index $\kappa(g)$. Translating the range (8.8) into an efficiently computed expression of the form $g = f(x)$ may be a simple or a challenging enterprise, depending on the functional form of $\kappa(g)$ that results from the grouping of row-indices.

8.4.2 A Sampler of Explicit APFs

Theorem 8.3 assures us that Procedure APF-Constructor produces a valid APF no matter how the copy-index $\kappa(g)$ grows as a function of the group-index g. However, the ease of computing the resulting APF, and its compactness, depend crucially on this growth rate. We now illustrate how one can use this growth rate as part of the design process, in order to stress either the ease of computing an APF or its compactness.

A. APFs that stress ease of computation. We first implement Procedure APF-Constructor with *equal-size groups*, i.e., with $\kappa(g) = constant$. For each $c \in \mathbb{N}^+$, let $\mathscr{T}^{\langle c \rangle}$ be the APF produced by the procedure with $\kappa^{\langle c \rangle}(g) \equiv c - 1$. One computes easily that

$$\mathscr{T}^{\langle c \rangle}(x, y) \stackrel{\text{def}}{=} 2^{\lfloor (x-1)/2^{c-1} \rfloor} [2^c (y - 1) + (2x - 1 \bmod 2^c)].$$

Lemma 8.4. *Each $\mathscr{T}^{\langle c \rangle}$ is a valid APF whose base row-entries and strides are given by*

$$B_x^{\langle c \rangle} \leq S_x^{\langle c \rangle} = 2^{\lfloor (x-1)/2^{c-1} \rfloor + c}. \tag{8.9}$$

Each $\mathscr{T}^{\langle c \rangle}$ is easy to compute but has base row-entries and strides that grow *exponentially* with row-indices. Increased values of c (= larger fixed group sizes) decrease the base of the growth exponential, at the expense of modest increase in computational complexity. Computing a few sample values illustrates how a larger value of c penalizes a few low-index rows but gives all others significantly smaller base row-entries and strides; cf. the top half of Figure 8.7.

x	g	$\mathscr{T}^{\langle 1 \rangle}(x,y)$				
14	13	8192	24576	40960	57344	73728 $\cdots$
15	14	16384	49152	81920	114688	147456 $\cdots$

x	g	$\mathscr{T}^{\langle 3 \rangle}(x,y)$				
14	3	24	88	152	216	280 $\cdots$
15	3	40	104	168	232	296 $\cdots$
$\vdots$	$\vdots$	$\vdots$	$\vdots$	$\vdots$	$\vdots$	$\vdots$
28	6	448	960	1472	1984	2496 $\cdots$
29	7	128	1152	2176	3200	4224 $\cdots$

x	g	$\mathscr{T}^{\#}(x,y)$				
28	4	400	912	1424	1936	2448 $\cdots$
29	4	432	944	1456	1968	2480 $\cdots$

x	g	$\mathscr{T}^{*}(x,y)$				
28	3	328	840	1352	1864	2376 $\cdots$
29	3	344	856	1368	1880	2392 $\cdots$
		$\vdots$	$\vdots$	$\vdots$	$\vdots$	$\vdots$

Fig. 8.7 *Sample values by several APFs.*

B. APFs that balance computation ease and compactness. The functional form of the exponent of 2 in (8.9) suggests that one can craft an APF whose base row-entries and strides grow subexponentially by allowing the parameter c to grow with x, in a way that (roughly) balances $x/2^c$ against c. This strategy leads us to consider the copy-index $\kappa^{\#}(g) = g$. When we implement Procedure APF-Constructor with copy-index $\kappa^{\#}$, we arrive at an APF $\mathscr{T}^{\#}$ that is rather easy to compute and whose base row-entries and strides grow only *quadratically* with row-indices. To wit: The copy-index $\kappa^{\#}(g) = g$ aggregates row-indices into groups of exponentially growing sizes. Each group g comprises row-indices $2^g, 2^g + 1, \ldots, 2^{g+1} - 1$. By (8.8), then, one computes easily that[4]

$$\kappa^{\#}(g) = g = \lfloor \log x \rfloor. \tag{8.10}$$

Instantiating (8.10) in the definitional scheme (8.6), we find that

---

[4] Throughout, all logarithms have base 2.

$$\mathscr{T}^{\#}(x,y) = 2^{\lfloor \log x \rfloor}\left(2^{1+\lfloor \log x \rfloor}(y-1) + (2x+1 \bmod 2^{1+\lfloor \log x \rfloor})\right). \qquad (8.11)$$

Lemma 8.5. *The function* $\mathscr{T}^{\#}$ *specified in (8.11) is a valid APF whose base row-entries and strides (as functions of x) are given by*

$$B_x^{\#} < \mathscr{S}_x^{\#} = 2^{1+2\lfloor \log x \rfloor} \le 2x^2,$$

hence grow quadratically with x.

Comparing $\mathscr{T}^{\#}$ **and the** $\mathscr{T}^{\langle c \rangle}$. For sufficiently large x, the (exponentially growing) strides of any of the APFs $\mathscr{T}^{\langle c \rangle}$ will be dramatically larger than the (quadratically growing) strides of the APF $\mathscr{T}^{\#}$. However, it takes a while for $\mathscr{T}^{\#}$'s superiority to manifest itself; for instance,

- it is not until $x = 5$ that $\mathscr{T}^{\langle 1 \rangle}$'s strides are always at least as large as $\mathscr{T}^{\#}$'s;
- the corresponding number for $\mathscr{T}^{\langle 2 \rangle}$ is $x = 11$;
- the corresponding number for $\mathscr{T}^{\langle 3 \rangle}$ is $x = 25$.

C. APFs that stress compactness. By choosing a copy-index $\kappa(g)$ that grows superlinearly with g, one can craft APFs whose base row-entries and strides grow subquadratically, thereby beating the compactness of $\mathscr{T}^{\#}$. But one must choose $\kappa(g)$'s growth rate judiciously, because faster growth need not enhance compactness.

Achieving subquadratic growth. Many copy-index growth rates yield APFs with subquadratic compactness. However, all of the APFs we know of that achieve this goal are rather difficult to compute and actually achieve the goal only asymptotically, hence are more likely of academic than practical interest.

Consider, for each $k \in \mathbb{N}^+$, the APF $\mathscr{T}^{[k]}$ specified by the copy-index $\kappa^{[k]}(g) = g^k$. By (8.8), the row-indices x belonging to group g now lie in the range

$$1 + 2 + 2^{2^k} + \cdots + 2^{(g-1)^k} < x \le 1 + 2 + 2^{2^k} + \cdots + 2^{g^k},$$

so that $g = (1 + o(1))\lceil (\log x)^{1/k} \rceil$. We actually use the simplified, albeit slightly inaccurate, expression $g = \lceil (\log x)^{1/k} \rceil$ in our asymptotic analyses of the $\mathscr{T}^{[k]}$, because the $o(1)$-quantity decreases very rapidly with growing x. Although closed-form expressions for $\mathscr{T}^{[k]}$ in terms of x have eluded us, we can verify that each $\mathscr{T}^{[k]}$ does indeed enjoy subquadratic stride growth.

Lemma 8.6. *Each function* $\mathscr{T}^{[k]}$ *produced by Procedure* APF-Constructor *from the copy-index* $\kappa^{[k]}(g) = g^k$ *is a valid APF whose base row-entries and strides (as functions of x) are given by*

$$B_x^{[k]} \le \mathscr{S}_x^{[k]} = 2^{O\left((\log x)^{1/k} + \log x\right)} = x2^{O\left((\log x)^{1/k}\right)}, \qquad (8.12)$$

hence grow subquadratically with x.

We illustrate a close relative of $\mathscr{T}^{[2]}$ that exhibits its subquadratic compactness at much smaller values of x than $\mathscr{T}^{[2]}$ does, namely, the APF $\mathscr{T}^{\star}$ that Procedure APF-Constructor produces from the copy-index

$$\kappa^\star(g) = \left\lceil \frac{1}{2} g^2 \right\rceil. \tag{8.13}$$

Mimicking the development with $\kappa^{[k]}$, we see that the value of g associated with this copy-index is $g = (1 + o(1))\lceil \sqrt{2\log x} \rceil + 1$, which we simplify for analysis to the slightly inaccurate expression

$$g = \left\lceil \sqrt{2\log x} \right\rceil + 1.$$

We can easily compute $\mathcal{T}^\star$ from (8.13), in the presence of (8.6), (8.8).

Lemma 8.7. *The base row-entries and strides of the APF $\mathcal{T}^\star$ satisfy*

$$B_x^\star \leq \mathcal{S}_x^\star = 2^{1+g+\kappa^\star(g)} \approx 8x4^{\sqrt{2\log x}}.$$

Comparing $\mathcal{T}^\star$ and $\mathcal{T}^\#$. Any function that grows quadratically with x will eventually produce significantly larger values than a function that grows only as $x4^{\sqrt{2\log x}}$. Therefore, $\mathcal{T}^\star$'s strides will eventually be dramatically smaller than $\mathcal{T}^\#$'s. Figure 8.7 indicates that this difference takes effect at about the same point as the exponential vs. quadratic one noted earlier, albeit at the cost of greater computational complexity.

The danger of excessively fast growing κ. If $\kappa(g)$ grows too fast with g, then the base row-entries and strides of the resulting APF grow *super*quadratically with the row-indices x, thereby confuting our goal of beating quadratic growth. We exemplify this fact by supplying Procedure APF-Constructor with the copy-index $\kappa(g) = 2^g$; the reader can readily supply other examples. By (8.8), we see that in this case, $g = \lfloor \log \log x \rfloor + O(1)$. Therefore, whenever x is the smallest row-index with a given group-index g (of course, infinitely many such x exist) we have

$$x = 2^{\kappa(0)} + 2^{\kappa(1)} + \cdots + 2^{\kappa(g-1)} + 1 \approx \sqrt{2^{\kappa(g)}},$$

while the stride associated with x is (cf. (8.7))

$$S_x = 2^{1+g+\kappa(g)} > 2^{\kappa(g)}\kappa(g) \approx x^2 \log x.$$

We do not yet know the growth rate at which faster growing $\kappa(g)$ starts hurting compactness. Finding this rate is an atractive research problem.

Chapter 9
Computability Theory

9.1 Introduction and History

Mathematics has been "practiced" for thousands of years, yet it was not until the nineteenth century that people began to attempt to crystalize/formalize the notion of *proof*.

> This attempt is an ongoing struggle: the advent of computers and computer-assisted proofs has led to a rethinking of the formal notions that were developed in the nineteenth and early twentieth centuries.

The formal notion of proof that led to the birth of the field of mathematical logic made a proof a kind of rewriting system. One started with a set of *axioms* ("statements or propositions that are regarded as being established, accepted, or self-evidently true"), which were automatically granted the status of *theorem* ("a truth established by means of accepted truths"). One then added a set of *rules of inference*, or rewriting rules, that allowed one to derive new theorems by selectively "rewriting" preexisting ones.

Inspired by this apparently "mechanistic" notion of "proving a theorem," at the very end of the nineteenth century, the great German mathematician David Hilbert challenged mathematicians to devise "automatic procedures"—what we would now call *algorithms*—that would either prove or refute purported theorems within the elementary theory of numbers.

> If you have encountered portions of the "elementary" theory of numbers that have not seemed elementary to you, do not despair! In the context of number theory, the term designates results that follow from first principles rather than from a long chain of other results.

As we know from our earlier discussions, the hope for any such procedure was dashed in 1931 by the famous incompleteness theorem of Kurt Gödel [30]. Informally, that theorem says that in any mathematical system—i.e., axioms plus rules of inference—that is powerful enough to "talk" about (or express) a quite simple repertoire of properties of the positive integers, the notion of "theorem" could never be powerful enough to encompass the notion of "true statement."

In rough terms, the system had only to be able to express the equality of expressions in which positive integers are combined via addition and multiplication. One such assertion might be "$x \times y = z + 2x$."

As we described in Section 7, Gödel's proof of the theorem built upon the mathematical tools that Georg Cantor developed for comparing the relative "sizes" of infinite sets. In the mid-1930s, Alan Turing adapted Gödel's logic-oriented framework to a computational setting [104], thereby initiating what we now call the theory of computability, or computability theory. Just as Gödel's 1931 paper turned on its ear the (mistaken) intuition that mathematics—even elementary number theory—could be mechanized, Turing's 1936 paper played a similar role in the realm of computation. In detail, Turing's work showed that there exist *specific* functions $f : \mathbb{N} \to \{0, 1\}$ that are quite simple to specify informally but that cannot be computed by any "reasonable" notion of digital computer.

Here are a couple of ways of thinking about Turing's landmark work. These are stated very informally, but they can be made precise and formal. In fact, *you* will be able to formalize these assertions by the end of this chapter!

- *There exist digital computers whose behaviors can be specified totally unambiguously that cannot be built.*

- *There exist processes whose behaviors can be specified totally unambiguously that cannot be programmed on any digital computer.*

In order to appreciate the ingenuity (and imagination) that Turing displayed in formalizing these assertions, you should recall that digital computers did not yet exist in the 1930s and that only rudimentary types of "programmed machines"—such as the Jacquard loom, invented by Joseph Marie Jacquard in 1801—had yet seen the light of day!

The "reasonable" notion of digital computer that Turing studied was the eponymous *Turing machine*, a variant of which we described in Section 3.3. This formal model was *so* simple in its structure and its per-step computing capabilities that despite Turing's demonstrations of the model's computing power, one could not avoid wondering whether more powerful "reasonable" notions existed, which would not fall prey to Turing's proof of uncomputability!

What, indeed, is a "reasonable" notion of digital computer?

A glib answer is that, paraphrasing U.S. Supreme Court justice Potter Stewart's remark about obscenity in his concurring opinion in *Jacobellis v. Ohio 378 U.S. 184* (1964), you know it when you see it!

A more careful answer is that any competent computer designer should agree that the candidate notion can indeed be built—and programmed—using existing (or foreseeable) technology.

More generally, while it is hard to write a short list of criteria that separate "reasonable" models from "unreasonable" ones, one can bolster intuition by looking at a few models that most people would consider *not* to be "reasonable." Because computability theory focuses on procedures that manipulate strings and numerals, any of the following capabilities would render a model "unreasonable":

- the ability to manipulate objects that *admit no finite representation*, e.g., general infinite series and real numbers;

- the ability to make *infinitesimal discriminations*, say, by having no bound on resolution (think of "resolution" as word size in a digital computer);
- the ability to perform *continuous operations*, as, say, an analog computer does.

Remarkably, none of the myriad attempts to outdo Turing produced a formal model that (*a*) was more powerful than the Turing machine and (*b*) was deemed "reasonable" by the computing community. The studies that invented these new models were certainly not wasted time, though: Many of the attempts gave rise to equally powerful, quite different, alternative formulations of computation theory, thereby supplying quite new insights into the intrinsic nature of digital computation. (Think of the conceptual axiom that introduces Chapter 2.) Many of the others produced models that, while not more powerful than the Turing machine, accomplished their assigned tasks much more efficiently than Turing's rudimentary instruction repertoire permitted. Acknowledging the many contributions of Turing's competitors, in addition to those of Turing himself, the theory has freely absorbed notions from competing theories based on many distinct formalisms. To name just a few of the most successful competitors, we have

- the *lambda calculus* of Alonzo Church [14] (which supplied the theoretical underpinnings of LISP and other functional programming languages; see, e.g., [31]),
- Stephen Cole Kleene's theory of *recursive functions* [49, 50],
- the *combinatory logic* of Haskell B. Curry, Robert Feys, William Craig, Moses Schönfinkel, and their collaborators [21, 22, 93],
- Noam Chomsky's *type-0 grammars* [12] (which had its greatest impact in the field of formal languages),
- Andrey A. Markov's *Markov algorithms* [61],

and on and on. The confluence of the theories that emerge from the many disparate attempts to formulate a theory of computability has led all mainstream computer scientists and mathematicians to accept, as an operating principle, the extramathematical *Church–Turing Thesis:*

The Church–Turing thesis. *The informal notion "computable by a digital computer" is equivalent to the formal notion "computable by a Turing machine."*

Notes.
- The Church–Turing thesis is "extramathematical," i.e., not subject to proof, because it asserts the equivalence of a formal notion and an informal one.
- Church and Turing are honored in the name of the Thesis because of their pioneering work on the theory.
- The reader should view the Church–Turing thesis in the light of more than seventy years of unsuccessful attempts to refute it (by devising a more powerful "reasonable" model).

The present chapter is devoted to developing the basic conceptual and technical tools of computability theory, leading up to a few of the theory's blockbuster theorems. The most famous of these theorems is, of course, Theorem 9.1, which identifies the decision function for the *halting problem*—given a program P and an input x for P, does P ever halt when started on input x?—as being uncomputable. Perhaps even more dramatic than Theorem 9.1, though, is the Rice–Myhill–Shapiro theorem

(Theorem 9.5), which—very informally—can be viewed as the sweeping statement that one cannot algorithmically decide anything about the dynamic behavior of a program P from a static description of P (say, as a list of instructions). (You'll have to read Section 9.5 in order to understand what Theorem 9.5 *really* says.) On the way to developing Theorems 9.1 and 9.5, we shall encounter some blockbuster concepts. Notable here are the notion of *reducing one computational problem to another* (Section 9.4) and the notion that a computational problem, A, is *complete* within a class of computational problems (Section 9.6)—meaning, quite informally, that A is the computationally "hardest" problem in the class. (Here again, you'll have to read Section 9.6 in order to understand what "complete" *really* means.) The "big ideas" underlying the preceding blockbuster concepts and results are not of just academic interest: They should be part of the intellectual toolkit of every person who is concerned with the technical aspects of computation—from the computation theorist to the serious applications programmer.

The present chapter is incredibly rich intellectually. It begins with a background section (Section 9.2) that continues the development from Chapter 2, but that focuses on concepts and tools that relate specifically to computability theory. While the material in this section may appear to be of only mathematical interest, it actually develops the main intellectual toolset for the entire theory of computability. Once we turn to the technical development of computability theory, beginning in Section 9.3, we employ the *model-independent* approach that seems to have originated in the now classical text of Hartley Rogers, Jr. [80]. In brief, this approach invites you to think about the process of computing in terms of whatever "reasonable" model you find congenial. (But do not forget that your model must be "reasonable" in the sense of the Church–Turing thesis!) In particular, this approach invites you to think about these concepts using (the virtual machines associated with) your own favorite (real!) programming language—anything from APL to BASIC to C to C++ to FORTRAN to Java to .... This freedom should allow you to employ intuitions that you have developed from your experience in programming and/or otherwise using real digital computers—and thereby to appreciate the relevance of computability theory to real computing. The development in Section 9.2 should convince you that the specific programming language that you choose as your "model" is irrelevant: *Every real programming language can be "encoded" as any another.*

> Perhaps the biggest advertisement for the model-independent approach that we inherit from [80] is the early textbook on computability theory by Martin Davis [23], which develops the theory entirely within the context of the "classical" Turing machine of [104]. While the development in [23] can be reassuring to the student, in that details that many texts leave to the reader are carefully derived, it may make it difficult for some to see "the big picture" because of the focus on details. Since it is usually a nontrivial exercise to translate intuition from one programming abstraction (i.e., model) to another, one may struggle to see the relevance of certain Turing machine–oriented constructs to the programming languages that one uses in daily life. This is a powerful argument for letting each reader use her own model.

The many benefits of a model-independent approach to the theory notwithstanding, it would be a mistake not to present at least some of the evidence for the Church–Turing thesis. We do this in Section 9.8, within the context of the online Turing machine

model that we introduced in Section 3.3. The exercise of considering explicit translations from one detailed model to another is also needed for our study of complexity
theory, in Chapter 13, because that theory cannot be developed with the same degree
of model independence that computability theory can: While complexity theory certainly does not demand the confines of a single "standard" model for algorithms and
digital computers, it does impose limitations on how different competing models can
be from one another before the complexity theory of one model diverges from that of
another. In particular, there is no thesis for complexity theory that has anything like
the sweep of the Church–Turing thesis.

9.2 Preliminaries

9.2.1 Representing Computational Problems as Formal Languages

In Section 2.4.2, we discussed briefly how to talk about a variety of computational
problems using the medium of formal languages. This unusual way of talking about
computation has left its tracks in the terminology we use in computation theory—
but so also have the other precursors of the theory, such as mathematical logic and
computer science. The reader should see the traces of some of these precursors in the
following list, which will recur throughout the remainder of the book:

A *set* (of integers or strings) is: $\left\{ \begin{array}{c} decidable \\ recursive \end{array} \right\}$ or $\left\{ \begin{array}{c} undecidable \\ nonrecursive \end{array} \right\}$

A *computational problem* is: solvable   or   unsolvable

A *property of a system* is: decidable   or   undecidable.

I really should be careful here to add a qualifier such as "computation-theoretically" or "in the
sense of computation theory" to the words "solvable and "unsolvable," because these words
are sometimes used in reference to other, quite different, types of problems, such as:

- finding positive integers x and y such that $x^2 = 2y^2$,
- finding a real number x such that $x^2 + 1 = 0$,
- finding the roots of a general quintic polynomial using radicals,
- finding positive integers x, y, and z such that $z^3 = x^3 + y^3$.

I take the risk of not using such qualifiers, because we shall use the words consistently in
their computation-theoretic senses.

The positive—i.e., desirable—notions, *decidable, recursive,* and *solvable,* are equivalent to one another, as are the negative—i.e., undesirable—notions, *undecidable,
nonrecursive,* and *unsolvable.* When referring to a set/language A, the assertion that
A is *decidable* is equivalent to the assertion that the characteristic function[1] κ_A of A
is *computable*; the assertion that A is *undecidable* is equivalent to the assertion that
κ_A is *not computable*. (We really need some standard model of computer in order to
make the preceding sentence precise. Any "reasonable" model will work.)

---

[1] Recall that $\kappa_A(x) = 1$ when $x \in A$ and $= 0$ when $x \notin A$.

In order to develop computability theory, we must extend the preceding notions in a direction that you may not have anticipated:

A *set* (of integers or strings) is: $\left\{ \begin{array}{c} \textit{semidecidable} \\ \textit{recursively enumerable} \end{array} \right\}$

A *computational problem* is: *partially solvable*

A *property of a system* is: *semidecidable*.

All of the preceding notions are equivalent to one another. When referring to a set/language A, the assertion that A is *semidecidable* is equivalent to the assertion that the semicharacteristic function,[2] κ'_A of A is *semicomputable*. (Here again, we can invoke any "reasonable" model to ground the notion.)

> Probably the best concrete way to think about semicharacteristic functions in a computational setting is as follows. If the semicharacteristic function κ'_A is computed by a program P, then P would halt and say "YES" when presented with an input that belongs to set A, and P would never halt when presented with an input that does not belong to set A. (Perhaps P would go into a tight loop in response to an input that is not a member of A.) To emphasize the possibility that P may not halt, we call the function κ'_A *semi*computable.

Even at this early stage of our study, we have access to an important result, which helps define the terrain that we are traversing, by relating the notions we have just been discussing. For simplicity, we mention only decidability and semidecidability; we could just as well focus on any of the equivalent notions.

Lemma 9.1. *A language $L \subseteq \Sigma^*$ is decidable if and only if both L and $\overline{L} = \Sigma^* \setminus L$ are semidecidable.*

Proof. Say first that L is decidable, so that its characteristic function κ_L is computable, say by the program P_L (which, recall, halts on all inputs). We can then semicompute the semicharacteristic functions of L and $\overline{L}$ via the following schematic programs. On input $x \in \Sigma^*$:

- $P'_L(x)$ computes $\kappa'_L(x)$ by simulating program P_L. If $P_L(x)$ halts and outputs 1 on input x, which means that $\kappa_L(x) = 1$, then P'_L halts and outputs 1; if $\kappa_L(x) = 0$, then P'_L enters a loop (hence, never halts).
- Similarly, $P'_{\overline{L}}(x)$ computes $\kappa'_{\overline{L}}(x)$ by simulating program P_L. If $P_L(x)$ halts and outputs 0 on input x, which means that $\kappa_L(x) = 0$, then $P'_{\overline{L}}$ halts and outputs 1; if $\kappa_L(x) = 1$, then $P'_{\overline{L}}$ enters a loop (hence, never halts).

The programs P'_L and $P'_{\overline{L}}$ exist because the function κ_L is total and computable. Moreover, P'_L clearly computes κ'_L, while $P'_{\overline{L}}$ clearly computes $\kappa'_{\overline{L}}$.

Say next that both L and $\overline{L}$ are semidecidable, so that κ'_L is semicomputable via some program P'_L, while $\kappa'_{\overline{L}}$ is semicomputable via some program $P'_{\overline{L}}$. The following program, call it P_L, computes κ_L. On input $x \in \Sigma^*$:

1. $P_L(x)$ simulates one step of $P'_L(x)$ and one step of $P'_{\overline{L}}(x)$. At most one of the simulated programs can have halted in this step. If either one halts, then $P_L(x)$ halts and gives the appropriate output: it outputs 1 if it was $P'_L(x)$ that halted, and 0 if it was $P'_{\overline{L}}(x)$ that halted.

---

[2] Recall that $\kappa'_A(x) = 1$ when $x \in A$.

2. $P_L(x)$ iterates this step until one of $P'_L(x)$ and $P'_{\overline{L}}(x)$ has halted.

If neither $P'_L(x)$ nor $P'_{\overline{L}}(x)$ has halted in the simulation thus far, then $P_L(x)$ simulates one more step of $P'_L(x)$ and one more step of $P'_{\overline{L}}(x)$. At most one of the simulated programs can have halted in this step. If either one halts, then $P_L(x)$ halts and gives the appropriate output: it outputs 1 if it was $P'_L(x)$ that halted, and 0 if it was $P'_{\overline{L}}(x)$ that halted.

Because $P'_L(x)$ computes $\kappa'_L(x)$, and $P'_{\overline{L}}(x)$ computes $\kappa'_{\overline{L}}(x)$, and because $L \cup \overline{L} = \Sigma^\star$, (precisely) one of the programs $P'_L(x)$, $P'_{\overline{L}}(x)$ will halt eventually. Therefore, $P_L(x)$ halts on all inputs, and its output tells whether that input belongs to L. In other words, P_L computes κ_L. $\square$

9.2.2 Functions and Partial Functions

This is a good time to review the material in Section 2.3.

Although one can map any set S to any set T via functions (cf. Section 2.3), we greatly simplify our exposition by focusing on only a few very restricted sets S and T as we develop computability and complexity theory. We accomplish this by selecting *a fixed countable "universal" set* U that will serve as a universe of discourse whenever we talk about computing functions. The encodings we presented when discussing countability in Chapter 7 give us an extremely broad range of choices for the set U, but we shall usually stick with just a few sets that have long histories in studies of computability theory and, more generally, of computation theory. These "standard" universal sets are:

- the set $\mathbb{N}$ of nonnegative integers;
- $\mathbb{N}$'s almost-twin, the set $\mathbb{N}^+$ of positive integers;
- the set $\Sigma^\star$ of finite-length strings over some finite alphabet $\Sigma = \{\sigma_1, \sigma_2, \dots, \sigma_n\}$. We do not care which Σ you choose to use—*as long as* Σ *contains both* 0 *and* 1. For simplicity, we shall often employ $\Sigma = \{0, 1\}$ as our universal set.
- $\Sigma^\star$'s specialization to binary strings: the set $\{0, 1\}^\star$.

With each topic we study, we shall choose a universal set U and always talk either about (partial) functions $f : U \to U$ or about (partial) functions $g : U \to \{0, 1\}$. (The latter class of functions is mandated by our focus on *languages*.) Thus, while we have almost free choice of the source set S, we constrain the target set T to be either S or $\{0, 1\}$. The qualifier "partial" that we used in describing the functions f and g emphasizes that a function $f : U \to U$ or $g : U \to \{0, 1\}$ may, in fact, be *defined*— i.e., produce a result—on only a proper subset of the set U; indeed, when g is the semicharacteristic function of a set S such that $S \subset \Sigma^\star$ (note the *proper* subset sign), then g is guaranteed to be defined on only a proper subset of the set U.

> Technically, the existence of *nontotal* functions means that the set U is the *source* of the function f, rather than its *domain*, which is, by definition, the subset of U where f is defined.

Section 7.1 and Chapter 8 tell us that there exist pairing functions for the just-enumerated universal sets that are not only computable, but even *efficiently* computable. Therefore, we never have to widen our focus explicitly in order to accommodate *multivariate* functions—functions of several variables. By using a pairing function to map $U \times U$ one-to-one onto U, where U is one of our universal sets, we ultimately reduce all computable functions to computable *univariate* (one-variable) functions. For instance, if we wish to discuss the bivariate function $h : U \times U \to U$, then we would select some easily computed pairing function $p : U \times U \leftrightarrow U$ and refocus our attention from the function h to the univariate function $q : U \to U$ defined by

$$q(u) \stackrel{\text{def}}{=} h(p^{-1}(u)).$$

Of course, the functions h and q are computationally equivalent, in the sense that $h(u_1, u_2) = q(p(u_1, u_2))$; moreover, if we have chosen p well, then h and q are also roughly equal in computational complexity.

> Our insistence on a fixed universe is mostly a happy decision, as it allows us to talk about functions and compositions of functions without worrying about questions of compatibility between sources and targets, or between domains and ranges. One unhappy consequence of this insistence, though, is that we have to develop computability theory and Computational Complexity Theory as theories of *partial* functions.

We reiterate from Section 2.3 the possibly unfortunate—but critically important—(historical) fact that *every function is, by default, a* partial *function—even the total ones;* in other words, "partial" is the generic term, with "total" being a special case.

Although we seldom talk explicitly about nontotal partial functions in everyday discourse, we in fact deal with such functions all the time, especially in our professional roles as students and/or practitioners of computing. We illustrate this fact with a few common sample nontotal functions, each having the set $\mathbb{N}$ of nonnegative integers as its source:

- the function $f(n) = \sqrt{n}$ is *partial*, being defined only when n is a perfect square;
- the function $g(n) = n/2$ is *partial*, being defined only when n is even;
- the function $h(n) = n - 1$ is *partial*, being defined only when n is positive.

Although it is not relevant to the point we are making here, it is worth noting that we often—but certainly not always—choose to simplify our lives by extending nontotal functions such as these to make them total.

- We replace $f(n)$ by either $\widehat{f}(n) = \lceil \sqrt{n} \rceil$ or $\check{f}(n) = \lfloor \sqrt{n} \rfloor$.
- We replace $g(n)$ by either $\widehat{g}(n) = \lceil n/2 \rceil$ or $\check{g}(n) = \lfloor n/2 \rfloor$.
- We replace $h(n)$ by $\widetilde{h}(n) = n \ominus 1$.

> In these extensions, $\lfloor x \rfloor$ denotes the *floor* of x (Section 2.6), $\lceil x \rceil$ denotes the *ceiling* of x (Section 2.6), and $\ominus$ denotes *positive subtraction* (Section 8.2.1).

9.2.3 Self-Referential Programs: Interpreters and Compilers

The major concepts of computability theory were developed before programmable computers existed. It is quite remarkable, therefore, to note that the people who developed the theory came up with the notion of an *interpreter*: a program P that

- takes two strings, x and y, as arguments,
- interprets string x as an encoding of a program in some predetermined language,
- interprets string y as an encoding of an input to program x,
- simulates program x step by step on input y.

As we remarked earlier, all programs and inputs ultimately get encoded as binary strings in a real computer, so the preceding scenario is not so far-fetched.

Here is where things take an interesting turn. There is no reason that the strings P, x, and y could not all be the same string! Were this the case, then the interpreter would be simulating itself operating on itself. Such *self reference* plays havoc with our intuitions, as you can see by pondering whether the following sentence is true: "This sentence is false." As we begin our excursion into the mysteries of computability theory, keep the notion "self reference" in mind. In some sense, it is the origin of many of the unpleasantnesses that the work of Gödel and Turing uncovered—namely, incompleteness and uncomputability. Like it or not, self reference and all of its by-products are part of our lives as mathematicians and computer scientists—and speakers of complex natural languages. The next section exposes in detail the first of these by-products.

9.3 The Halting Problem: The "Oldest" Unsolvable Problem

This is a good time to review the material in Section 7, which contains the mathematically "purest" version of the mathematical tools we shall be invoking from now on.

This section is devoted to exposing the fundamental nature of the *halting problem*, one of the first computational problems to have been identified as "unsolvable," in the computation-theoretic use of the word.

> You may want to refer back to Section 1.1.2, more specifically to the comment there by Andrew Pitts, for a bit more historical detail on the "original" computation-theoretically unsolvable problem.

Once we understand the halting problem, we shall be able to lay the foundation for understanding the problem's far-reaching computational consequences.

The *halting problem* (HP, for short) is the following set of ordered pairs of strings:

$$\text{HP} \stackrel{\text{def}}{=} \{\langle x, y \rangle \mid \text{Program } x \text{ halts when presented with input } y\}.$$

By using a pairing function, we turn HP into a language. The *diagonal* halting problem (DHP, for short) is the set of all programs that halt when supplied with their own descriptions as input. Symbolically,

$$\text{DHP} \stackrel{\text{def}}{=} \{x \mid \text{Program } x \text{ halts when presented with input } x\}.$$

(*a*) Note the *self reference* inherent in DHP: each string $x \in$ DHP is both a program and an input to that program!

(*b*) If you look back at our discussion of *un*countability in Section 7.2, you should understand the adjective "diagonal" in DHP's name.

9.3.1 The Halting Problem Is Semisolvable but Not Solvable

Hopefully, we have built up the suspense sufficiently that you *really* want to understand why HP and DHP are not solvable. Let's relieve the suspense.

Theorem 9.1. *The diagonal halting problem is not solvable. In other words: the set DHP is not decidable. Hence, the same is true for the halting problem HP.*

Proof. We focus only on proving that DHP is unsolvable, because trivially, if HP were solvable, then so also would be DHP. To wit: for any string x, we have $x \in$ DHP if and only if $\langle x,x \rangle \in$ HP. Therefore, if we could decide the truth or falsity of the sentence "$\langle x,x \rangle \in$ HP," then we could also decide the truth or falsity of the sentence "$x \in$ DHP."

> We shall see in Section 9.4 that the preceding argument actually shows that DHP is *mapping-reducible* to HP; i.e., instances of DHP can be *encoded* as instances of DHP, under a strong notion of encoding called a *mapping reduction*. The argument is thus a very simple illustration of an extremely powerful tool for analyzing the logical interrelationships among computational problems. The uses of the tool abound in computability and complexity theories.

Assume, for contradiction, that DHP were decidable. There would, then, be a program—call it P—that operates on strings and that behaves as follows.

On input x, program P outputs:
> 1 if string x, interpreted as a program, halts when presented
>   with string x as an input,
> 0 if string x, interpreted as a program, does not halt when presented
>   with string x as an input.

(Program P, if it existed, would compute the characteristic function κ_{DHP} of DHP.)

For convenience, let us henceforth apply a shorthand to programs such as P, by rewriting P in the following way.

On input x, program P outputs:
> 1 if program x halts on input x,
> 0 if program x does not halt on input x.

I now want to draw on your experience writing programs. You should agree that if you were presented with program P, then you could modify it to obtain a program P' that behaves as follows.

On input x, program P' outputs:

$$\begin{array}{ll} 1 & \text{if program } x \text{ } \textit{does not halt} \text{ on input } x, \\ \begin{array}{l}\text{LOOP} \\ \text{FOREVER}\end{array} & \text{if program } x \text{ halts on input } x. \end{array}$$

(Program P', if it existed, would compute the *semi*characteristic function κ'_{DHP} of DHP.)

It is worth spending a moment to make sure that we are "on the same page." How would one construct program P' from program P? You would take program P and apply input x to it, and then wait to see what output program P emits. *Note that we are assuming that program P halts on all inputs!* If program P outputs 0 when it halts, then we would have program P' halt and output 1; if program P outputs 1 when it halts, then we would have program P' loop forever, via a statement such as

FOO: **goto** FOO

Now, there is no need for you to waste *your* time running program P on input x. You could, instead, invoke an interpreter for program P to do this for you. Program P' would then have a form something like the following. (This is the format we shall use henceforth for "high-level" programs.)

Program P'	
Input	x
if	program x outputs 0 on input x
then	output 1
else	loop forever

Back to our argument. We have placed no restriction on the input to either program P or program P'. In particular, this input could be the string P' itself. (Here is the self reference!) How does program P' respond to being presented with its own description? The following sequence of biconditionals ("if-and-only-if" statements) tells the story. (The highlighted sentences are explanatory notes, not part of the story.)

program P' *halts* when presented with input P'

if and only if

program P' outputs 1 when presented with input P'

/*By definition, P' outputs 1 *if it halts, and (of course) it halts if it outputs* 1. */

if and only if

program P' *does not halt* when presented with input P'

/*This is how program P' is specified! */

You can "play" this sequence of statements in either direction—biconditionals point in both directions. In either case, the final statement that you arrive at contradicts the first statement.

What can be wrong here? The contradictions that we derive by traversing our sequence of biconditionals in both directions tell us incontrovertably that there is something wrong somewhere in our argument.

1. *If* DHP were solvable, *then* we could write a version of program P that halted on every input.

2. *If* we could write a version of program P that halted on every input, *then* we could write a version of program P' that behaves as claimed.

But our problem with the sequence of biconditionals tells us that we *cannot* write a version of program P' that behaves as claimed. As we have just seen, the only reason that program P' cannot exist is that *program P cannot exist*. This means that DHP cannot be solvable, which proves the theorem. □

Although Theorem 9.1 shows that problem HP is "hard," this bad news is somewhat moderated by the fact that the problem is *partially* solvable, in the following sense.

Theorem 9.2. *The halting problem HP is partially solvable; that is, as a set HP is semidecidable.*

Proof. To semidecide if the input $\langle x, y \rangle \in HP$, construct a program P that behaves as follows:

- Program P simulates program x on input y.
- If program x ever halts and gives an output, then program P halts and gives the same output.

Note that we have given no indication of what program P does if program x never halts. It will be our standard practice to have such unspecified behavior betoken a program's entering a tight loop, hence never halting. This is fine when we are *semi*deciding a language, as we are doing here with HP; cf. Section 9.2.1. □

9.3.2 Why We Care about the Halting Problem—An Example

The reader can legitimately question why the halting problem (by which I mean both HP and DHP) is so important. After all, the question whether a program halts on a specific input does not arise very often in "real" programming (although one does often program as though such halting were guaranteed). In fact, the real importance of HP and DHP in "real" computing contexts is immense, albeit indirect. Myriad problems that *are* a central focus in "real" programming turn out to be unsolvable precisely because HP and DHP are! While the preceding claim must be justified gradually over the next three sections, we now present one simple example that hints at these myriad consequences of Theorem 9.1.

Imagine that you have written a program P, and you are worried that there may exist inputs that will send P into an infinite loop. It would be very comforting to have access to a "metaprogram" $P^\star$ that would "look at" P and reassure you that your worries are groundless. Regrettably, program $P^\star$ cannot exist in general—although versions of $P^\star$ may exist for very constrained classes of programs P. Moreover, the fact that no general "magic bullet" such as program $P^\star$ can exist follows from the undecidability of HP and DHP! Here is the mathematical formulation of the preceding problem and the proof of its unsolvability.

The "magic bullet" program P^* would compute the characteristic function κ_{TOT} of the set/language

$$TOT = \{x \mid \text{program } x \text{ halts on all inputs}\}.$$

The fact that P^* cannot exists thus takes the following form.

Theorem 9.3. *The set TOT is undecidable.*

Proof. As with most theorems that assert the undecidability of a set/language S, this theorem is proved by assuming the decidability of S and the derivation of a contradiction from that assumption.

Assume, for contradiction, that the set TOT were decidable—or equivalently, that TOT's characteristic function κ_{TOT} were computable. Let *ONE* denote a program that computes the (total) constant function $f(n) \equiv 1$; for definiteness, we note that *ONE* could look as follows:

Program *ONE*
Input n
output 1

Consider the following infinite family of programs that are indexed by all of the strings in $\{0,1\}^*$. For each $x \in \{0,1\}^*$, the program associated with string x is:

Program *ONE$_x$*
Input n
if   program x halts on input x
then   simulate program *ONE* on input n
else   loop forever

You should be able to verify that the function F_{ONE_x} computed by program *ONE$_x$* satisfies the following:

$$F_{ONE_x}(n) = \begin{cases} F_{ONE}(n) & \text{if } x \in DHP, \\ \text{undefined} & \text{otherwise.} \end{cases}$$

In particular, we find the following chain of biconditionals that expose the behavior of program *ONE$_x$*:

Program *ONE$_x$* halts on *all* inputs n

if and only if

program x halts on input x

if and only if

program x belongs to the set/language DHP; i.e., $x \in$ DHP.

This chain of biconditionals boils down to the following crucial one.

$$[ONE_x \in TOT] \iff [x \in DHP].$$

The preceding biconditional means that if the set TOT were decidable, then we could use its (*computable!*) characteristic function to decide the undecidable set DHP— by deciding whether $ONE_x \in$ TOT. This fact can be expressed symbolically via the following equation.

$$\kappa_{TOT}(ONE_x) = \kappa_{DHP}(x).$$

Because we know that κ_{DHP} is not computable, we infer that κ_{TOT} is also not computable, which means, of course, that TOT is not decidable.   □

Of course, we could have used a program for any specific total computable function in place of *ONE* in the preceding proof. Some simple candidates: a program for the identity function on $\mathbb{N}$, $f(n) = n$; or a program for the "square" function on $\mathbb{N}$, $f(n) = n^2$; or a program for the function that reverses (say, binary) strings, $f(x) = x^R$. You should check your understanding of the proof by making sure that you understand how to replace *ONE* by any of these other functions.

The proof of Theorem 9.3 shows that one can "reduce" the halting problem DHP to the problem TOT, in the (mathematical) vernacular sense of that verb. We now turn to the task of formalizing the notion of reduction, via a special form of *encoding* of one problem as another. (In fact, there are many notions of "reduction" in computability theory—cf. [80]—and even more in complexity theory—cf. [18, 48] for early examples.) In the next section, we begin to develop one of the most common notions of the reduction of one problem to another, one that is particularly easy to interpret as an "encoding." As the remainder of the book unfolds, you will see how powerful the following simple notion of "reduction" is.

9.4 Mapping Reducibility

At an intuitive level, the ability to "reduce" one computational problem, A, to another computational problem, B, means that we can use the ability to solve an instance of problem B to "help" us produce a solution to an instance of problem A. Referring back to Section 9.3.2, we were able to decide whether program ONE_x computed a total function—i.e., whether ONE_x belonged to the language TOT—by deciding whether program x halted when presented with input x—i.e., whether x belonged to the language DHP.

> The specific question, "Does ONE_x compute a total function?" is an *instance* of problem TOT; the set/problem TOT can be viewed as the totality of such instances.

The major source of informality in the preceding description is the meaning of the word "help." In the context of computability theory, we use the word to mean that "one" can convert any program that decides language B into a program that decides language A. The quotes around "one" are intended to suggest that we have not yet resolved all the informality with our opening intuitive definition. Specifically, we are not willing to allow human intervention in the helping process: we want the process of producing a solution to an instance of A from a solution to an instance of B to be accomplished by a program! More specifically, we want to be able to write a program $P_{B \to A}$ that produces an instance of problem B from each instance of problem A, in such a way that, for all strings x,

$$[x \in A] \quad \text{if and only if} \quad [P_{B \to A}(x) \in B].$$

Note that program $P_{B \to A}$ *must always work*; i.e., it must compute a *total* function.

The fact that $P_{B \to A} \in \text{TOT}$ is crucial for its role in reducing problem A to problem B. It is, however, a coincidence that our first encounter with such a function occurs as we are discussing a reduction to problem TOT.

When we study complexity theory, in Chapter 13, we add some notion of *efficiency* to the requirements for program $P_{B \to A}$. But let's stick with computability theory for now.

Let's start to get formal. For convenience, let us fix on a specific (but arbitrary) finite alphabet Σ and say that all the languages of interest are subsets of $\Sigma^\star$. The following is one of the most important concepts in all of computation theory.

Language $A \subseteq \Sigma^\star$ is *mapping-reducible* (*m-reducible*, for short) to language $B \subseteq \Sigma^\star$, written

$$A \leq_m B,$$

if and only if there exists a *total computable function* $f : \Sigma^\star \to \Sigma^\star$ such that for all $x \in \Sigma^\star$:

$$[x \in A] \quad \text{if and only if} \quad [f(x) \in B]. \tag{9.1}$$

We call the function f that encodes instances of language A as instances of language B a *reduction function*.

It is fruitful to view the function f as a mechanism for *encoding* instances of problem A as instances of problem B, so that is the terminology we shall use most of the time. We shall see now how such encoding can "help" one decide or semidecide a set/language.

We focus solely on mapping reducibility throughout our introduction to computability theory and complexity theory. The reader should realize, though, that there are many other important notions of reducibility that enrich at least computability theory, and possibly complexity theory also. The interested reader should consult an advanced source such as [80] to see the range of significant notions studies within computability theory. At the "weakest" end of such notions, one encounters a version of m-reducibility, called *one-one reducibility*, that employs only reduction functions that are *one-to-one*. At the "strongest" end of such notions, one encounters the notion of *Turing reducibility*, which allows the computation that decides language B to "help" the computation that decides language A *throughout* the latter computation (by "answering" questions about the computation thus far), not just at its inception (by encoding one input as another as m-reducibility does). The description of notions of reducibility as "weaker" or "stronger" is intended to indicate that if language A is reducible to language B under some specific notion of reducibility, then A is reducible to B also under any "stronger" notion. (The relation "stronger" is the converse of the relation "weaker.") For the three notions we have mentioned explicitly, for instance, we find that one-one reducibility is "weaker" than m-reducibility, which, in turn, is "weaker" than Turing reducibility.

The reason that we focus solely on mapping reducibility is its combination of mathematical simplicity and intuitive appeal.

- Mapping reducibility admits a very appealing informal interpretation via the notion of *encoding* instances of language A as instances of language B.

- Sources such as [80] describe the *structure* that the various notions of reducibility give to the decidable and semidecidable languages, by studying which decidable or semidecidable languages are reducible to which others. If one surveys such results, one finds that the reductions produced under m-reducibility are the most uniformly consistent with one's intuition, if one interprets "reducible to" as "helped by."

9.4.1 Basic Properties of m-Reducibility

The preceding section laid out an intuitive goal for m-reducibility. (We shall generally use this shortened name in place of "mapping reducibility.") Specifically, if one set/language A is m-reducible to another set/language B, than the ability to decide or semidecide B should "help" one do the same process for A. It turns out to be easy to find a formal sense in which m-reducibility plays the desired role.

Lemma 9.2. *Let A and B be languages over the alphabet Σ, and say that $A \leq_m B$.*

(a) *If language B is semidecidable (resp., decidable), then language A is semidecidable (resp., decidable).*

(b) *Contrapositively, if language A is* not *semidecidable (resp., not decidable), then language B is* not *semidecidable (resp., not decidable).*

Proof. We need prove only part (a) explicitly. Let f be the total computable function that m-reduces A to B, and let φ be a program that computes f. Let's review precisely what this means.

If language B is *decidable,* then there is a program P that always halts such that when presented with a string $x \in \Sigma^\star$, program P outputs 1 when $x \in B$ and 0 when $x \notin B$. If language B is *semidecidable,* then there is a program P' that, when presented with a string $x \in \Sigma^\star$, halts and outputs 1 precisely when $x \in B$; program P' loops forever if $x \notin B$.

In either case, we can use program φ as a preprocessor to either program P or program P'. Now, by definition, φ converts any input $y \in \Sigma^\star$ whose membership in language A is of interest to an input $f(y) \in \Sigma^\star$ that belongs to language B if and only if y belongs to A. Therefore, the composite program φ-then-P is a decider for language A; and, the composite program φ-then-P' is a semidecider for language A. Here, more explicitly, is what the composite program φ-then-P looks like:

Program φ-then-P
Input $x \in \Sigma^\star$
Compute program φ on input x: $y \leftarrow \varphi(x)$
Compute program P on input y

.

The composite program φ-then-P' differs only in the presence of P' on the last line, instead of P. □

It is of great importance for the development of the theory that the m-reducibility relation is transitive: for any three languages $A, B, C \subseteq \Sigma^\star$, if $A \leq_m B$ and $B \leq_m C$, then $A \leq_m C$.

Lemma 9.3. *The relation "is mapping-reducible to,"* $\leq_m$, *is transitive.*

Proof. Focus on three arbitrary languages, $A, B, C \subseteq \Sigma^\star$. Say that:

- $A \leq_m B$ via the total computable function f; i.e.,

$$(\forall x \in \Sigma^\star) \, [x \in A] \Longleftrightarrow [f(x) \in B].$$

- $B \leq_m C$ via the total computable function g; i.e.,

$$(\forall x \in \Sigma^\star) \, [x \in B] \Longleftrightarrow [g(x) \in C].$$

Then it is easily seen that

$$(\forall x \in \Sigma^\star) \, [x \in A] \Longleftrightarrow [g(f(x)) \in C]. \tag{9.2}$$

Since the composition of total computable functions is another total computable function, condition (9.2) is equivalent to saying that $A \leq_m C$. □

9.4.2 The s-m-n Theorem: Where Does One Find Encodings?

We now present a result from [50] that allows us to formalize the technique that we used in Section 9.3.2 to encode one problem as another. Although this result is quite transparent to anyone familiar with computers, we should never lose sight of the fact that in common with much of the foundational work in computability theory, this result was discovered and proved by people who had never seen a programmable computer!

The result is known within the computability theory community as the s-*m-n theorem*. The m and n in the name of the theorem actually are *variables* that range over $\mathbb{N}$, so that specific invidual instantiations of the s-*m-n* theorem have names such as "the s-1-2 theorem." (The "s" in the name is an uninterpreted letter, but it probably meant "substitute" originally. You'll see why imminently). Our applications of the theorem will typically have $m \in \{1, 2\}$ and $n = 1$, although these restrictions are decidedly not inherent in the result.

In preparation for stating the s-*m-n* theorem, let us revisit the proof of Theorem 9.3. A more systematic rendition of what we did in the proof is the following.

First, we converted the single-input program *ONE*, which computes the constant function $f(n) \equiv 1$ to the two-input program $ONE^{(2)}$:

Program $ONE^{(2)}$
Input $\quad x\ (x \in \{0,1\}^\star)$
Input $\quad n\ (n \in \mathbb{N})$
if $\qquad$ program x halts on input x
then $\quad\;$ simulate program ONE on input n
else $\qquad$ loop forever

Next, we converted program $ONE^{(2)}$ into the infinite family of programs ONE_x, where x varies over $\{0,1\}^\star$, which is depicted in the proof of Theorem 9.3. It was this infinite family that yielded the contradiction that establishes Theorem 9.3.

What the s-m-n theorem (really the s-1-1 theorem) asserts is that one can *automate* the process of producing the family $\{ONE_x\}$ of indexed programs from the original program ONE. Specifically, one can write a single-input program P *that always halts* and that in response to any input $x_0 \in \{0,1\}^\star$ will produce the string that is program ONE_{x_0}. One can view program P as actually producing the following variant, ONE'_x, of program ONE_x.

Program ONE'_x
Input $\quad n\ (n \in \mathbb{N})$
$x := x_0$
if $\qquad$ program x halts on input x
then $\quad\;$ simulate program ONE on input n
else $\qquad$ loop forever

Note that in essence, all program P does in response to input x_0 is transform program $ONE^{(2)}$ by replacing the input statement "**Input** x" with the assignment statement "$x := x_0$" (which is often articulated "x gets x_0"). Clearly, then, one can write a version of program P that halts on all inputs.

The type of transformation we have just described is easily generalized to produce the s-m-n theorem. We state the theorem in its general form, but we leave its proof, which just generalizes our description of program P, to the reader.

Theorem 9.4. (The s-m-n theorem) *Let us be given the program Ψ that has $m + n$ input variables, $X_1, X_2, \ldots, X_m$ and $Y_1, Y_2, \ldots, Y_n$, that is depicted schematically in Figure 9.1(a). There exists a program P that has m inputs: $X_1, X_2, \ldots, X_m$, such that:*

- *Program P halts on all inputs.*
- *In response to inputs $x_1, x_2, \ldots, x_m$, program P converts program Ψ to the program Ψ' that has n input variables, $Y_1, Y_2, \ldots, Y_n$, that is depicted schematically in Figure 9.1(b).*

The next section exercises Theorem 9.4 vigorously, so be sure that you understand the proof before proceeding.

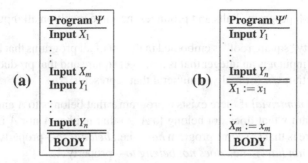

Fig. 9.1 (a) *The program* Ψ *and* (b) *the program* Ψ' *of Theorem 9.4.*

9.5 The Rice–Myhill–Shapiro Theorem

The theorem we are about to prove states—informally!—that there is nothing "nontrivial" that one can determine about the function computed by a program from the program's static description. The word "nontrivial" here precludes behavioral properties that are true either of no program or of every program. Now, on to the formal statement of the theorem.

A set/language $A \subseteq \Sigma^\star$ (whose constituent strings are interpreted as programs) is a *property of functions* (*PoF*, for short) if the following is true. (We say it in three distinct but equivalent ways for emphasis.)

> *If the programs x and y compute the same function (say, for definiteness, from $\Sigma^\star$ to $\Sigma^\star$), then either both x and y belong to A, or neither belongs to A.*

Equivalently,

> *If a program x belongs to A, then so also do all other programs that compute the same function as x.*

Equivalently,

> *All programs that compute the same function lie on the same side of the metaphorical line that separates the set/language A from its complement $\overline{A} = \Sigma^\star \setminus A$.*

PoFs are our formal mechanism for talking about functions within computability theory: we identify a "property of functions" with the set of all programs that compute functions that enjoy the desired property. A few examples:

- The property "total function" is embodied in the set of all programs that halt on every input (hence, compute functions that are total). This is our (by now) old friend TOT $= \{x \mid \text{program}_x$ halts on all inputs$\}$.
- The property "empty function" is embodied in the set EMPTY of all programs that never halt on any input: EMPTY $= \{x \mid \text{program}_x$ never halts on any input$\}$.
- The property "constant function" is embodied in the set of all programs that halt and produce the same answer, no matter what the input is:

$\{x \mid \text{program}_x \text{ halts and produces the same result on all inputs}\}$.

- The property "square root" is embodied in the set of all programs that halt precisely when their input n is an integer that is a perfect square and that produce, when they halt, the output $\sqrt{n}$ (or really, a numeral that represents $\sqrt{n}$).

A PoF A is *nontrivial* if there exists a program x that belongs to A and there exists another program y that does not belong to A, so that neither A nor $\overline{A}$ is empty. The main point here is that (a) some program *has*—i.e., *belongs to*—property A, and some program *does not have*—i.e., *does not belong to*—property A.

Amazingly, all we need to know is that a set A is a nontrivial PoF in order to know that A is not decidable.

Theorem 9.5. (The Rice–Myhill–Shapiro theorem) *Every nontrivial PoF is undecidable. In other words: If a language A is a nontrivial property of functions, then A is not decidable. Furthermore, if a program for the empty function EMPTY belongs to PoF A, then A is not semidecidable.*

An alternative statement of Theorem 9.5. *Every problem that corresponds to a nontrivial PoF is unsolvable.*

Proof. Let us concentrate, for definiteness, on the alphabet $\Sigma = \{0, 1\}$ and on programs that compute (partial) functions from $\{0, 1\}^*$ to $\{0, 1\}^*$. As we now know, this is really no restriction because of our ability to encode any finite set as any other finite set (say, by using pairing functions).

Let us denote by "program x" the *program specified by the string x* and by F_x the *function computed by program x*. In particular, let e be a string such that program e loops forever on every input, so that F_e is the empty (i.e., nowhere defined) function; using our earlier notation, $e \in \text{EMPTY}$.

It should be clear how to supply the details necessary to turn the following pseudoprogram into a real interpreter program, in a real programming language (of your choice):

Program Simulate.1	
Inputs	x, y, z
if	program x halts on input x
then	simulate program y on input z
else	loop forever

You should be able to verify that

$$F_{\text{Simulate.1}}(x, y, z) = \begin{cases} F_y(z) & \text{if } x \in \text{DHP}, \\ F_e(z) & \text{if } x \notin \text{DHP}. \end{cases}$$

If you are comfortable with the development thus far, then you should agree that we can replace the input y in **Program** Simulate.1 by a specific string—call it y_0—that is fixed once and for all. (Note that formally, we are applying the s-1-1 theorem of Section 9.4.2 in order to effect this replacement.) We thereby obtain the following pseudoprogram, which again you can convert into a real interpreter program, in a real programming language:

Program Simulate.2	
Inputs x, z	
if	program x halts on input x
then	simulate program y_0 on input z
else	loop forever

You should now be able to verify that

$$F_{\text{Simulate.2}}(x, z) = \begin{cases} F_{y_0}(z) & \text{if } x \in \text{DHP}, \\ F_e(z) & \text{if } x \notin \text{DHP}. \end{cases}$$

Now, the dependence of **Program** Simulate.2 on input x is so formulaic that we could actually supply x to a *preprocessor* for our interpreter program that automatically inserts a value for x into **Program** Simulate.2. Indeed, we can design this preprocessor so that in response to any string x, it produces the following program, which will be the input to our interpreter program:

Program Simulate.3	
Input z	
if	program x halts on input x
then	simulate program y_0 on input z
else	loop forever

In more detail, the preprocessor is specified by the following program.

Program Preprocessor	
Input x	
	"**Input** z
Output	**if** program x halts on input x
	then simulate program y_0 on input z
	else loop forever"

Note that **Program** Preprocessor just outputs a string that is fixed except for the indicated inclusion of the input x. This string is **Program** Simulate.3 with the appropriate value of x in the indicated place. Thus, **Program** Preprocessor *always halts* on every input x; i.e., it computes a *total computable* function, $\mathscr{F} : \{0,1\}^* \to \{0,1\}^*$. You should now be able to verify that

$$F_{\mathscr{F}(x)}(z) = F_{\text{Simulate.3}}(z) = \begin{cases} F_{y_0}(z) & \text{if } x \in \text{DHP}, \\ F_e(z) & \text{if } x \notin \text{DHP}. \end{cases} \tag{9.3}$$

Let us now shift gears and start thinking about an arbitrary but fixed nontrivial PoF A. Now, the program e for the empty function belongs either to A or to $\overline{A}$.

Let us assume that $e \notin A$. We return to the alternative assumption after deducing the consequences of this assumption. We infer the following chain of properties about A.

1. Because A is a *property of functions,* we know that *no program* $y \in A$ *is equivalent to program* e. This is because by definition, if A contained such a program, then it would have to contain program e also.
2. Because of fact 1, we know that *every program* $y \in A$ *halts on* some *input*. This is because any program that violated this would be equivalent to program e.
3. Because A is *nontrivial,* there must be some program that belongs to A. Let y_0 be a program that belongs to A.

Let's see what happens when we let the program y_0 of fact 3 serve as the program y_0 mentioned in **Program** Simulate.3. When this happens, we can infer from (9.3) that

$$[x \in \text{DHP}] \iff [\mathscr{F}(x) \in A].   (9.4)$$

Why is this true? There are two alternatives we must consider.

$x \in \text{DHP}.$ If this is true, then $F_{\mathscr{F}(x)} \equiv F_{y_0}$, as functions. This means that program $\mathscr{F}(x)$ and program y_0 compute the same function. Because $y_0 \in A$ (by hypothesis) *and* because A is a PoF, this means that $\mathscr{F}(x) \in A$.

$x \notin \text{DHP}.$ If this is true, then $F_{\mathscr{F}(x)} \equiv F_e$, as functions. This means that program $\mathscr{F}(x)$ and program e compute the same function (the empty function in this case). Because $e \in \overline{A}$, *and* because $\overline{A}$ is a PoF, this means that $\mathscr{F}(x) \in \overline{A}$.

When analyzing the case "$x \notin \text{DHP}$," we used the following fact, which you should verify: *The set A is a PoF if and only if its complement $\overline{A}$ is a PoF.*

The preceding alternatives verify (9.4).

What have we shown here? Looking at (9.4) and comparing it to the "formula" (9.1) for mapping reductions, we find that we have proved the following.

For any nontrivial PoF A that does not contain *program e, we have* $\text{DHP} \leq_m A$.
By Lemma 9.2, this means that any such set/language A is undecidable.

Finally, what happens to the preceding reasoning when $e \in A$? You will be asked as an exercise to make the changes in our argument occasioned by this change of assumption. By making these changes, you should end up with an argument that proves the following.

For any nontrivial PoF A that does contain *program e, we have* $\overline{\text{DHP}} \leq_m A$.
In this case, Lemma 9.2 tells us that the set A is not semidecidable!

This completes the proof.   □

The proof of Theorem 9.3 is essentially an instantiation of the proof of Theorem 9.5 for a specific language A—in this case, language TOT—that does not contain program e. Let us close the current section with an instantiation of the proof of Theorem 9.5 for a specific language A that *does* contain program e. Let's use the language EMPTY as our example.

Corollary 9.1 *The set EMPTY is not semidecidable.*

Proof. We shall refer lavishly to the proof of Theorem 9.3 in order to avoid needless repetition.

Assume, for contradiction, that the set EMPTY were semidecidable—or, equivalently, that EMPTY's semicharacteristic function κ'_{EMPTY} were partially computable. Let *ONE* denote the program that computes the (total) constant function $f(n) \equiv 1$, as in the proof of Theorem 9.3. Consider the following infinite family of programs that are indexed by all of the strings in $\{0,1\}^*$. For each $x \in \{0,1\}^*$, the program associated with string x is program ONE_x (from the proof of Theorem 9.3 in Section 9.3.2). You should be able to verify that the function F_{ONE_x} computed by program ONE_x satisfies the following:

$$F_{ONE_x}(n) = \begin{cases} F_{ONE}(n) & \text{if } x \in \text{DHP}, \\ F_e(n) & \text{if } x \notin \text{DHP} \text{ (i.e., if } x \in \overline{\text{DHP}}). \end{cases}$$

Note that saying "$F_{ONE_x}(n) = F_e(n)$" is equivalent to saying "$F_{ONE_x}(n)$ is undefined."

In particular, we find the following chain of biconditionals that expose the behavior of program ONE_x:

Program ONE_x fails to halt on *the single* input n
if and only if
program x fails to halt on input x
if and only if
program x belongs to the set/language $\overline{\text{DHP}}$; i.e., $x \notin \text{DHP}$.

This chain of biconditionals boils down to the following crucial one.

$$[ONE_x \in \text{EMPTY}] \iff [x \in \overline{\text{DHP}}].$$

The preceding biconditional means that if the set EMPTY were semidecidable, then we could use its (*partially computable!*) characteristic function to semidecide the nonsemidecidable set $\overline{\text{DHP}}$—by deciding whether $ONE_x \in \text{EMPTY}$. This fact can be expressed symbolically via the following equation:

$$\kappa'_{\text{EMPTY}}(ONE_x) = \kappa'_{\overline{\text{DHP}}}(x).$$

Now, we know that $\kappa'_{\overline{\text{DHP}}}$ is *not* partially computable, or else, by Lemma 9.1, κ_{DHP} would be computable—which it's not (Theorem 9.1). We conclude therefore that κ'_{EMPTY} is also not partially computable, which means, of course, that EMPTY is not semidecidable. $\square$

9.6 Complete, or "Hardest," Semidecidable Problems

One of the most exciting features of mapping-reducibility is that there exist semidecidable problems/languages that in a precise, formal sense are the *hardest* semidecidable problems. We call these hard problems *m-complete*, where, as in the term "m-reducible," the "m" stands for "mapping."

> From this point until the end of the book, we consistently capitalize the words "Complete" and "Hard" when used in the technical sense of this section, so the the reader will recognize easily when the words are being used in a technical sense and when in the vernacular.

A problem $A \subseteq \mathbb{N}$ is m-*Complete* precisely when:

1. A is semidecidable.
2. Every semidecidable problem B is mapping-reducible to A; symbolically: $B \leq_m A$.

The reason that we call m-Complete problems the "Hardest" semidecidable problems is that (by Lemma 9.2): *If any semidecidable problem were decidable, then* all *semidecidable problems would be decidable.*

An informal—but not imprecise—reading of the definition of "m-Complete" indicates that *every semidecidable problem can be "encoded" as any m-Complete problem.* This is an extremely strong property, so strong, in fact, that it is not clear a priori that there exist m-Complete problems! In fact, though, we have been dealing with two of them throughout this chapter. If you had any doubts about the importance (and the relevance) of the halting problem, the following two results should allay your concerns!

Theorem 9.6. *The set HP (the halting problem) is m-Complete.*

Proof. The semidecidability of HP having been established in Theorem 9.2, we concentrate only on the fact that every semidecidable problem m-reduces to HP.

Let A be an arbitrary semidecidable problem. By definition—see Section 9.2.1—this means that the semicharacteristic function κ'_A of the set A is semicomputable. More formally, there is a program P_A such that for all strings x,

$P_A(x)$ *halts precisely when* $x \in A$.

If y is a string (or integer, if you prefer) that is actually a "name" of program P_A—hence, of set A—then the preceding condition can be rewritten as

$\langle y, x \rangle \in$ HP *if and only if* $x \in A$.

Clearly, this last assertion implies that $A \leq_m$ HP via the total computable function f_y defined by $f_y(x) = \langle y, x \rangle$.

> If you choose to develop computability theory in terms of (nonnegative) integers, then the function f_y is a pairing function of the sort developed in Section 7.1; we exhibited several computable ones there. If you choose to use strings (as we have done most of the time), then f_y would be some kind of string encoding of the sequence (left-angle-brace, string y, comma, string x, right-angle-brace). The development in Section 7.1 shows that there are myriad computable options here.

Because A was an arbitrary semidecidable problem, the Theorem follows. □

Showing that the set DHP is m-Complete takes a bit more work, because it is not transparent how to embed a reference to the encoded set A within DHP's single unstructured constituent strings.

Corollary 9.2 *The set DHP (the diagonal halting problem) is m-Complete.*

Proof. The semidecidability of DHP having been established in Theorem 9.2, we concentrate only on the fact that every semidecidable problem m-reduces to DHP. Further, because mapping reducibility is transitive (Lemma 9.3), it suffices to show that HP $\leq_m$ DHP and then invoke Theorem 9.6, which shows that HP is m-Complete.

We begin our demonstration that HP $\leq_m$ DHP by revisiting Section 7.1 in the light of what we now know about computability. Specifically, we recall that there exist *computable injections* $F : \Sigma^* \times \Sigma^* \to \Sigma^*$ for any finite Σ. (We leave the easy verification to the reader.) This means that we can *computably* deal with any ordered pair of strings $\langle x, y \rangle$ as though it were a single string $F(x, y)$.

Consider now the following program.

Program Simulate.1
Input x
Input y
Input z
if   program x halts on input y
then   output z
else   loop forever

Note that whenever the first two inputs to **Program** Simulate.1 form a pair $\langle x, y \rangle$ that belongs to the language HP, the program computes the identity function on its third input. Hence, in this case, **Program** Simulate.1 halts for every value of the third input. When the first two inputs form a pair $\langle x, y \rangle$ that belongs to $\overline{\text{HP}}$ (because program x does not halt on input y) then **Program** Simulate.1 computes *the empty function*. Hence, in this case, **Program** Simulate.1 never halts for any value of the third input.

As we did in the proof of the Rice–Myhill–Shapiro theorem (Theorem 9.5) we can replace **Program** Simulate.1 by an infinite family of programs—one for each ordered pair of strings $\langle x, y \rangle$. (Note that, formally, we are applying the s-2-1 theorem of Section 9.4.2 in order to effect this replacement.) The family member that corresponds to the specific ordered pair $\langle x_0, y_0 \rangle$ reads as follows:

Program Simulate.2:$\langle x_0, y_0 \rangle$
Input z
if   program x_0 halts on input y_0
then   output z
else   loop forever

Note that each program **Program** Simulate.2:$\langle x_0, y_0 \rangle$ in this family either computes the identity function, *which is total*, or the empty function, *which is nowhere defined*; the former occurs when $\langle x_0, y_0 \rangle \in$ HP; the latter occurs when $\langle x_0, y_0 \rangle \in \overline{\text{HP}}$.

Once again, we invoke our knowledge of how interpreters work to assert that we can write a program P that always halts—hence, computes a total computable function—and that on any input pair $\langle x, y \rangle$ produces the string that is the program **Program** Simulate.2:$\langle x, y \rangle$. In other words, this program is the value $P(x, y)$.

We can now ask, for any pair of strings $\langle x, y \rangle$, whether the string $P(x, y)$ belongs to DHP, i.e., whether program $P(x, y)$ halts when it is run with a copy of itself as input. From what we have said earlier, $P(x, y) \in$ DHP if and only if program $P(x, y)$ halts on all inputs—and this happens if and only if $\langle x, y \rangle \in$ HP.

We have thus shown that for all pairs of strings $\langle x, y \rangle \in \Sigma^\star \times \Sigma^\star$, $[\langle x, y \rangle \in$ HP$]$ iff $[P(x, y) \in$ DHP$]$. By definition, then, we have shown that HP $\leq_m$ DHP. It follows that the latter language is m-Complete. □

9.7 Some Important Limitations of Computability

Almost all of our comments about computability theory to this point have extolled the theory's power—as manifest, say, in the Rice–Myhill–Shapiro theorem—and broad applicability—as manifest in the Church–Turing thesis. In order to make the theory part of one's professional life, it is fully as important that one understand its limitations as its strengths. We now briefly describe two of these limitations, one relating to the "negative" assertions that the theory makes and one to its "positive" assertions. **What does "uncomputable/undecidable/unsolvable" mean?** We have demonstrated the unsolvability of a number of problems in this chapter. I want to discuss two of them in a bit more detail. You should be able to extrapolate this discussion to other problems quite easily.

The problem of deciding whether a given program halts on a given input (the Halting Problem, HP) is unsolvable, as is the problem of deciding whether a program halts on all inputs (the set TOT). Yet we clearly encounter program-input pairs for which the halting problem *is* solvable all the time, and there are many programs that we *can* prove halt on all inputs. Indeed, we have discussed many such program–input pairs and programs in this chapter! So what is computability theory trying to tell us? Informally, computability theory is telling us that

- One cannot *automate the process* of deciding whether a given program halts on a given input or whether a given program halts on all inputs.
- *There exist* specific programs for which one cannot decide halting behavior. (The infinite family of programs Program Simulate.2:$\langle x_0, y_0 \rangle$ of the preceding section provide examples.)

Thus, computability theory is really telling us what can be achieved *automatically* and *in general*. The tricky part is that because of the power of encodings, one can never be certain that one's apparently innocent program is not a computable encoding of some troublesome program such as a bad instance of Program Simulate.2:$\langle x_0, y_0 \rangle$. **What does "computable" mean?** It is not surprising that computability theory's "negative" assertions must be read with care, but how can a "positive" assertion of a function's computability create problems? The problem arises from the so-called *law of excluded middle*, a logical principle that posits the truth of the assertion

$$A_1 \vee A_2 \vee \cdots \vee A_n \vee \overline{(A_1 \vee A_2 \vee \cdots \vee A_n)},$$

no matter what the disjuncts $A_1, \ldots, A_n$ are. This logical "law"—which is usually stated in the two-alternative case (the case $n = 1$)—tells us that we can infer the truth of an exhaustive disjunction of alternatives *without knowing which of the alternatives is actually true!* This "law" is, for example, the infrastructure for every proof by contradiction! In such a proof, we show that assertion A leads to an absurdity, so we (usually implicitly) invoke the law of excluded middle to infer that $\overline{A}$ must be true.

How can this "law" that we invoke all the time lead to problems? Consider the following (total) *run-of-7's function* $f : \mathbb{N} \to \mathbb{N}$. Letting π denote, as usual, the ratio of the circumference of a circle to its diameter, define the function f as follows: For each $n \in \mathbb{N}$,

$$f(n) = \begin{cases} 1 & \text{if there is a run of } \geq n \text{ instances of the digit 7 in the decimal} \\ & \text{expansion of } \pi, \\ 0 & \text{otherwise.} \end{cases}$$

I claim that the function f is computable—even though I have no idea how to compute it, and as far as I know, no one else knows how to compute it either! So why is f computable? Consider the alternatives.

1. For every $n \in \mathbb{N}$, the decimal expansion of π contains a run of instances of the digit 7 whose length exceeds n.
2. There exists an $n_0 \in \mathbb{N}$ such that the decimal expansion of π contains a run of n_0 instances of the digit 7, but it does not contain any run of length longer than n_0.

By the law of excluded middle, one of the preceding alternatives is true. Note that there are really infinitely many alternatives here: there exists no longest run; for each $n \in \mathbb{N}$, n is the length of the longest run.

On the one hand, if alternative 1 is true—there is no longest run—then the function f is computed by the program *ONE* described in the proof of Theorem 9.3.

On the other hand, if alternative 2 is true, and n_0 is the length of the longest run, then the function f is a so-called *step function*

$$f(n) = \begin{cases} 1 & \text{if } n \leq n_0, \\ 0 & \text{if } n > n_0. \end{cases}$$

In this case, the following program computes f:

Program ONE-until-n_0	
Input	n
if	$n \leq n_0$
then	output 1
else	output 0

Here is where the footing gets slippery! We have just listed (with the aid of parameterization) an infinite set of programs. By the law of excluded middle, *one of them* computes the function f. This means that there exists a program that computes the function f. By definition, then, f is a computable function. The only stumbling block is that we do not know *which one* of our infinitely many candidate programs actually computes f. But computability theory requires only that we show that *there exists* a program that computes f. The theory does not require us to be able to point to the program!

I am willing to call the preceding problem a "limitation" of computability theory, perhaps even a shortcoming of the theory. If there is any branch of mathematics that should be "constructive," one would expect computability theory to be such a branch—but it's not!

> If the preceding story disturbs you, be assured that you are not alone. There have been several vibrant schools of *constructive* mathematics over the past century or more. (The opposing school is said to do *classical* mathematics.) All of the constructive schools require proofs of existence to be accompanied by explicit demonstrations of an object that satisfies the desired condition. Practitioners of constructive mathematics would, in particular, reject our "proof" that the function f is computable—precisely because we cannot identify the program that computes f.
>
> It is hard to give up tools such as the law of excluded middle that, even when not really needed, tend to make arguments shorter and simpler. For this reason, even some mathematicians who sympathize at some level with the constructive agenda persist in using classical arguments.
>
> If you are interested in seeing a sophisticated attempt to put much of the math you have studied in high school and college on a constructive footing, you should thumb through the fascinating book by Erret Bishop [6].

9.8 (Online) Turing Machines and the Church–Turing Thesis

The current section has three mutually supporting goals that center on exposing the reader to a number of variants of the Turing machine model. We hope thereby to:

1. give the reader some intuition for why people accept the Church–Turing thesis as a working hypothesis.

   We have asserted earlier that all efforts to find a "reasonable" model of "digital computer" that is more powerful than the Turing machine have failed. We want to give some technical meat to these assertions: (*a*) by discussing a number of the proposed augmentations of the TM model; (*b*) by describing how an ordinary TM can simulate each of these proposals.

2. help the reader gain a better understanding of the TM model, particularly the online Turing machine (OTM).

   It is easy to underestimate the capabilities of the OTM because of its primitive resources for data storage and manipulation. Our study thus far has spent ample time exposing the limitations of TMs and their variants, via the notion of undecidability and its even more limiting kin (such as unsemidecidability). It is time to restore the balance a bit.

3. present a sampler of variants of the TM model that one might expect to be *less* powerful than the Turing machine—but are not!
   In order to fully appreciate the capabilities and limitations of digital computation, one must understand how *weak* one can make a computing repertoire—instruction set, data structures—yet still enable the full power of digital computation, as well as how *strong* one can make a computing repertoire without augmenting computing power beyond that of a TM.

The first of these goals is the primary motivation of this section, for all of computability theory and complexity theory depend on the Church–Turing Thesis for the depth of their impact, which results from (the belief in) the breadth of their implications. We therefore organize this section as a series of subsections devoted to individual competitors to the OTM model. We hope that the menu of competitors that we consider is adequate in number and in disparate computing repertoires to enhance the reader's appreciation for the thesis and to supply the reader with the technical tools needed to add to our menu.

> Studying even a small fraction of the massive number of competitors for the TM that have been proposed over the decades is beyond the scope of this book. But we urge the interested reader to follow the pointers sprinkled throughout our discussion in Section 9.1 and elsewhere. Each new competitor model that one considers will enhance one's understanding of the capabilities and limitations of digital computation—and one's appreciation for the Church–Turing thesis.

Regarding our other two goals for this section: Understanding the TM model via its competitors better helps one understand why the apparently simplistic OTM model remains relevant to "real" computing to this very day. Indeed, embellished versions of the OTM can provide a basis for an algorithmic theory of data-structure topology. Specifically, for many types of computations, one can expose the algorithmic impact of data-structure topology by comparing the efficiencies—say in computation time—of competing genres of OTM whose worktapes have competing topologies. We thus abstract the control portion of an algorithm down to a finite state-transition system (the competing OTMs' finite-state controls), and use the OTMs' worktapes to model access to data structures. Indeed, the development in Section 5.5 can be viewed as a valuable test study of this point of view. We urge the reader to reread that section from the described vantage point. For the aspiring researcher, variants on the classical TM are a wonderful source of algorithmic problems about data structures!

Throughout this section, we employ the OTM of Figure 3.5 as our "base" model. In Section 9.8.1, we show how a variety of ways of apparently weakening the OTM model can, in fact, efficiently simulate the OTM. In Section 9.8.2, we show how a variety of ways of apparently strengthening the OTM model can, in fact, be simulated efficiently by the OTM. Because this section is devoted to supplying intuition, we supply highly descriptive sketches of proofs, rather than highly detailed formal proofs.

9.8.1 Simplifying an OTM without Diminishing Its Power

This section is devoted to five modifications of the OTM model that one might expect to deprive the model of some computing power. For each of the five, we show that the new model can simulate arbitrary computations by an unmodified OTM, hence is no weaker than an OTM. The first four of our apparently weaker OTMs can actually simulate an OTM rather efficiently—specifically, with only polynomial slowdown— a fact that played a significant role in the early development of complexity theory, as we discuss in Chapter 13.

A. An OTM with a *One-Ended Tape*

An OTM M can extend its worktape either to the left or the right as a computation proceeds. This two-way extendibility has the potential of complicating analyses of M's behavior, by sometimes requiring an awkward indexing of M's tape squares using both positive and negative integers. Consider, for instance, the (imaginary) snapshot of a computation by M in Figure 9.2(left). In the figure, we imagine that we have

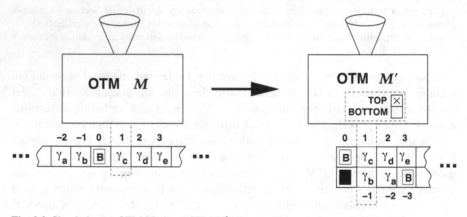

Fig. 9.2 Simulating an OTM M via an OTM M' whose worktape is *one-ended*.

labeled as "square 0" the tape square where M's read/write head resided at the start of some computation and that we have labeled all other squares relative to this base square. As we suggest in Figure 9.2(right), at the cost of complicating M's worktape alphabet a bit, we can rewrite M's program so that M will never extend its worktape leftward from the base square. This allows us to index all squares with *nonnegative* integers—which sometimes leads to simpler analyses. We now flesh out the suggestion of Figure 9.2 to a simulation algorithm.

Proposition 9.1 *For every OTM M, there exists an equivalent OTM M' whose worktape is* one-ended, *i.e., is never extended to the left. M' can simulate M step for step: it executes any t-step computation by M in t steps.*

Proof (Sketch). We produce the one-ended OTM M' from the arbitrary OTM M with the help of an algorithmic device that has broad applications in the worlds of both TMs and data structures: structuring a worktape into *tracks*. Refer to Figure 9.2(right) as we describe the concept and implementation of tracks.

Having a two-track worktape amounts, logically, to splitting each square of the worktape into an upper half and a lower half. Collectively, the sequence of upper halves of the squares forms the *top track* of the tape, and the sequence of lower halves forms the *bottom track*. Extending this idea to $k > 2$ tracks requires merely clerical changes in our description; e.g., one obtains three tracks by splitting each tape square into a top third, a middle third, and a bottom third. The formal mechanism for algorithmically implementing the track concept is via a *product-structured* worktape alphabet. For instance, in order to algorithmically implement a two-track tape, each of whose tracks employs the worktape alphabet Γ, one endows the full tape with the worktape alphabet $\Gamma \times \Gamma$; for three tracks, one would use $\Gamma \times \Gamma \times \Gamma$; and so on.

> There is a hair that needs to be split here. In our intuitive description of tracks, and in Figure 9.2(right), the two tracks of the tape are depicted as being one on top of the other. Our suggested formal implementation via product-structured worktape alphabets views the tracks as implemented by having symbols be ordered pairs (for two tracks), ordered triples (for three tracks), and so on. This inconsistency is dictated by our employing a "notation" that accommodates the different strengths of the textual and graphic media. Once forewarned, the reader should not be misled by the inconsistent conventions.

We apply tape-tracking to the problem at hand in the following way. Let us begin with an OTM M that has the freedom of extending its worktape in either direction; let Γ be M's worktape alphabet. As indicated in the paragraph preceding the proposition, we view M's trajectory on its worktape as inducing a labeling of the squares of the tape with integers. *This labeling is for our convenience in analyzing M's behavior; M does not have access to it.* We view the square where M begins its journey on the tape as getting the label 0. The labels of other squares are determined from the label of this base square: each square's label is 1 greater than the label of its left-hand neighbor and 1 less than the label of its right-hand neighbor. These logical labels, which are provided in Figure 9.2(left), give us a convenient way to describe how we replace M with the desired equivalent OTM whose worktape can be extended only to the right. We obtain a snapshot of M''s tape by "folding" the corresponding snapshot of M's tape, in the manner indicated in Figure 9.2(right). Note that each of M's tape squares that has a positive label k is paired via our folding with the square of M's tape that has label $-k$. Essentially, M' will now be able to simulate M step for step, by

- mimicking M's moves exactly when M is in the positively labeled region of its worktape; M' moves left when M does, and it moves right when M does;
- "flipping" M's moves when M is in the negatively labeled region of its worktape; M' moves left when M moves right, and it moves right when M moves left.

The only complication to this simple step-for-step simulation occurs when M moves onto its square 0 and continues moving in the same direction. The problem is that under our simulation strategy, this sequence of moves requires M' to switch from one track of its worktape to the other. We must endow it with the resources needed to make this switch. We do so via two further slight additions to our description of M'.

1. It is important that M' know when it is at the left end of its worktape, so that as it simulates a move of M wherein M leaves its square 0 toward the left, M' does not try to extend its tape leftward nor to move left from its square 0 and thereby "fall off" the tape. We rather want M' to simulate this leftward move by M by switching from the top track of its tape to the bottom track. The mechanism that we institute to identify square 0 for M' is illustrated in Figure 9.2(right): We do *not* pair square 0 of M's tape with another square of M's tape as we craft square 0 of M''s tape. Instead, we pair square 0 of M's tape with a special symbol $\blacksquare$ that is distinct from all letters in Γ, i.e., $\blacksquare \notin \Gamma$. M' places $\blacksquare$ in the bottom track of the tape square where it starts a computation, which is square 0 of M''s tape; it never writes $\blacksquare$ anywhere else. Thus, we endow M' with the worktape alphabet

$$\Gamma \times \left(\Gamma \cup \{\blacksquare\} \right)$$

rather than just the product $\Gamma \times \Gamma$.

2. As M' simulates moves of M, it must know when to take its current worktape symbol from the top track of its tape and when from the bottom. This is an easy determination, because M':

   - knows that it starts its journey on square 0;
   - can tell when it returns to square 0, by the presence of the symbol $\blacksquare$ on the bottom track of the square;
   - can tell that it is in
     - "positive" territory when M's most recent departure from its square 0 was via a rightward move,
     - "negative" territory when M's most recent departure from its square 0 was via a leftward move.

   We simplify the formal specification of M''s track selection by explicitly implanting a TOP/BOTTOM toggle in M''s state-set. Formally this toggle is implemented by replacing M''s state-set Q by the set $Q \times \{\text{TOP, BOTTOM}\}$.

   The informal description we have provided, augmented by several formal hints, should allow the reader to complete the details needed to turn our detailed descripton into a formal proof of the proposition. $\square$

There is another approach to constructing the one-ended OTM M' from M. One can endow M' with a one-track one-ended worktape and have M' allocate the odd-labeled squares (resp., the even-labeled squares) of this tape, in order, to the squares of M's tape that have nonnegative (resp., negative) labels. We leave the details to the reader.

We have chosen our track-based simulation strategy for two reasons. (1) The idea of endowing a TM's worktape with tracks is useful in a large variety of algorithmic applications. (Indeed, we shall employ tracks again in the next section.) (2) At the cost of endowing M' with a worktape alphabet whose size is roughly the square of M's, the track-based strategy yields *a step-for-step simulation:* M' simulates t steps of a computation by M in exactly t steps. The parity-based simulation strategy allows M' to have an alphabet that is only roughly twice the size of M's, but the simulation incurs a factor-of-2 slowdown.

B. An OTM with *Two Stacks* instead of a Worktape

A *stack* can be viewed as a tape whose contents are accessed and manipulated in a very constrained way. In the following description, we refer repeatedly to Figure 9.3(right), which depicts an OTM $M^{(\text{stack})}$ that has two stacks, which it employs in place of a standard worktape. A stack has a *top*, where all access to data takes place.

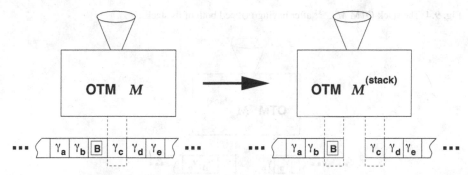

Fig. 9.3 Simulating an OTM M via an OTM $M^{(\text{stack})}$ that uses *two stacks* in place of a worktape.

To conserve precious resources, we have drawn Figure 9.3(right) and its two successor figures with $M^{(\text{stack})}$'s two stacks *on their sides*. Thus, in the figures, the "top" of the lefthand stack is, in fact, the tape's rightmost square, and the "top" of the right-hand stack is, in fact, the tape's leftmost square.

A stack is *read* via the POP operation, which removes the stack's top square. In the figure, when $M^{(\text{stack})}$ POPs its left-hand stack, it thereby reads the symbol $\boxed{\text{B}}$, simultaneously removing the symbol from the stack; a similar operation on its right-hand stack produces the symbol γ_c. The operation POP is destructive: the old top square (which has been POPped) is no longer on the stack. The double-POP just described thus produces the configuration depicted in Figure 9.4. A stack is *written* via the PUSH operation, which places a new top square "on top of" the previous one. Note that the PUSH operation is not destructive: the old top square is still in the stack; it is just the second-from-top square now. Figure 9.5 illustrates how the configuration depicted in Figure 9.4 changes when $M^{(\text{stack})}$ PUSHes the symbol γ_f onto its right-hand stack.

We say that a stack is *empty* if it does not contain any symbols other than the blank symbol $\boxed{\text{B}}$. Accordingly, when one POPs an empty stack, one receives, in response, an instance of $\boxed{\text{B}}$. If, as is convenient in many computations, one wants to detect the "bottom" of the stack—i.e., the symbol that was at the top when the computation began—then one must initially PUSH onto the stack a designated marker for this "bottom."

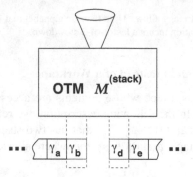

Fig. 9.4 The stack-OTM $M^{(\text{stack})}$ after having POPped both of its stacks.

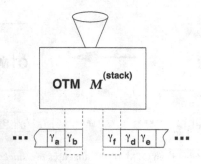

Fig. 9.5 The stack-OTM $M^{(\text{stack})}$ after having PUSHed a new symbol, γ_f onto its right-hand stack.

The computing device $M^{(\text{stack})}$ depicted in Figure 9.3(right)—and in Figures 9.4 and 9.5—is a 2-*stack OTM* (2-*STM*, for short). $M^{(\text{stack})}$ looks much like an OTM, except that two *stacks* jointly form the OTM's only unbounded data structure. Every computation begins with both of $M^{(\text{stack})}$'s stacks "empty," meaning that each has just one square, and that square contains the blank symbol $\boxed{\text{B}}$. The reader should be able to flesh out how $M^{(\text{stack})}$ computes, based on our discussions of the stack data structure and of the semantics of OTMs (in Section 3.3). Of course, we could easily extend the STM model by endowing an OTM with $k > 2$ stacks, but the following result illustrates that the 2-STM already has the power of an OTM, hence fills the needs of this section.

Proposition 9.2 *For every OTM M, there exists an equivalent 2-STM $M^{(\text{stack})}$. $M^{(\text{stack})}$ can simulate any t-step computation by M in $O(t)$ steps.*

Proof (Sketch). We design the 2-STM $M^{(\text{stack})}$ to simulate a given OTM M as follows. Say that $M^{(\text{stack})}$ has a *left-hand* stack and a *right-hand* stack, as in Figure 9.3.

Overall setup

- M begins a computation with a blank worktape, and $M^{(\text{stack})}$ begins with two blank stacks.

- Whenever M polls its input port, $M^{(\text{stack})}$ does likewise.
- Whenever M halts and either accepts or rejects, $M^{(\text{stack})}$ does likewise.
- The inductive correspondence between M's configuration and $M^{(\text{stack})}$'s is as follows:
  - the contents of $M^{(\text{stack})}$'s left-hand stack (bottom to top) are identical to the contents of that portion of M's worktape that lie to the left of M's read/write head;
  - the top symbol on $M^{(\text{stack})}$'s right-hand stack is identical to the symbol currently under scan by M's read/write head;
  - the contents of $M^{(\text{stack})}$'s right-hand stack (top to bottom) that lie *below* the topmost symbol are identical to the contents of that portion of M's worktape that lie to the right of M's read/write head.

Operation

- When M reads the symbol currently under scan on its worktape, $M^{(\text{stack})}$ POPs *both* of its stacks.
  If M reads the symbol γ from its worktape, then, by the inductive hypothesis, γ is the symbol that $M^{(\text{stack})}$ reads from its *right-hand* stack. Let $\overline{\gamma}$ be the symbol that $M^{(\text{stack})}$ reads from its left-hand stack, simultaneously with its reading γ from its right-hand stack.
- Say that M rewrites symbol γ as symbol $\overline{\gamma}'$.

  - If M *stays stationary* on its worktape at this step, then $M^{(\text{stack})}$:
    1. PUSHes $\overline{\gamma}'$ onto its right-hand stack;
    2. PUSHes $\overline{\gamma}$ onto its left-hand stack.
  - If M *moves left* on its worktape at this step, then $M^{(\text{stack})}$:
    1. PUSHes $\overline{\gamma}'$ onto its right-hand stack;
    2. PUSHes $\overline{\gamma}$ onto its right-hand stack.
  - If M *moves right* on its worktape at this step, then $M^{(\text{stack})}$:
    1. PUSHes $\overline{\gamma}$ onto its left-hand stack;
    2. PUSHes $\overline{\gamma}'$ onto its left-hand stack.

$M^{(\text{stack})}$ thus performs $O(1)$ elementary steps for each elementary step by M. Moreover, after $M^{(\text{stack})}$ has performed its elementary steps, the inductive situation is reestablished. This completes the proof. $\square$

It is not difficult to show that a 1-STM is quite weak computationally—certainly not nearly in the same league as an OTM. The easiest path to this insight has three steps. (1) We invoke Chomsky's early demonstration, in [12], that a 1-STM accepts only context-free languages. (2) We recall, from Section 6.1.3, that the language $L = \{a^n b^n c^n \mid n \in \mathbb{N}\}$ is not context-free. (3) We show quite easily how an OTM can accept the language L: the OTM can write the block that consists of occurrences of a on its worktape and then match that string's length against the length of the block that consists of occurrences of b and the block that consists of occurrences of c.

C. An OTM with a *FIFO Queue* instead of a Worktape

A *FIFO queue* can be viewed as a tape whose contents are accessed and manipulated in a very constrained way. (Usually the qualifier "FIFO" is understood and not stated explicitly; we shall continue this tradition.) In the following description, we refer repeatedly to Figure 9.6, which depicts an OTM $M^{(\text{queue})}$ that has a queue, which it employs in place of a standard worktape. A queue has two ends that we call its IN

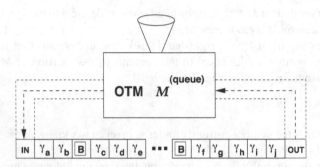

Fig. 9.6 An OTM $M^{(\text{queue})}$ that uses *a FIFO queue* in place of a worktape.

and OUT ports. In a single step, one can DEQUEUE a symbol *from* the OUT port of a queue—thereby (destructively) reading from the queue—and/or ENQUEUE a symbol *into* the IN port of the queue—thereby writing onto the queue. In Figure 9.6, a DE-QUEUE would remove/read the symbol γ_j; an ENQUEUE of a symbol γ_k would place that symbol to the left of γ_a in the figure's depiction of the queue.

The computing device $M^{(\text{queue})}$ depicted in Figure 9.6 is a *queue OTM* (*QTM*, for short). $M^{(\text{queue})}$ operates much like an OTM, except that it has *a single FIFO queue* as its only unbounded data structure. As expected, $M^{(\text{queue})}$ reads its worktape by DE-QUEUING the symbol at its queue's OUT port; it writes its worktape by ENQUEUING a new symbol at its queue's INport. Initially, $M^{(\text{queue})}$'s queue is empty, meaning that it contains just one symbol, $\boxed{\text{B}}$, and that its IN and OUT ports "coincide," in the sense that the pointers, IN and OUT, that represent the QTM's ports point to the same tape square.

> Because a queue is a *logical* data structure rather than a physical device, its ports are implemented using pointers—so having ports coincide now and then, but not always, presents neither problems nor contradictions.

The reader should be able to flesh out how $M^{(\text{queue})}$ computes, based on our discussions of the queue data structure and of the semantics of OTMs (in Section 3.3).

Proposition 9.3 *For every OTM M, there exists an equivalent QTM $M^{(\text{queue})}$, which can simulate any t-step computation by M in $O(t^2)$ steps.*

Proof (Sketch). The way that the QTM $M^{(\text{queue})}$ simulates a given OTM M differs from our previous simulations, in that $M^{(\text{queue})}$ does not just manipulate an encoding of M's worktape within its queue. In fact, $M^{(\text{queue})}$ will employ a simulation strategy that we have not encountered before, but that we shall reencounter a number of times in Chapter 13: $M^{(\text{queue})}$ will reproduce M's computation, as a sequence of *total states* of M. (Recall the notion *total state* from Section 3.3.)

Overall setup

- Letting Γ be M's worktape alphabet and Q its state-set (which, as always, is disjoint from Γ), the worktape alphabet of $M^{(\text{queue})}$ is $\Gamma^{(\text{queue})} = \Gamma \cup Q$.
- M begins a computation with a blank worktape, and $M^{(\text{queue})}$ begins with a queue that contains only M's initial state q_0, followed by special symbol $\boxtimes$ that it uses to identify the boundary between total states of M that are successive in the computation being simulated; see Figure 9.7.

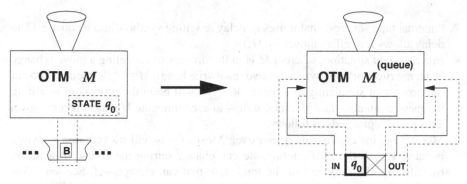

Fig. 9.7 Beginning the simulation of an OTM M via a QTM $M^{(\text{queue})}$. The queue square that contains state q of M is made very bold to highlight it.

- Whenever M polls its input port, $M^{(\text{queue})}$ does likewise.
- Whenever M halts and either accepts or rejects, $M^{(\text{queue})}$ does likewise.

Operation

The inductive correspondence between M's total states and $M^{(\text{queue})}$'s queue configurations during the course of a simulation is as follows. Consult Figures 9.8, and 9.9 as we proceed. Focus on a moment in a computation when M is in state $q \in Q$ and has the string

$$\gamma_a \gamma_b \boxed{B} \gamma_c \gamma_d \gamma_e \gamma_f \gamma_g$$

on its worktape, with its read/write head on symbol γ_c.

- In a *stable* situation, wherein M is in the process of polling its worktape (and maybe its input port also), $M^{(\text{queue})}$'s queue will contain a copy of M's total state

$$\gamma_a \gamma_b \boxed{B} q \gamma_c \gamma_d \gamma_e \gamma_f \gamma_g.$$

This is the situation depicted in Figure 9.8. $M^{(\text{queue})}$ recognizes that this is a stable situation because the delimiter symbol $\boxtimes$ is in the rightmost of the three

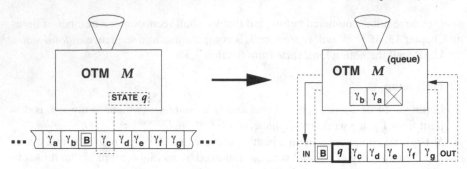

Fig. 9.8 A *stable* moment during the simulation of an OTM M via a QTM $M^{(\text{queue})}$. The queue square that contains state q of M is made very bold to highlight it.

"internal tape squares" that it uses to delay rewriting symbols into its queue. (The delay allows it to effect moves by M.)

- In a *transient* situation, wherein M is in the process of executing a move (change state, rewrite worktape symbol, move read/write head), $M^{(\text{queue})}$'s queue will be in the process of simulating M's move. $M^{(\text{queue})}$ will be in the process of rewriting its queue contents—i.e., M's total state—to accommodate M's most recent move. This process proceeds as follows.

$M^{(\text{queue})}$ begins copying its queue over. Most of this will be verbatim copying, because very little of M's total state can change during the course of a move. Specifically, the only part of the total state that can change—cf. Section 3.3— involves the 3-symbol sequence within the total state that contains M's current state q at its center. In Figure 9.8, this sequence is

$$\boxed{B}\, q\gamma_c.$$

Say that at this step, M rewrites the symbol γ_c to symbol $\overline{\gamma}_c$, moves one square to the right on its worktape, and changes state to $\overline{q}$. The relevant 3-symbol sequence now becomes

$$\boxed{B}\, \overline{\gamma}_c \overline{q}.$$

As $M^{(\text{queue})}$ rewrites its queue contents, it uses its internal (finite-state) memory in order to "read ahead" three symbols, so that when it commits itself to writing the active 3-symbol sequence, it can write the updated version. Figure 9.9 depicts a moment within the transient situation created by the described move by M:

$$\boxed{B}\, q\gamma_c \;\longrightarrow\; \boxed{B}\, \overline{\gamma}_c \overline{q}.$$

Analysis

Focus on a moment when M has k symbols on its worktape, and it executes a step. Before the step, $M^{(\text{queue})}$'s queue contains $k+1$ symbols: the symbols from M's tape, plus M's state symbol. As $M^{(\text{queue})}$ updates its queue in order to simulate M's move, it performs $O(k)$ copy-plus-update operations. This reckoning accounts quite

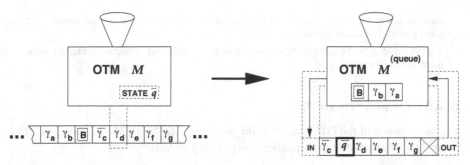

Fig. 9.9 A *transient* moment during the simulation of an OTM M via a QTM $M^{(\text{queue})}$. The queue square that contains state $\overline{q}$ of M is made very bold to highlight it.

conservatively for the possibility that $M^{(\text{queue})}$ must add an additional symbol to its queue (because M has extended its tape), as well as for the fact that $M^{(\text{queue})}$ must stagger its updating a bit in order to have time to change the active 3-symbol sequence from M's total state before the change.

Because M can add no more than one symbol to its worktape during each of its steps, it follows that $M^{(\text{queue})}$ can simulate a t-step computation by $M^{(\text{queue})}$ within

$$O\left(\sum_{i=0}^{t} i\right) = O(t^2)$$

steps.

Note the *big* assumption in the preceding summation. We have pulled the big-O from the *inside* of the summation to the *outside*. We are justified in doing this, because the constant factor hidden within the big-O is uniform across all steps of the simulation. (You should verify this crucial fact.) Were this not the case, then we could theoretically have constants that grow with the number of steps that M has executed, and our alleged time bound would be totally bogus. Thankfully, this is *not* "not the case," so we are quite justified in pulling the big-O outside the summation, and the result holds. (Note the use of the law of excluded middle (cf. Section 9.7) in the preceding sentence!)

This completes the proof. □

When coupled with the weakness we noted in paragraph C regarding OTMs whose only unbounded data structure is a single stack, this paragraph's result about QTMs can be viewed as demonstrating a sense in which a queue is a more powerful data structure than a stack.

Is it true in general that queues are stronger than stacks? The answer is somewhat interesting.

In 1972, Robert E. Tarjan studied the problem of sorting sequences of integers using networks of stacks or of queues [101]. He discovered that stacks and queues were in some sense dual to one another in power, in the sense that the smallest network of stacks that would sort the sequence equaled the number of increasing runs that make up the input sequence, while smallest number of queues equaled the number of decreasing runs.

In 1992, Lenwood S. Heath, F. Thomson Leighton, and the author studied the problem of using networks of stacks or of queues to lay out the nodes of a graph (in a manner consistent with certain circuit-layout problems) [37]. They uncovered graphs that needed *exponentially more* queues to lay out than stacks. Within this context, then, one could say that stacks are a more powerful structure than are queues.

The relationship between the computational powers of stacks and of queues is thus far from a cut-and-dried topic.

D. An OTM with *"Paper"* Tape

Our next variant of the OTM model is the *paper-tape OTM* (*PTM*, for short). A PTM is an OTM whose worktape alphabet Γ is *partially ordered*.

A *(strict) total order* on a set S is a transitive binary relation R on the set, under which, given any two distinct elements $a, b \in S$, we have either aRb or bRa. A familiar example is the relation "less than" on the natural numbers: for any two distinct integers $m, n \in \mathbb{N}$, either $m < n$ or $n < m$.

A *partial order* on a set S is similar to a total order, but it lacks the insistence that every pair of distinct elements of S be related (in one direction or the other). One familiar partial order arises naturally with the set $\mathbb{N} \times \mathbb{N}$ of ordered pairs of natural numbers. Under this order, one says that $\langle x_1, y_1 \rangle < \langle x_2, y_2 \rangle$ iff $x_1 < x_2$ and $y_1 < y_2$. Note that pairs such as $\langle 1, 2 \rangle$ and $\langle 2, 1 \rangle$ are just not related under this relation "$<$."

The partial order "$<$" that we posit for the worktape alphabets of PTMs are intended to be induced by the holes in a paper tape. Intuitively, if $\gamma_1 \in \Gamma$ has holes in the same configuration as $\gamma_2 \in \Gamma$, but γ_1 has *more holes* than γ_2, then $\gamma_2 < \gamma_1$ under the partial order on Γ; otherwise γ_1 and γ_2 are not related under "$<$."

A PTM M operates much like an OTM, except that when M overwrites a symbol γ on its worktape, *it must do so with some symbol $\bar{\gamma}$ that is greater than γ in the partial order*. (Intuitively, a PTM can "add holes" to the representation of γ, but it cannot "remove holes.")

Interestingly, despite the apparent disadvantage of having "paper tape," a PTM can simulate an OTM. Moreover, the simulating PTM can be made to operate in time polynomial in the computation time of the OTM. We now prove this assertion indirectly, by proving that a PTM can simulate a QTM with only polynomial slowdown.

Proposition 9.4 *For every QTM $M^{(\text{queue})}$, there exists an equivalent PTM. By transitivity, therefore, for every OTM M, there exists an equivalent PTM $M^{(\text{paper})}$. $M^{(\text{paper})}$ can simulate any t-step computation by $M^{(\text{queue})}$ in $O(t^3)$ steps; hence, it can simulate any t-step computation by M in $O(t^5)$ steps.*

Proof (Sketch). Let us be given a QTM $M^{(\text{queue})}$. We design a PTM $M^{(\text{paper})}$ that will simulate any computation by $M^{(\text{queue})}$. The overall strategy of the simulation will be for $M^{(\text{paper})}$ to always maintain a *fresh* copy of $M^{(\text{queue})}$'s queue. Every time that $M^{(\text{queue})}$ updates its queue, by dequeuing one symbol and enqueuing one symbol, $M^{(\text{paper})}$ copies its fresh copy to a new, as-yet unused, portion of its tape, making the required updates as it proceeds. $M^{(\text{paper})}$ orchestrates this copying by making extra holes in tape squares, to indicate whether they are *new* (one extra hole), *old but still valid* (two extra holes), or *obsolete* (three extra holes). During its lifetime, every used square of $M^{(\text{paper})}$'s worktape thus contains a symbol γ from $M^{(\text{queue})}$'s worktape alphabet—which will never be rewritten—plus either one, two, or three extra holes; the lifetime thus progresses as follows, using an ordered-pair notation of the form $\langle \gamma, \bullet\bullet \rangle$ to denote the worktape symbol γ embellished with extra holes (each denoted by $\bullet$):

$$(\text{unused}) \longrightarrow \langle \gamma, \boxed{\bullet} \rangle \longrightarrow \langle \gamma, \boxed{\bullet\bullet} \rangle \longrightarrow \langle \gamma, \boxed{\bullet\bullet\bullet} \rangle.$$

Figure 9.10(right) contains a version of this notation that is adapted for legibility by

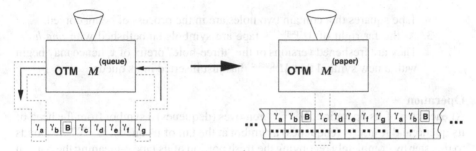

Fig. 9.10 Simulating a QTM $M^{(\text{queue})}$ via an PTM $M^{(\text{paper})}$ whose worktape is "made of paper."

omitting the brackets and boxes.

Overall setup

- The worktape alphabet $\Gamma^{(\text{paper})}$ of $M^{(\text{paper})}$ is $\Gamma \times \{\boxed{\bullet}, \boxed{\bullet\bullet}, \boxed{\bullet\bullet\bullet}\}$, where Γ is $M^{(\text{queue})}$'s worktape alphabet and where the instances of "$\bullet$" denote holes that embellish the elements of Γ.
- $M^{(\text{queue})}$ begins a computation with an empty queue, and $M^{(\text{paper})}$ begins with a blank worktape.
- Whenever $M^{(\text{queue})}$ polls its input port, $M^{(\text{paper})}$ does likewise.
- Whenever $M^{(\text{queue})}$ halts and either accepts or rejects, $M^{(\text{paper})}$ does likewise.
- The inductive correspondence between $M^{(\text{queue})}$'s configuration and $M^{(\text{paper})}$'s is as follows. Consult Figure 9.10 as we proceed.

  - Consider, for illustration, string x that forms the contents of $M^{(\text{queue})}$'s queue captured in Figure 9.10:

    $$x = \gamma_a \gamma_b \boxed{B} \gamma_c \gamma_d \gamma_e \gamma_f \gamma_g.$$

In any such situation, the contents of $M^{(\text{paper})}$'s tape will have the following form

1. The entire left end of the tape will contain symbols embellished with *three holes*. The rightmost symbols with three holes will be a (possibly null) prefix of the string x.

In Figure 9.10(right) the "three-hole" prefix of x is the string

$$\gamma_a\, \gamma_b\, \boxed{B}\, \gamma_c.$$

Tape squares that contain three holes have completed their useful lives and will never be revisited.

2. Immediately following the string of squares with three holes will be the remainder of x, with each tape square embellished with *two holes*.

   In Figure 9.10(right) the "two-hole" suffix of x is the string

$$\gamma_d\, \gamma_e\, \gamma_f\, \gamma_g.$$

   Tape squares that contain two holes are in the process of being copied.

3. At the far right of $M^{(\mathrm{paper})}$'s tape are symbols embellished with *one hole*. They are freshened versions of the "three-hole" prefix of x, hence may begin with a new symbol that $M^{(\mathrm{queue})}$ has just inserted in its queue.

Operation

At each step of its operation, $M^{(\mathrm{queue})}$ removes (dequeues) a symbol from the head of its queue, and it inserts (enqueues) a symbol at the tail of its queue. $M^{(\mathrm{paper})}$ responds to this step by completely recopying the fresh portion of its tape—meaning the portion that does not yet have three holes. It accomplishes this via the following sequence of operations.

1. $M^{(\mathrm{paper})}$ goes left on its tape until it encounters a three-hole square.
2. $M^{(\mathrm{paper})}$ "picks up" (using its finite-state memory) a copy of the symbol that resides immediately to the right of the three-hole square. This square should have two holes; $M^{(\mathrm{paper})}$ immediately gives it a third hole (thereby terminating its useful life).
3. $M^{(\mathrm{paper})}$ carries the newly acquired symbol rightward, until it encounters a new, unused square. It deposits the symbol there, embellished with one hole.

Analysis

Focus on a moment when $M^{(\mathrm{queue})}$ has k symbols in its queue, and it executes a step. As $M^{(\mathrm{paper})}$ freshens its copy of $M^{(\mathrm{queue})}$'s queue, it traverses an $O(k)$-symbol segment of its tape k times. We must assess length $O(k)$ for the traversed segment, rather than k, because $M^{(\mathrm{queue})}$ may have lengthened its queue (by one symbol) at this step. It follows that, in aggregate, it takes $M^{(\mathrm{paper})}$ $O(k^2)$ steps to simulate this one step by $M^{(\mathrm{queue})}$.

Because $M^{(\mathrm{queue})}$ can add no more than one symbol to its queue during each of its steps, it follows that $M^{(\mathrm{paper})}$ can simulate a t-step computation by $M^{(\mathrm{queue})}$ in

$$O\left(\sum_{i=0}^{t} i^2\right) = O(t^3)$$

steps.

The implications of the preceding analysis for the speed with which $M^{(\text{paper})}$ can simulate a t-step computation by an OTM follow via an invocation of Proposition 9.3. This completes the proof. □

E. An OTM with *Registers* instead of a Worktape

The final "simple" model that we consider is a variant of the OTM model that has, as its unbounded storage medium, some *fixed number*, $k \in \mathbb{N}$, of *registers*, denoted by $R_1, R_2, \ldots, R_k$. Each register R_i is capable of holding any nonnegative integer $m \in \mathbb{N}$ (no matter how large). We call the computational model that we study in this section a *k-register machine* (*k-RM*, for short). A k-RM M can be viewed as an OTM that has k registers in place of a worktape. (In particular, M polls its input port in exactly the way that an OTM does, via polling and autonomous states.) Whereas an OTM interacts with its worktape by reading a single square, respecifying the contents of that square, and possibly transferring its attention to an adjacent square, the k-RM M interacts with its registers by means of the following constrained set of interactions. At each step of a computation, M performs the following operations independently on each of its registers. Let us denote the *contents* of a register R_i, i.e., the integer that resides in R_i, by "$\underline{R_i}$."

- *Test register R_i for 0*
  At each step, M tests each register R_i independently to determine whether $\underline{R_i} = 0$. This is how M polls (or, "reads") its storage medium.
- *"Increment" a register.*
  At each step, M alters the contents of each register R_i independently, by adding $+1$ or -1 or 0 to the number that the register holds. We denote this alteration by the assignment

$$\underline{R_i} \leftarrow \underline{R_i} + \alpha,$$

  where $\alpha \in \{-1, 0, +1\}$.
  This is how M alters (or "writes to") its storage medium.

Thus, the "syntax" of the (one-step) transition function of a k-RM is

$$\delta : \left((Q_{\text{poll}} \times \Sigma) \cup Q_{\text{aut}} \right) \times \{\text{zero, nonzero}\}^k \longrightarrow Q \times \{-1, 0, +1\}^k.$$

(Compare this with the analogous form for OTMs (3.4).) The notions of *computation*, *acceptance*, and *rejection* by a k-RM can easily be inferred from the analogous notions for an arbitrary OA (see Section 3.1). You should be able to supply details.

A k-RM thus has no explicit list-processing capability—although it can simulate such a capability surprisingly simply. We now use the STM model as an intermediary in showing that RMs having two or more registers can simulate OTMs rather simply, albeit not very efficiently.

Before proceeding with the proposition, we endow RMs with another atomic operation on registers, one that is not part of the model's traditional repertoire. For the sake of expositional convenience, at a cost of no more than a small constant factor in performance, *we allow an RM to transfer the contents of one of its registers, say*

R_i, *to another of its registers, say* R_j, in a single step. The small Program Register-Transfer indicates that this transfer instruction, which we denote symbolically by

$$R_j \leftarrow R_i,$$

increases the speed, but not the computing power, of an RM.

Program Register-Transfer R_i, R_j
/*Implement the operation $R_j \leftarrow R_i$*/
Inputs R_i, R_j
$R_j \leftarrow 0$
do until $R_i = 0$
$\quad R_i \leftarrow R_i - 1$
$\quad R_j \leftarrow R_j + 1$
enddo

Proposition 9.5 *1. a. For every 2-STM M, there is an equivalent 3-RM that can simulate any t-step computation by M in $2^{O(t)}$ steps.*

b. For every OTM M', there is an equivalent 3-RM that can simulate any t-step computation by M' in $2^{O(t)}$ steps.

2. a. For every 2-STM M, there is an equivalent 2-RM that can simulate any t-step computation by M in $2^{2^{O(t)}}$ steps.

b. For every OTM M', there is an equivalent 2-RM that can simulate any t-step computation by M' in $2^{2^{O(t)}}$ steps.

Proof (Sketch). We present an explicit proof only for the "*a*" parts of the proposition, relying on Proposition 9.2 plus the proof of the "*a*" parts to prove the "*b*" parts of the proposition.

1. A 3-RM $M^{(3\text{register})}$ that simulates a 2-STM M

Overall setup

Let Γ be the stack alphabet of the 2-STM M. We employ the following strategy as we construct $M^{(3 \text{ register})}$.

- *Register usage*

  – We use the integer in register R_1 (resp., register R_2) to "simulate" stack S_1 (resp., stack S_2) of M.

  – We use register R_3 as an auxiliary register that allows $M^{(3 \text{ register})}$ to manipulate registers R_1 and R_2 as it simulates stack-moves by M.

- *Encoding of stack contents*

– We assume that the elements of Γ are the integers $\{0, 1, \ldots, |\Gamma| - 1\}$, with the blank symbol $\boxed{B}$ playing the role of the digit 0. If necessary, we relabel the elements of Γ to make this true.

– We view each string that appears in one of M's stacks as a *numeral* in base $g \stackrel{\text{def}}{=} |\Gamma| - 1$, with the digit at the top of the stack being the low-order digit.

Thus, if one of M's stacks contains the string

$$x = \gamma_n \gamma_{n-1} \cdots \gamma_1 \gamma_0, \tag{9.5}$$

where each $\gamma_i \in \Gamma$, and where γ_0 resides at the top of the stack, then the register of $M^{(3 \text{ register})}$ that represents this stack in the simulation will contain the integer

$$m = \sum_{i=0}^{n} \gamma_i g^i,$$

that is, the base-g value of numeral x.

Under the preceding setup, $M^{(3 \text{ register})}$ can simulate M's PUSH and POP operations by simple arithmetic analogues, which we call A-PUSH and A-POP, respectively. Our implementations of these arithmetic analogues are motivated by the following facts. Let the string x of (9.5) reside on stack S_i of M, and let m, the base-g value of numeral x, reside in register R_i of $M^{(3 \text{ register})}$. Then:

- For every $\gamma \in \Gamma$, define the effect of the operation

$$\text{A-PUSH } \gamma \text{ onto } R_i$$

via the assignment

$$\underline{R_i} \leftarrow (\underline{R_i} \cdot g) + \gamma.$$

This definition preserves the encoding we have established, because after the operation "A-PUSH γ onto R_i" is executed, the base-g numeral for the number in register R_i is

$$\gamma_n \gamma_{n-1} \cdots \gamma_1 \gamma_0 \gamma,$$

which is the result of the M-operation

$$\text{PUSH } \gamma \text{ onto } S_i.$$

- For every $\gamma \in \Gamma$, define the effect of the operation

$$m \leftarrow \text{A-POP } R_i$$

via the pair of assignments

$$\underline{R_3} \leftarrow \lfloor \underline{R_i} \div g \rfloor,$$
$$m \leftarrow \underline{R_i} - \underline{R_3}.$$

This definition preserves the encoding we have established, because after the operation "$m \leftarrow$ A-POP R_i" is executed:

– variable m has the numerical value γ_0, which corresponds to the value of variable m that is the result of the M-operation

$$m \leftarrow \text{POP } S_i$$

– the residual string

$$\gamma_n \gamma_{n-1} \cdots \gamma_1$$

is a base-g numeral for the residual number in register R_i.

Thus, these definitions of the arithmetic analogues of PUSH and POP propagate the encoding of M's stack contents that $M^{(3 \text{ register})}$ uses throughout its simulation of M.

Note the benefits from having $\boxed{\text{B}}$ play the role of the digit 0. (a) To simulate one of M's stacks being empty, such as at the beginning of a computation, $M^{(3 \text{ register})}$ need only set the corresponding one of its registers to 0. (b) If M POPs an empty stack, it receives a copy of $\boxed{\text{B}}$ in response; if $M^{(3 \text{ register})}$ A-POPs a register that contains 0, it receives 0 as a result—which is the "code" for $\boxed{\text{B}}$.

In more detail, the arithmetic analogues of PUSH and POP are implemented by the procedures specified in Program **Stack-Register Operations**.

Analysis

Focus on M while it is computing. By step t of its computation, (at least) one of M's stacks, say S_i, could contain a string of length t, but no longer. Under our encoding, register R_i (which $M^{(3 \text{ register})}$ uses to simulate stack S_i) could contain a number of magnitude as large as 2^t, but no larger.

For M, each of its stack-updating operations, PUSH and POP, takes a single step. As $M^{(3 \text{ register})}$ simulates instances of these operations via its arithmetic analogues, A-PUSH and A-POP, the 3-RM requires $\leq g \cdot 2^t$ steps to simulate M's single-step PUSH or POP.

It follows that after M has been computing for t steps, the number of steps that $M^{(3 \text{ register})}$ will have executed while simulating M is no greater than

$$\sum_{i=0}^{t} g \cdot 2^i = g \cdot \sum_{i=0}^{t} 2^i \leq 2g \cdot 2^t.$$

Since g is a fixed constant, the proof of part 1 is complete.

2. A 2-RM $M^{(2 \text{ register})}$ that simulates a 3-RM $M^{(3 \text{ register})}$

We design the 2-RM $M^{(2 \text{ register})}$ by modifying our design of the 3-RM $M^{(3 \text{ register})}$. We shall therefore establish part 2 of the proposition by describing only how to map register configurations of $M^{(3 \text{ register})}$ onto register configurations of $M^{(2 \text{ register})}$

Program Stack-Register Operations	
/*Implement the arithmetic analogues of the stack operations*/	
Operation by M	**Operation by** $M^{(3 \text{ register})}$
PUSH γ onto S_i	A-PUSH γ onto R_i
	$R_3 \leftarrow 0$ **do until** $R_i = 0$ $\quad \overline{R_i} \leftarrow R_i - 1$ $\quad$ **do** g times $\qquad \overline{R_3} \leftarrow R_3 + 1$ $\quad$ **enddo** **enddo** /*$\underline{R_3} = R_i \cdot g$*/ $R_i \leftarrow \overline{R_3}$ /*Our first use of register exchange to simplify exposition*/ **do** $\quad \gamma$ times $\quad \overline{R_i} \leftarrow R_i + 1$ **enddo** /*$\underline{R_3} = (R_i \cdot g) + \gamma$*/
$m \leftarrow$ POP S_i	$m \leftarrow$ A-POP R_i
	$R_3 \leftarrow 0$ **do until** $R_i < g$ $\quad$ **do** g times $\qquad \overline{R_i} \leftarrow R_i - 1$ $\quad$ **enddo** $\quad \overline{R_3} \leftarrow R_3 + 1$ **enddo** /*$\underline{R_i} = \lfloor R_i \div g \rfloor$*/ $m \leftarrow R_i$ $R_i \leftarrow \overline{R_3}$

and to manipulate the latter configurations. We retain the notation of part 1, wherein R_1, R_2, and R_3 are the three registers of $M^{(3 \text{ register})}$. In order to avoid confusion, we call the two registers of $M^{(2 \text{ register})}$ P_1 and P_2.

Overall setup

- During a *stable* moment—i.e., a moment when $M^{(2 \text{ register})}$ *is not* in the process of updating its representation of $M^{(3 \text{ register})}$'s registers—$M^{(2 \text{ register})}$ will encode the *triple* of integers

$$\langle R_1, R_2, R_3 \rangle \tag{9.6}$$

contained in $M^{(3 \text{ register})}$'s registers by the *pair* of integers

$$\langle P_1, P_2 \rangle, \tag{9.7}$$

where

$$P_1 = 2^{R_1} 3^{R_2} 5^{R_3},$$

$$\underline{P_2} = 0.$$

- During an *unstable* moment—i.e., a moment when $M^{(2 \text{ register})}$ *is* in the process of updating its representation of $M^{(3 \text{ register})}$'s registers—$M^{(2 \text{ register})}$ will encode the *triple* of integers (9.6) by a pair (9.7), where $\underline{P_1}$ and $\underline{P_2}$ have the following forms for some $c \in \mathbb{N}$:

$$\underline{P_1} = 2^{\underline{R_1}}3^{\underline{R_2}}5^{\underline{R_3}} - \alpha c \quad (\alpha \in \{1,2,3,5\}),$$
$$\underline{P_2} = \beta c \qquad\qquad\qquad (\beta \in \{1,2,3,5\}).$$

The reader can firm up the values of α and β from the coming description of the details of $M^{(2 \text{ register})}$'s operation.

Operation

$M^{(2 \text{ register})}$ must simulate the following register operations by $M^{(3 \text{ register})}$. $M^{(2 \text{ register})}$ begins all of these operations in a *stable* configuration.

We focus only on how $M^{(2 \text{ register})}$ simulates operations by $M^{(3 \text{ register})}$ on register R_1. In order to shift the focus to register R_2 or to register R_3 instead of register R_1, one just substitutes, respectively, the integer 3 or the integer 5 for the integer 2 in the following description.

- *Test register for* 0
  Say that $M^{(3 \text{ register})}$ tests the condition

$$\underline{R_1} = 0?$$

This is *equivalent* to having $M^{(2 \text{ register})}$ test the condition

Is $\underline{P_1}$ *not* divisible by 2?

In order to perform this test, $M^{(2 \text{ register})}$ attempts to divide $\underline{P_1}$ by 2. If this attempted division leaves a remainder, then $\underline{P_1}$ *is not* even, so that $\underline{R_1}$ *is* 0; if the attempted division *does not* leave a remainder, then $\underline{P_1}$ *is* even, so that $\underline{R_1}$ *is not* 0.

The attempted division proceeds as follows. $M^{(2 \text{ register})}$ repeatedly subtracts 2 from $\underline{P_1}$ (using two successive subtractions of 1) and adds 1 to $\underline{P_2}$ until $\underline{P_1}$ is either 0 or 1. If $\underline{P_1}$ ends up as 0, then the number it contained before the division was even; if it ends up as 1, then the predivision number was odd. $M^{(2 \text{ register})}$ now knows the "answer" to the test. To regain a *stable* configuration, it need only restore the value of $\underline{P_1}$, which it can easily do by reversing the preceding steps, i.e., by repeatedly subtracting 1 from $\underline{P_2}$ and adding 2 to $\underline{P_1}$ (using two successive additions of 1) until $\underline{P_2}$ reaches 0.

- *Increment a register*
  $M^{(2 \text{ register})}$ simulates $M^{(3 \text{ register})}$'s incrementing (resp., decrementing) register R_1 by multiplying (resp., dividing) $\underline{P_1}$ by 2. (For simplicity, we assume that

$M^{(3\ \text{register})}$ will never attempt to decrement a register that contains 0; this just means that the decrement is preceded by a test for 0 that has a *negative* outcome.)

A multiplication proceeds as follows: $M^{(2\ \text{register})}$ repeatedly subtracts 1 from $\underline{P_1}$ and adds 2 to $\underline{P_2}$ (using two successive additions of 1) until $\underline{P_1}$ reaches 0.

A division proceeds as follows: $M^{(2\ \text{register})}$ repeatedly subtracts 2 from $\underline{P_1}$ (using two successive subtractions of 1) and adds 1 to $\underline{P_2}$ until $\underline{P_1}$ reaches 0.

$M^{(2\ \text{register})}$ regains a *stable* configuration after a multiplication or division by transferring the contents of register P_2 to register P_1.

Analysis

The correctness of our strategy for having $M^{(2\ \text{register})}$ simulate $M^{(3\ \text{register})}$ is a consequence of the fundamental theorem of arithmetic. The complexity of our simulation follows by adapting the corresponding analysis for part 1, in light of the fact that $M^{(2\ \text{register})}$ manipulates integers that are exponentially larger than those manipulated by $M^{(3\ \text{register})}$, because we use a prime-power pairing function. □

What about the case $k = 1$? It is a straightforward exercise to show that 1-RMs *cannot* simulate arbitrary OTMs. One can build a proof by showing that a 1-STM can simulate a 1-RM by using a *tally code* in its stack, i.e., by representing each integer m by a string of m 1's. It then follows from Section 6.1.3 that no 1-RM can accept the language $L = \{a^n b^n c^n \mid n \in \mathbb{N}\}$ which is easily accepted by an OTM (as we showed in subsection B of the current section).

9.8.2 Augmented TMs That Are No More Powerful Than OTMs

This section is devoted to three modifications of the OTM model that one might expect to enhance the computing power of the model. For each of the modifications, we show that an unmodified OTM can simulate the new model on arbitrary computations, with only polynomial slowdown. As we discuss in Chapter 13, the efficiency of these simulations played a significant role in the early development of complexity theory.

A. An OTM with *Several Worktapes*

For any integer $k > 1$, a *k-tape OTM M* operates much as does an ordinary (1-tape) OTM, but it has k worktapes that it interacts with—reads from, writes to, moves on—independently. As you might expect by this point in our journey, the single-step operation of M is specified by a state-transition function with the following form:

$$\delta : \left((Q_{\text{poll}} \times \Sigma) \cup Q_{\text{aut}} \right) \times \Gamma^k \longrightarrow Q \times \Gamma^k \times \{N, L, R\}^k. \qquad (9.8)$$

(Compare this with its one-tape analogue in (3.4).)

A mathematical aside: Note how the use of the kth powers of the sets Γ and $\{N,L,R\}$ in (9.8) automatically carries with it the fact that M operates *independently* on its k worktapes.

The notions of *computation*, *acceptance*, and *rejection* by a k-tape OTM can easily be inferred from the analogous notions for an arbitrary OA (see Section 3.1). You should be able to supply the details.

It is not too surprising that an ordinary (1-tape) OTM can simulate arbitrary computations by any k-tape OTM. It is not so obvious, though, that the simulations can be rather efficient.

Proposition 9.6 *For every k-tape OTM $M^{(k)}$, there is an equivalent 1-tape OTM M that can simulate any t-step computation by $M^{(k)}$ in $O(t^2)$ steps.*

Proof (Sketch). Let the k-tape OTM $M^{(k)}$ have worktape alphabet Γ.

Overall setup

Refer to Figure 9.11 as we describe our design of an ordinary OTM M that can simulate $M^{(k)}$ on arbitrary computations. As the figure suggests, we endow our 1-

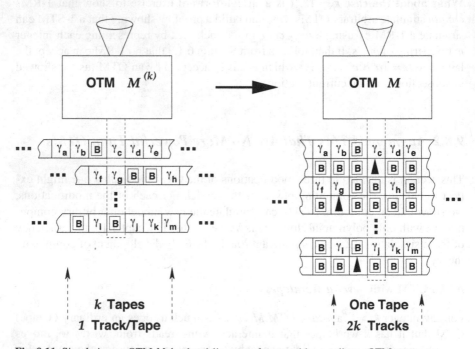

Fig. 9.11 Simulating an OTM M that has k linear worktapes with an ordinary OTM.

tape OTM M with a worktape that has $2k$ tracks. M uses each *odd-numbered* track $2i-1$ of its tape, where $i \in [1,k]$, to simulate tape i of $M^{(k)}$; hence, each square of each odd-numbered track of M's tape can hold any symbol from Γ. M uses each *even-numbered* track $2i$ of its tape, where $i \in [1,k]$, to keep track of the position of

$M^{(k)}$'s read/write head on tape i. To this end, each square of each even-numbered track of M's tape can hold either of the two symbols $\boxed{B}$ and $\blacktriangle$. In summation, then, M's worktape alphabet is the k-fold product

$$\left(\Gamma \times \left\{\boxed{B}, \blacktriangle\right\}\right) \times \left(\Gamma \times \left\{\boxed{B}, \blacktriangle\right\}\right) \times \cdots \times \left(\Gamma \times \left\{\boxed{B}, \blacktriangle\right\}\right).$$

At every instant, *precisely one* square of each even-numbered track $2i$ of M's tape will hold an instance of $\blacktriangle$, thereby indicating where $M^{(k)}$'s read/write head resides on tape i; all other squares of track $2i$ will hold instances of $\boxed{B}$. Figure 9.11 depicts a generic configuration of $M^{(k)}$'s k tapes and the corresponding configuration of M's $2k$-track tape.

Operation

M simulates a single move by $M^{(k)}$ by executing the following protocol. As we describe the protocol, keep in mind that M is designed specifically to simulate $M^{(k)}$. Therefore, in what follows, k is a fixed constant. (This is important because it allows us to design M so that it can store $O(k)$ items in its internal memory.)

1. M assembles in its internal memory a k-place vector

$$\langle \gamma_1, \gamma_2, \ldots, \gamma_k \rangle \in \Gamma^k \tag{9.9}$$

that specifies the k symbols that $M^{(k)}$ is reading on its multiple tapes at this step. The intention of our notation is that $M^{(k)}$ is reading $\gamma_i \in \Gamma$ on its ith tape. M assembles this vector as follows.

- M goes to the leftmost non-$\boxed{B}$ symbol on its worktape.
- Starting there, M traverses the extent of its worktape until it encounters the rightmost non-$\boxed{B}$ symbol.
- As M encounters each instance of the symbol $\blacktriangle$ on an even-numbered track $2i$ of its tape, it stores the symbol from Γ that resides in the corresponding square of track $2i - 1$. M "keeps track" in its internal memory of the fact that it now knows what $M^{(k)}$ is reading on its ith tape at the step being simulated. This memory allows M to stop its traversal as soon as it knows all k entries of the vector (9.9).

In Figure 9.11, the vector (9.9) is

$$\langle \gamma_c, \gamma_g, \ldots, \boxed{B} \rangle.$$

2. M uses the vector (9.9)—together with $M^{(k)}$'s program, plus the symbol at its input port if $M^{(k)}$ is currently in a polling state—in order to determine the move that $M^{(k)}$ would make at this step. This includes how $M^{(k)}$ would rewrite each of the k tape symbols it is currently scanning and where it would move each of its k read/write heads. M records this information in a $2k$-place vector of the following form in its internal memory:

$$\langle\langle\overline{\gamma}_1, D_1\rangle, \langle\langle\overline{\gamma}_2, D_2\rangle, \dots, \langle\langle\overline{\gamma}_k, D_k\rangle\rangle.$$

3. M makes the necessary updates on its tape by means of another complete sweep through the nonblank portion of the tape. In detail:

   - M goes to the leftmost non-$\boxed{\text{B}}$ symbol on its worktape.
   - Starting there, M traverses the extent of its worktape until it encounters the rightmost non-$\boxed{\text{B}}$ symbol.
   - As M encounters each instance of the symbol $\blacktriangle$ on an even-numbered track $2i$ of its tape, it
     - rewrites the symbol from Γ that resides in the corresponding square of track $2i - 1$, in accord with $M^{(k)}$'s program; this replaces symbol γ_i by symbol $\overline{\gamma}_i$.
     - moves the $\blacktriangle$ on track $2i$ one square in direction D_i, placing an instance of $\boxed{\text{B}}$ in the vacated square.

This completes M's simulation of a single step by $M^{(k)}$.

Analysis

Our analysis is quite similar to that employed in the proof of Proposition 9.3.

Focus on a moment wherein $M^{(k)}$ has just executed the tth step of a computation and on the corresponding moment after M has just simulated the tth step of this computation by $M^{(k)}$. Because $M^{(k)}$ can add no more than one new tape square to each of its k worktapes in a single step, we know that at the moment we are considering, none of its worktapes can exceed t squares in length. Under our simulation strategy, therefore, at the corresponding moment, M's (single) worktape cannot exceed kt squares in length.

Now let us observe M simulating the next step of $M^{(k)}$'s computation.

1. M goes to the leftmost non-$\boxed{\text{B}}$ square on its tape.
2. M makes a complete sweep across its tape, gathering information about the k symbols that $M^{(k)}$ is reading on its worktapes.
3. M does an internal computation to decide how $M^{(k)}$ would react to the k symbols it has read.
4. M makes a complete sweep across its tape, making the changes needed to update the tape contents, as mandated by $M^{(k)}$'s program.

Steps 1, 2, and 4 of this accounting each takes $\leq kt$ steps by M; step 3, being internal, is "instantaneous" (because state transitions are "instantaneous"). In aggregate, then, simulating the $(t + 1)$th step of $M^{(k)}$'s computation takes M no more than $3kt$ steps. Recalling yet again that k is a fixed constant here, this reckoning shows that M can simulate the first t steps of $M^{(k)}$'s computation in no more than

$$\sum_{i=1}^{t} 3ki = O(t^2)$$

steps.

This completes the proof. ☐

The timing of the simulation strategy of Proposition 9.6 changes by only constant factors as the single-tape OTM M simulates OTMs having different numbers of worktapes. In particular, the strategy takes quadratic time to simulate even a 2-tape OTM $M^{(2)}$. In [39], Fred C. Hennie and Richard E. Stearns develop a more sophisticated simulation algorithm, via which a 1-tape OTM can simulate t steps by a 2-tape OTM in $O(t \log t)$ steps.

B. An OTM with a *Multidimensional Worktape*

This section deals with OTMs whose worktapes are 2-*dimensional;* see Figure 9.12. We recommend that as readers follow our description of this model, they extrapo-

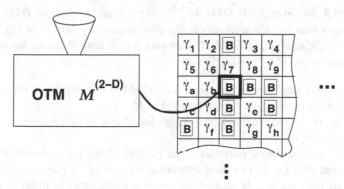

Fig. 9.12 An OTM $M^{(2-D)}$ with a two-dimensional worktape.

late the model's details to OTMs whose worktapes are k-*dimensional* for values of k greater than 2 and, more ambitiously, to OTMs whose worktapes have the structure of fixed-degree *rooted trees*.

Let $M^{(2-D)}$ be an OTM with a 2-dimensional worktape, as depicted in Figure 9.12. We call $M^{(2-D)}$ a 2-*dimensional OTM*, or, for short, a 2-D *OTM*. The squares of $M^{(2-D)}$'s tape are indexed by $\mathbb{N}^+ \times \mathbb{N}^+$, with the *origin square*, whose label is $\langle 1, 1 \rangle$, being the place where $M^{(2-D)}$'s read/write head begins every computation. A step by $M^{(2-D)}$ proceeds exactly as does a step by an ordinary OTM, except that $M^{(2-D)}$ can move its read/write head one square in any of the *four* compass directions:

$$\text{northward:} \quad \langle i, j \rangle \rightarrow \langle i-1, j \rangle \quad \text{if } i > 1,$$
$$\text{eastward:} \quad \langle i, j \rangle \rightarrow \langle i, j+1 \rangle,$$
$$\text{southward:} \quad \langle i, j \rangle \rightarrow \langle i+1, j \rangle,$$
$$\text{westward:} \quad \langle i, j \rangle \rightarrow \langle i, j-1 \rangle \quad \text{if } j > 1.$$

The conditions on northward and westward moves prevent $M^{(2-D)}$ from moving its head in a way that causes the head to "fall off" the tape. The notions of *computation*, *acceptance*, and *rejection* by a 2-D OTM can easily be inferred from the analogous notions for an arbitrary OA (see Section 3.1).

2-D OTMs are often defined with worktapes that can extend without limit in all four compass directions (so "falling off" is not a danger). One then simplifies the formal development by proving a two-dimensional analogue of Proposition 9.1 that converts the "full" two-dimensional tape to the quadrant-structured tape that we have posited. The proof of this analogue does not require any ideas that are not already present in our proof of Proposition 9.1—except, of course, that we now must fold the tape over twice, resulting in four tracks; therefore, we build this simplification into our model and leave details to the reader.

We show now that although having two-dimensional tapes can enhance the efficiency of an OTM: (*a*) it does not enhance the raw computing power of the model, and (*b*) it can enhance efficiency precisely because it can pack more squares close to one another. You may want to review Section 5.5 before you embark on the following result.

Proposition 9.7 *For every* 2-*D OTM* $M^{(2\text{-}D)}$, *there is an equivalent OTM M with two linear tapes that can simulate any t-step computation by* $M^{(2\text{-}D)}$ *in* $O(t^2)$ *steps. Hence, there is an equivalent OTM* M' *with one linear tape that can simulate any t-step computation by* $M^{(2\text{-}D)}$ *in* $O(t^4)$ *steps.*

Proof (Sketch). We shall design only the 2-tape OTM M, relying on an invocation of Proposition 9.6 for the design of the 1-tape OTM M'.

Refer to Figure 9.13 as we describe a design for M. For simplicity, we assume that M has *one-ended* tapes.

We have M poll its input port and make (accept or reject) decisions about the input it has read thus far in a manner consistent with $M^{(2\text{-}D)}$'s program. Therefore, we concentrate only on how M manipulates its two linear tapes in order to simulate $M^{(2\text{-}D)}$'s manipulation of its single 2-dimensional tape.

Overall setup

When $M^{(2\text{-}D)}$'s tape—call it $T^{(2\text{-}D)}$—has the configuration illustrated in Figure 9.13 (top), then:

- M's first tape, T_1, contains a linearization of $T^{(2\text{-}D)}$ that is obtained using the diagonal pairing function $\mathscr{D}(x,y)$ specified in (8.1). As we discussed in Section 8.2.1, $\mathscr{D}(x,y)$ linearizes $\mathbb{N}^+ \times \mathbb{N}^+$ along "diagonal shells." The thin curve that is superimposed upon $T^{(2\text{-}D)}$ in Figure 9.13(top) indicates how $\mathscr{D}(x,y)$ would order these tape squares. This is precisely the order in which these tape squares occur on T_1; cf. Figure 9.13(bottom).
  When the square being placed on T_1 corresponds to a square of $T^{(2\text{-}D)}$ that $M^{(2\text{-}D)}$ has already visited, then the square contains a symbol that $M^{(2\text{-}D)}$ has explicitly written there. However, when the square corresponds to a square of $T^{(2\text{-}D)}$ that $M^{(2\text{-}D)}$ has *not* yet visited, then *as a default*, M writes the symbol $\boxed{\text{B}}$ on the square. This action does not impair M's simulation of $M^{(2\text{-}D)}$, because if $M^{(2\text{-}D)}$ ever visits this square in the course of its computation, then the square will, indeed, contain $\boxed{\text{B}}$, because that is how the TM model treats squares that are visited for the first time.
- M's second tape, T_2, contains three numerals, one each for:

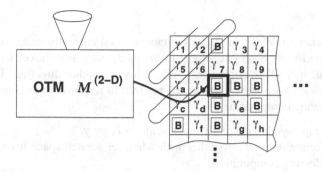

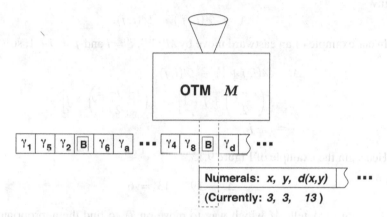

Fig. 9.13 Toward simulating a two-dimensional worktape with a linear worktape.

- the (integer) value x that indicates the index of the *row* of $T^{(2\text{-}D)}$ where $M^{(2\text{-}D)}$'s read/write head resides;
- the (integer) value y that indicates the index of the *column* of $T^{(2\text{-}D)}$ where $M^{(2\text{-}D)}$'s read/write head resides;
- the (integer) value $\mathscr{D}(x,y)$ that indicates the index, under the diagonal pairing function, of the *square* $\langle x,y \rangle$ of $T^{(2\text{-}D)}$ where $M^{(2\text{-}D)}$'s read/write head resides.

In Figure 9.13(bottom), $x = y = 3$, so T_2 contains two instances of a numeral for the value 3 and one instance of a numeral for the value $\mathscr{D}(3,3) = 13$.

Additionally, M will use some of the blank space on T_2 for scratch space as it updates the three numerals.

- M's first read/write head resides on the square of T_1 that corresponds to the square of $T^{(2\text{-}D)}$ where $M^{(2\text{-}D)}$'s read/write head currently resides.

  M's second read/write head resides on the leftmost square of T_2.

Operation

Focus on a step wherein $M^{(2\text{-D})}$ replaces the tape symbol currently under scan by γ and moves its read/write head one square to the *east*—so that it moves from the current square, call it $\langle i, j \rangle$, to square $\langle i, j+1 \rangle$. (Moves in other directions lead to similar computations, so we leave them to the reader.) In response to this move, M performs the following computation.

1. M replaces the tape symbol currently under scan on T_1 by γ.
2. M uses the information on T_2, together with whatever scratch space it needs, to perform the following computation.

   a. M computes, for the current square $\langle i, j \rangle$ and the target square $\langle i', j' \rangle$, the quantity

   $$\Delta = \mathscr{D}(i', j') - \mathscr{D}(i, j).$$

   In our example of an eastward move by $M^{(2\text{-D})}$, $i' = i$ and $j' = j+1$, so

   $$\Delta = \mathscr{D}(i, j+1) - \mathscr{D}(i, j)$$
   $$= \left[\binom{i+j}{2} + j + 1 \right] - \left[\binom{i+j-1}{2} + j \right]$$
   $$= i + j.$$

   Hence, in the example of Figure 9.13,

   $$\Delta = 19 - 13 = 6.$$

   b. The sign of Δ tells M which way to move on T_1 to find the appropriate new square: $(\Delta > 0)$ means "move right"; $(\Delta = 0)$ means "don't move"; $(\Delta < 0)$ means "move left." The magnitude (i.e., unsigned value) of Δ tells M how many squares to move. In our example of an eastward move by $M^{(2\text{-D})}$, M should move rightward $i + j$ squares, so in the example of Figure 9.13, M should move rightward six squares.

   M counts down from Δ to 0 on T_2 as it moves on T_1 in order to know how many squares to move on T_1.

3. M updates the three numerals on T_2 to their appropriate new values. In our example of an eastward move by $M^{(2\text{-D})}$, M leaves the x numeral unchanged, adds $+1$ to the y numeral, and replaces the $\mathscr{D}(x, y)$ numeral by a numeral for $\mathscr{D}(x, y+1)$ (which it has computed in the course of computing Δ).

This completes M's simulation of a single step by $M^{(2\text{-D})}$.

Analysis

The dominant components of M's simulation of the tth step of a computation by $M^{(2\text{-D})}$ are as follows. This analysis works whether M represents Δ using a tally encoding or a positional number system.

- Computing the quantity Δ.

  This computation can be accomplished in time proportional to the lengths of the representations of x and y. This is clearly $O(t)$.

- Orchestrating and executing M's move along T_1.

  Orchestrating the move involves counting down the quantity Δ on T_2. It is obvious that this can be accomplished in $O(t)$ steps when M represents Δ using a tally encoding. For positional number systems, this calculation can be derived using the techniques in, e.g., Chapter 1 of [53].

  Executing M's move along T_1 takes $O(t)$ steps, because the source and target squares in this move reside, on $T^{(2\text{-}D)}$, in adjacent diagonal shells. One verifies easily that this means that the function $\mathscr{D}(x,y)$ assigns these squares values that are only $O(t)$ apart.

Because M can thus simulate the tth step of $M^{(2\text{-}D)}$'s computation in $O(t)$ steps, it follows via techniques we have used several times before in this section that M can simulate an entire t-step computation by $M^{(2\text{-}D)}$ in $O(t^2)$ steps.

This completes the proof. $\square$

C. An OTM with a *"Random-Access"* Worktape

The final model that we discuss in our brief survey attempts to inject a modicum of realism into an OTM-based model. A *random-access OTM* $M^{(\text{RA})}$ (*RA-OTM*, for short) has both a linear *address* worktape $T^{(A)}$ and a two-dimensional *storage* worktape $T^{(S)}$. Refer to Figure 9.14 as we describe $M^{(\text{RA})}$'s operation. As suggested in

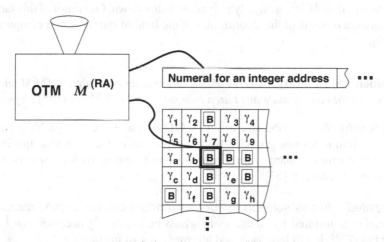

Fig. 9.14 A random-access OTM $M^{(\text{RA})}$. The square of $T^{(S)}$ that is currently being accessed is highlighted.

the figure, we assume, for simplicity, that $T^{(A)}$ is *one-ended* and that $T^{(S)}$ has the structure of a *quadrant* (as do the two-dimensional tapes of the preceding section).

An RA-OTM $M^{(RA)}$ behaves like all OTMs with regard to polling its input port and making decisions regarding acceptance and rejection of input strings. It differs from the other OTM models we have discussed in its very powerful mechanism for interacting with its storage worktape $T^{(S)}$. After every interaction with its worktapes—which includes rewriting the symbol currently under scan on $T^{(S)}$—$M^{(RA)}$ can:

- ignore its address worktape $T^{(A)}$ and move the read/write head on $T^{(S)}$ as though it were a 2-D OTM.
  After this action,

  - the read/write head on $T^{(S)}$ will have moved at most one square to the north, east, south, or west;
  - $T^{(A)}$ will be unchanged, with respect to both content and the position of its read/write head.

- write a numeral x on $T^{(A)}$.
  After this action,

  - the read/write head on $T^{(S)}$ will have moved to the leftmost square of row x of $T^{(S)}$, i.e., the row whose index is the number specified by numeral x;
  - $T^{(A)}$ will contain the new numeral x, and its read/write head will reside on the leftmost square of the tape.

The second of the preceding ways of moving on $T^{(S)}$ motivates our using the phrase "random access" to describe $M^{(RA)}$.

By this time, it will probably come as no surprise that any RA-OTM $M^{(RA)}$ can be simulated by an ordinary OTM. It may not be obvious, though, that an ordinary OTM can simulate $M^{(RA)}$ with only polynomial slowdown. Our proof of this fact will be reminiscent of some of the algorithmics of the field of *sparse matrix* computations [33, 103].

Proposition 9.8 *For every RA-OTM $M^{(RA)}$, there is an equivalent OTM M with two linear tapes that can simulate any t-step computation by $M^{(RA)}$ in $O(t^2)$ steps.*

Proof (Sketch). We describe two ordinary OTMs that solve this problem because of the algorithmic lessons provided by a comparison of the solutions. Indeed, this comparison mirrors in many ways a comparison of algorithms for *dense* vs. *sparse* matrix computations; cf. [33].

A *computed-address* solution for *dense* random-access computations. Our first solution is inspired by situations in which before $M^{(RA)}$ accesses row k of its storage tape $T^{(S)}$, it will have accessed all rows $i < k$ of the tape.

For this situation, we employ the OTM M in Figure 9.13(bottom) as our simulating OTM. The only change required to the simulation algorithm described in the proof of Proposition 9.7 regards M's response to a "random access" by $M^{(RA)}$ to $T^{(S)}$. If $M^{(RA)}$ writes a numeral for the integer k on its address tape $T^{(A)}$, then this mandates moving the read/write head of $T^{(S)}$ from the square, $\langle i, j \rangle$, that it occupies at this step to square $\langle k, 1 \rangle$ for the next step. M simulates this move exactly as in the proof of Proposition 9.7, but the computation of the "address computer" Δ is a bit more complicated, because

$$\Delta = \mathscr{D}(k,1) - \mathscr{D}(i,j)$$

need no longer have a simple form as in that proof. Importantly, though, when M is simulating the tth step of a computation by $M^{(RA)}$, the time-complexity of performing this computation and executing the resulting move on M's tape T_1 remains $O(t)$. We leave this verification—*which depends on the fact that $M^{(RA)}$'s computation is dense*—as an exercise. (*Hint:* How far away from $\mathscr{D}(i,j)$ can $\mathscr{D}(k,1)$ be on M's tape T_1?)

A *table-lookup* solution for *sparse* random-access computations. The OTM M that we designed for the computed-address solution can be dramatically inefficient if the computation by $M^{(RA)}$ utilizes its storage tape in a very sparse manner. To illustrate this point, while supplying intuition for the qualifier "sparse," consider the rather extreme—but eminently possible—scenario wherein $M^{(RA)}$ uses only row $2^{2^{10}}$ of its storage tape $T^{(S)}$. (The programmer of $M^{(RA)}$ has a sense of humor?) In order to simulate the step in which $M^{(RA)}$ accesses $T^{(S)}$ for the first time, the computed-address simulator M must traverse roughly $\mathscr{D}(2^{2^{10}},1)$ squares of its storage tape T_1 *no matter how many computation steps $M^{(RA)}$ has executed.* In some sense, therefore, M's simulation is unboundedly inefficient.

If we employ a *table-lookup* approach to linearizing $T^{(S)}$, then we can replace the unboundedly inefficient computed-address simulator M by a simulator M' that suffers only polynomial-time simulation slowdown. Extending the way that M uses its second tape, T_2, M' uses tape T_2 to store (numerals for)

- the coordinates $\langle x, y \rangle$ of the square of $T^{(S)}$ that $M^{(RA)}$ is currently accessing;
- the coordinates $\langle x', y' \rangle$ of the square of $T^{(S)}$ that $M^{(RA)}$ will access next.
  These coordinates can be specified in one of two ways.

  - If $M^{(RA)}$ decides to make a "two-dimensional-tape" move, by moving to a neighbor of the current square in one of the four compass directions, then M' computes the coordinates of the target square and places them on tape T_2. For instance, a northward move by $M^{(RA)}$ would lead M' to specify $\langle x', y' \rangle$ as $\langle x-1, y \rangle$.
  - If $M^{(RA)}$ decides to make a "random-access" move, by explicitly specifying a new row x_0 of $T^{(S)}$, then M' will specify $\langle x', y' \rangle$ as $\langle x_0, 1 \rangle$.

(M' has no need of the computed address $\mathscr{D}(x,y)$, so it will not store this value on T_2.) M' uses the coordinate numerals mostly as uninterpreted strings, for the purpose of pattern matching; it uses them as numerals only in computations such as "$x' \leftarrow x - 1$" that are occasioned by "two-dimensional-tape" moves by $M^{(RA)}$. A more pronounced difference between M and M' is visible in the respective organizations of their storage tapes T_1. Recall that M uses T_1 to store $M^{(RA)}$'s storage tape $T^{(S)}$, linearized according to the diagonal pairing function $\mathscr{D}(x,y)$. In contrast, M' stores on T_1 a *list* of those squares of $T^{(S)}$ that $M^{(RA)}$ has accessed thus far in its computation. This list is stored in the following format. If the square $\langle x, y \rangle$ of $T^{(S)}$ has been accessed by $M^{(RA)}$, and if it currently contains the symbol γ, then T_1 will contain the

following list entry:

$$x, y : \gamma;$$

The delimiter symbols "," and ":" and ";" play very specific roles in a list entry:

- Each comma (,) separates a pair of coordinate-strings x and y from one another.
- Each colon (:) separates a pair of coordinate-strings "x, y" from a worktape symbol γ.
- Each semicolon (;) terminates a list entry, hence separates each list entry from its successor.

Of course, this usage demands that the delimiter symbols *not* belong to either $M^{(RA)}$'s worktape alphabet or the alphabet used to encode the numerals x and y.

Say that M' has to simulate a move by $M^{(RA)}$ wherein $\langle x', y' \rangle$ are the coordinates of the sought new square of $T^{(S)}$. M' scans along tape T_1, to find a list entry that begins with the string "$x', y' :$".

- If no such list entry is found, then M' appends a new entry

$$x', y' : \boxed{B};$$

at the end of its list. The fact that this list entry did not exist means that $M^{(RA)}$ has not previously visited square $\langle x', y' \rangle$ of $T^{(S)}$; hence, on its first visit to the square, the symbol there will be $\boxed{B}$.
- If the sought list entry

$$x', y' : \gamma;$$

is found, then M' reads the tape symbol γ and consults (in its internal memory) $M^{(RA)}$'s program, in order to decide what $M^{(RA)}$ would do in the current situation, having read symbol γ on $T^{(S)}$.

How has this table-lookup approach avoided the unbounded—and unboundable— delays of the computed-access approach? The answer has three components.

1. What is the smallest number of steps that $M^{(RA)}$ could have been computing to this point if the coordinate-string $\langle x', y' \rangle$ appears on M''s tape T_2? We claim that $M^{(RA)}$ must have been computing for at least $\ell(x) + \ell(y)$ steps.

   - If $M^{(RA)}$ wandered along $T^{(S)}$ via a sequence of "two-dimensional-tape" moves to a square that is adjacent to square $\langle x', y' \rangle$, then $M^{(RA)}$ must have been computing for at least a number of steps proportional to the *number* $x + y$. This number certainly exceeds the combined length $\ell(x) + \ell(y)$ of the numerals x and y.
   - If $M^{(RA)}$ specified the coordinate-string during a "random-access" move, then $M^{(RA)}$ must have been computing for at least $\ell(x) + \ell(y)$ steps, just in order to write the string!

2. As M' proceeds along tape T_1, the number of list entries that it encounters cannot exceed the number of squares of $T^{(S)}$ that $M^{(RA)}$ has visited thus far. The number of list entries is thus a lower bound on the number of steps that $M^{(RA)}$ has executed thus far.
3. Say that M' is seeking a list entry that begins with the coordinate string $\langle x',y'\rangle$. (This means, in particular, that this coordinate string is written on tape T_2.) Consider how M' processes each list entry on tape T_1 by focusing on one that begins with the coordinate string $\langle x'',y''\rangle$. M' moves its two read/write heads to the left ends of $\langle x',y'\rangle$ (on T_2) and $\langle x'',y''\rangle$ (on T_1). It then moves its two read/write heads in tandem to check whether the two coordinate strings are identical. It can make this determination within

$$\min\big(\ell(x')+\ell(y'),\ \ell(x'')+\ell(y'')\big)$$

steps. (Either it gets interrupted by a mismatch in the two coordinate strings, or it gets all the way across.)
This reckoning makes it clear that M''s search for the coordinate string $\langle x',y'\rangle$ takes time no longer than twice the length of the word currently written on T_1. (The extra factor of 2 results from the fact that M' backs up to the left end of the string $\langle x',y'\rangle$ as it encounters each new list entry.)

Adding up all of the processes involved in the table-lookup approach to the simulation, one finds that M' spends $O(t)$ steps to simulate the tth step of an arbitrary computation by $M^{(RA)}$. The bound on simulation time follows.

A final question: Why bother with the computed-access simulation for dense computations if the table-lookup simulation achieves the same performance bound? The answer resides in the constant factors. When the computed-access simulation works well, it works very well, with small constant factors. While the table-lookup simulation never suffers the disastrous unbounded slowdown that can plague the computed-access simulation, it does incur larger constant factors. □

The "bottom line" is that even this most ambitious of our augmentations of the OTM model can be simulated by an ordinary OTM with only polynomial slowdown.

PART IV
NONDETERMINISM

Your bill is $495.75.
That is $0.75 for turning the screw, and
$495.00 for knowing which screw to turn.
(punch line of old joke)

We turn finally to the third of our pillars, *nondeterminism*, a somewhat unntuitive computational concept that plays a variety of fundamental roles in computation theory. We defer our (brief) discussion of these roles until after Chapter 10, because it is difficult to appreciate the computational benefits of nondeterminism without having seen it.

Chapter 10
Nondeterministic Online Automata

We saw in Section 3.1 that one can view OAs as abstract representations of actual circuits or machines or programs. In contrast, the generalization of OAs that we present now is a mathematical abstraction that cannot be realized directly from conventional hardware or software elements. It is best to view this model either as a pure mathematical convenience—whose utility we shall see imminently—or as a "computational strategy" that we shall try to realize via sophisticated transformation of a program.

10.1 Nondeterministic OAs

Informal development. Here is how the abstraction works. One can view an OA—deterministic or not—as "making a decision" regarding the choice of the next state whenever it receives an input symbol. Nondeterminism endows an OA with the ability to "hedge its bets" in this decision-making process. One way to look at this hedging is that in the new abstraction, an OA can create "alternative universes," making a (possibly) distinct choice of next state in each universe. We noted earlier (Section 3.1) that a computation by a (deterministic) OA M on a string $\sigma_0 \sigma_1 \sigma_2 \cdots \sigma_k$ can be viewed as a linear sequence

$$q_0 \xrightarrow{\sigma_0} q_1 \xrightarrow{\sigma_1} q_2 \xrightarrow{\sigma_2} \cdots \xrightarrow{\sigma_k} q_{k+1}.$$

The interpretation of the preceding sequence is that the OA M starts out in state q_0; in response to input symbol σ_0, it moves to state q_1; thence, in response to input symbol σ_1, it moves to state q_2; and so on.

In contrast, in the more general, nondeterministic, setting, a "computation" must be viewed as a *forest*, in order to allow the nondeterministic OA to split universes at each step. A nondeterministic analogue of the computation depicted in (3.1), wherein for convenience of presentation we always split universes in *two*, now has a form such as the following:

A.L. Rosenberg, *The Pillars of Computation Theory*, Universitext,
DOI 10.1007/978-0-387-09639-1_10, © Springer Science+Business Media, LLC 2010

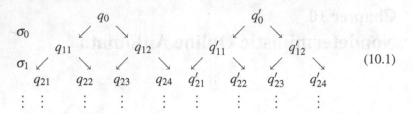

$$(10.1)$$

The computation depicted in (10.1) is a *forest*, rather than a tree, because we allow the OA to "hedge its bets" even before the computation starts—by beginning its computation in a set of start states, rather than in a single start state.

In order to flesh out the generalized model, we must, of course, indicate when a "nondeterministic" OA "accepts" a string. In the deterministic setting, acceptance resides in the fact that the terminal state, q_{k+1}, in the computation depicted in (3.1) is an accepting state (an element of F). In the generalized setting, after reading an input string $\sigma_0 \sigma_1 \sigma_2 \cdots \sigma_k$, the nondeterministic OA may be in different terminal states in different universes: in the computation depicted in (10.1), for instance, after reading $\sigma_0 \sigma_1$, the nondeterministic OA is in up to eight states, $q_{21}, q_{22}, q_{23}, q_{24}, q'_{21}, q'_{22}, q'_{23}, q'_{24}$ (which need not all be distinct) in its eight terminal universes. By convention, we say that the OA accepts an input string if *at least one* of the states that the string leads to is an accepting state. *Nondeterminism as thus construed is inherently built around an existential quantifier—"there exists a path to an accepting state."*

Formal development. Formalizing the preceding discussion, a *nondeterministic online automaton* (*NOA*, for short) is a system $M = (Q, \Sigma, \delta, Q_0, F)$ where:

1. Q, Σ, and F play the same roles as with a *deterministic* OA (henceforth called a *DOA*, for short);
2. Q_0 is M's *set* of initial states;
3. M's state transitions take sets of states to sets of states.

We elaborate on the third of these points. We begin with

$$\delta : Q \times \Sigma \to \mathscr{P}(Q),$$

where, as usual, $\mathscr{P}(Q)$ denotes the power set of Q. We extend δ to sets of states, i.e., to a function

$$\delta : \mathscr{P}(Q) \times \Sigma \to \mathscr{P}(Q),$$

in the natural way, via unions: For any subset $Q' \subseteq Q$ and $\sigma \in \Sigma$,

$$\delta(Q', \sigma) = \bigcup_{q \in Q'} \delta(q, \sigma).$$

There is a natural inductive extension of δ to $\mathscr{P}(Q) \times \Sigma^\star$. For any subset $Q' \subseteq Q$,

$$\delta(Q', \varepsilon) = Q',$$

and for all $\sigma \in \Sigma$ and $x \in \Sigma^*$,

$$\delta(Q', \sigma x) = \bigcup_{p \in \delta(Q', \sigma)} \delta(\{p\}, x).$$

We have just extended the state-transition function δ of an NOA twice, once to have it act on *sets of states*, rather than individual states, and once to have it act on *strings*, rather than individual symbols; each time we overloaded the symbol "δ" to accommodate the extended domain. The reader should verify, using the same type of careful reasoning we went through early in Section 3.1 (when extending δ to act on strings), that we have not jeopardized our firm technical footing via these extensions and overloadings.

Acceptance of a string by an NOA is formalized by the following condition:

$$L(M) = \{x \in \Sigma^* \mid \delta(Q_0, x) \cap F \neq \emptyset\}.$$

You should make sure that you see the correspondence between the formal setting of an NOA and its language, as just described, and our intuitive description preceding the formalism.

10.2 Nondeterminism as Unbounded Search, 1

In this section and Section 12.3, we present two results that expose the inherent nature of nondeterminism in two ways that appear to be quite distinct, yet in fact are just different ways of characterizing any nondeterministic "computation" as a deterministic computation that is accompanied by an unbounded search. It is traditional and convenient—but certainly not necessary—to have the search *precede* the deterministic computation. Our first characterization—in this section—is *algorithmic*, in the sense that it incorporates the search into the computation performed by an NOA. Our second characterization—in Section 12.3—is *logical*, in the sense that it incorporates the search into the *specification* (in logical notation) of the NOA's computation.

An algorithmic view of nondeterministic search. In the strict sense formalized by the proof of the following theorem, nondeterminism does not enhance computation theory at all, at least not in terms of what can be computed. Nondeterminism does, however, have profound effects in terms of various measures of efficiency, as well as in other respects that we discuss in Section 10.3.

Theorem 10.1. *Every language L that is accepted by an NOA M is also accepted by a DOA M' whose structure is determined by M's.*

Proof. Consider the language L that is accepted by the NOA $M = (Q, \Sigma, \delta, Q_0, F)$. Our proof relies on the following intuition, which is discernible in an NOA's acceptance criterion. Focus on a moment when M has thus far read the string $x \in \Sigma^*$ (and has branched into some collection of independent universes). If we want to determine how having read x will influence M's subsequent behavior as it reads possible new input symbols, all we need to know is the *set* of states $\delta(Q_0, x)$ that M is in within its various universes. Specifically, the number k of occurrences of a state q in this set is immaterial—as long as $k \neq 0$. It follows that we can deterministically simulate the (nondeterministic) computation of M on any string z just by keeping track of the successive sets of states that the successive symbols of z lead M to. The generality of the OA model allows us to accomplish this with a *deterministic* OA $M' = (Q', \Sigma, \delta', q'_0, F')$, which we construct as follows. Recall that our goal is for the DOA M' to simulate *all* of M's computational universes simultaneously, so that $L(M') = L(M)$.

Because M' need only keep track of the sequence of *sets of* states that an input string would lead M through, we can construct M' via the following so-called *subset construction*.

We repeat our earlier *caveat* about nondeterminstic "algorithms." In its full generality, the following construction is not an algorithm: it gives no hint about how to represent and manipulate the sets of M's states. We shall see in Chapter 11 that when the OA is *finite*, then this construction is easily converted into an algorithm.

We specify the various defining components of the DOA M':

- $Q' = \mathscr{P}(Q)$.
  This says, informally, that M' keeps track of sets of M's states.
- For all $R \in \mathscr{P}(Q)$ (which is another way of saying for all $R \subseteq Q$) and all $\sigma \in \Sigma$,

$$\delta'(R, \sigma) = \bigcup_{r \in R} \delta(r, \sigma).$$

  Thus, M' follows M from one set of states S (which personifies one set of simultaneous universes) to the set S's successor-set under input σ, as specified by M's state-transition function δ.
- $q'_0 = Q_0$.
  Thus M' begins correctly, by simulating M's set of start states, Q_0.
- $F' = \{R \in \mathscr{P}(Q) \mid R \cap F \neq \emptyset\}$.
  This definition captures the fact that M accepts a string iff that string leads M to a set of states that contains *one or more* (accepting) states of F.

Our intuitive justifications for each component of M' can be turned into a simple inductive proof that $L(M') = L(M)$, which is left to the reader.   □

We shall illustrate the subset-construction conversion algorithm in Chapter 11 (right after Theorem 11.1), because it is easier to appreciate how and why the construction works when the NOA in question is finite.

It is worth reflecting on what the proof of Theorem 10.1 tells us about the nature of nondeterminism. In essence, the DOA M' that replaces a given NOA M generates, for each (nondeterministic) "computation" by M, a *search tree* whose structure embodies the repeated nondeterministic ramifications by M. Indeed, the entire process of simulating M deterministically involves searching the various levels of this tree for the existence of accepting states. Of course, if we were to actually implement M'

algorithmically, we would have to craft some regimen for performing the searches through the search trees generated by M in response to the various input strings. We thus see nondeterminism as a kind of shorthand for unbounded searches. This insight will remain valid as we begin to study nondeterminism formally—especially in terms of its impact on computational complexity.

10.3 An Overview of Nondeterminism in Computation Theory

Having seen nondeterminism in a very abstract guise in the preceding section, we can begin to discuss the central role that nondeterminism plays in computation theory. In the case of both the weakest and most powerful classes of OAs, namely finite automata and Turing machines (or their equivalents), nondeterminism is just a convenience. Deterministic FAs can simulate nondeterministic ones (Section 11.1)—albeit at the cost of increased size—and deterministic TMs can simulate nondeterministic ones (Section 12.2)—albeit at the *apparent* cost of increased computing time. (Nobody knows for sure whether the slowdown is inevitable.) Even in these cases, though, we shall see that nondeterminism gives rise to simplified design algorithms and succinct specifications for automata (and the algorithms they represent).

With many classes of OAs whose computing power is intermediate between that of FAs and (unrestricted) TMs—whether the limitations on power arise from limitations on computational resources, such as time or memory, or because of restricted access to resources, as with so-called pushdown automata (see, e.g., [41])—it is best to view nondeterminism either as a purely mathematical convenience or as a "strategy" that we shall try to realize algorithmically via some sophisticated algorithmic transformation. The most important examples of this role of nondeterminism are found in formal language theory—where they were discovered by Chomsky [12, 13]—and in complexity theory—where they were discovered by Cook [18].

Within the domain of formal language theory, several families of languages that are defined by related syntactic schemas (which are kin to the formal grammars of Section 6.1.3) can actually be characterized as precisely the languages accepted by certain special classes of *nondeterministic* OAs. To cite just the two most important such families—important because of their roles in the formal study of both programming languages and natural languages: *context-free languages* are precisely the languages accepted by *nondeterministic* pushdown automata; and *context-sensitive languages* are precisely the languages accepted by *nondeterministic* linear-bounded automata. (A linear-bounded automaton is a Turing machine that, when computing on an input word x, is not allowed to access more than $\ell(x)$ squares of its worktape; see Section 13.2.2.) The preceding equivalences between families of languages and classes of automata originated in [12, 13]; texts such as [41] give a treatment that is more accessible to the modern reader.

As we shall see in Chapter 13, within the domain of complexity theory, nondeterminism is one of several concepts that people have used to enrich deterministic computation in a way that leads to significant, apparently new, classes of languages.

(For most classes, it is not known whether the associated deterministic and nondeterministic classes are the same!) Most typically, the classes are defined by restricting the resources—mainly time and memory—available to the computational model, usually as a function of the length of the input word being processed (as with the just defined linear-bounded automaton). Also, most typically, some fixed computational model—often Turing machines—is used as the anchor for the resulting theory. Within this theme, one of the—perhaps *the*—central questions in complexity theory asks how much computational resource is needed to simulate nondeterminism deterministically. The most famous instance of the preceding question is the **P**-vs.-**NP** problem that originated in [18] and that dominates our discussion of complexity theory in Chapter 13.

Chapter 11
Nondeterministic FAs

11.1 Nondeterministic FAs vs. Deterministic FAs

The material in this section refines the development in Chapter 10 by restricting attention to NOAs whose set Q of states is *finite*. We call such an NOA a *nondeterministic
finite automaton* (*NFA*, for short).

We noted in Chapter 4 that one can view FAs as abstract representations of actual circuits or machines or programs. However, when the FA model is generalized
to allow nondeterminism, the resulting NFA model is a mathematical abstraction
that cannot be realized directly from conventional hardware or software elements—
although its intended behavior can be mimicked via simulation. In the theories of
finite automata and regular languages, nondeterminism was invented independently
and roughly contemporaneously by two pairs of researchers, Gene H. Ott and Neil
H. Feinstein [75] and Michael O. Rabin and Dana Scott [79], as a simplifying conceptual and algorithmic tool for proving the Kleene–Myhill theorem (which we develop
in Section 11.2).

> Nondeterminism is indeed just a convenience! The original proof of the Kleene–Myhill the-
> orem, in [51], did not employ nondeterminism. It is worth looking at this version in order to
> appreciate the significance of nondeterminism in making the theorem's proof accessible.

11.1.1 NFAs Are No More Powerful Than DFAs

We now verify formally that nondeterminism is no more than a convenience within
finite automata theory, by showing that it affords NFAs no more computing power that
DFAs already have. Specifically, we show that NFAs accept only regular languages.

Theorem 11.1. *Every language accepted by an NFA is regular, i.e., is accepted by
some DFA.*

Proof. This result is really just a corollary of Theorem 10.1, via the insight that if
a given NOA $M = (Q, \Sigma, \delta, Q_0, F)$ is an *NFA*—because the set Q is finite—then

A.L. Rosenberg, *The Pillars of Computation Theory*, Universitext,   217
DOI 10.1007/978-0-387-09639-1_11, © Springer Science+Business Media, LLC 2010

the power set, $\mathscr{P}(Q)$, of Q is also finite. This means that the subset-construction algorithm in the proof of Theorem 10.1 actually produces a *DFA M'* that accepts the same language, $L(M)$, as M does. $\square$

We now present a couple of examples to illustrate the construction of Theorem 11.1.

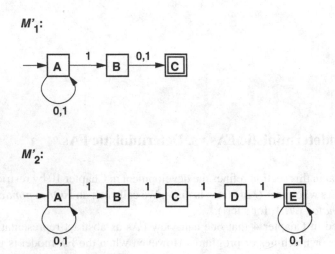

Fig. 11.1 Graph-theoretic representations of two simple NFAs.

M'_1	0	1
→ A	{A}	{A,B}
B	{C}	{C}
C	∅	∅

M'_2	0	1
→ A	{A}	{A,B}
B	∅	{C}
C	∅	{D}
D	∅	{E}
E	{E}	{E}

Table 11.1 Tabular representations of the NFAs of Figure 11.1.

Example 1. It is clear intuitively that the NFA M'_1 of Figure 11.1(top) and Table 11.1(left) accepts the language L of binary strings that have a 1 as the next-to-last symbol. Applying the subset construction of Theorem 11.1 to M'_1 produces the DFA M''_1 of Figure 11.2(right). An easy application of the state-minimization algorithm of Section 5.2 shows that M''_1 is minimal in number of states.

Example 2. It is clear intuitively that the NFA M'_2 of Figure 11.1(bottom) and Table 11.1(right) accepts the language L of all binary strings that contain a run of at least four consecutive 1's. This is the language $L(M_2)$, where M_2 is the DFA depicted in Figure 3.2 and Table 3.1. Applying the subset construction of Theorem 11.1

M'_1: M''_1:

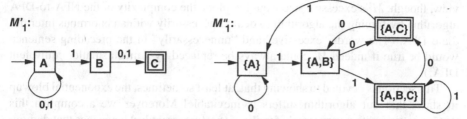

Fig. 11.2 The DFA M''_1 that Theorem 11.1 produces from the NFA M'_1 of Figure 11.1.

M'_2:

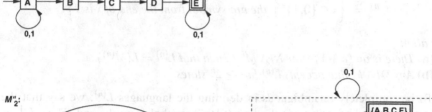

M''_2:

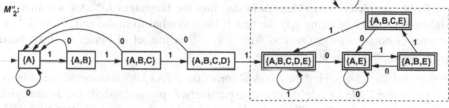

Fig. 11.3 The DFA M''_2 that Theorem 11.1 produces from the NFA M'_2 of Figure 11.1.

to M'_2 produces the DFA M''_2 of Figure 11.3(bottom). When the state-set of M''_2 is reduced using the state-minimization algorithm of Section 5.2, one finds that all of the accepting states of M''_2 can be merged—as we have done in the dashed box in the figure—thereby producing a DFA that is identical to M_2, aside from the different names for the states.

11.1.2 Does the Subset Construction Waste DFA States?

Our proof of Theorem 11.1 uses the subset construction as the basis of an algorithm that converts a given NFA to a DFA that accepts the same language. Because the power set $\mathscr{P}(S)$ of a finite set S is exponentially larger than S—specifically, $|\mathscr{P}(S)| = 2^{|S|}$—this proof strategy raises the specter that the DFA M' produced by our algorithm to replace an NFA M may have many more states than it needs—perhaps exponentially more! Now, from one point of view, such profligacy would not matter in the end, because one could employ the state-minimization algorithm of Section 5.2 to replace M' by an equivalent DFA M'' of the proper size. From a more practical point of view, though, M''s excessive size would explode the complexity of the NFA-to-DFA algorithm by forcing the algorithm to deal unnecessarily with an enormous intermediate DFA. (The words "excessive" and "unnecessarily" in the preceding sentence would be true if indeed M' could always be replaced by a much smaller equivalent DFA.)

This section is devoted to showing that, at least sometimes, the exponential blowup in states that our algorithm suffers is inevitable! Moreover, we accomplish this demonstration with a very simple family of (perforce, regular) languages that demonstrate the inevitability of the blowup for NFAs of all possible sizes.

Proposition 11.1 *For each positive integer n, define the language*

$$L^{(n)} = \left\{ x \in \{0,1\}^* \mid \text{the } n\text{th symbol from the end of } x \text{ is } 1 \right\}.$$

For all n:
   (a) *There is an $(n+1)$-state NFA $M^{(n)}$ such that $L^{(n)} = L(M^{(n)})$.*
   (b) *Any DFA M that accepts $L^{(n)}$ has $\geq 2^n$ states.*

To be completely unambiguous in defining the languages $L^{(n)}$: we say that the rightmost symbol of string x is the first ("1th") symbol from the end. Thus, $L^{(1)}$ is the set of binary strings that end with a 1, $L^{(2)}$ is the set of binary strings whose penultimate symbol is a 1, and so on.

Proof. **(a)** The idea behind the construction of the NFAs $M^{(n)}$ is discernible in Figure 11.1(top) and Table 11.1(left), which, respectively, depict and specify the 3-state NFA $M^{(2)}$. We extrapolate from this example to derive the generic $(n+1)$-state NFA $M^{(n)}$ via the specification in Table 11.2. The start-state A_0 of $M^{(n)}$ "temporizes" before (nondeterministically) "identifying" (via a guess) the input position that will be the nth symbol from the end of the input. Once having "identified" this position, $M^{(n)}$ checks that the symbol at this position is a 1 and then uses the rest of its states to verify (by counting) that the position it has "identified" is indeed the nth from the end. The easy details are left to the reader.

$M^{(n)}$		0	1
$\rightarrow$	A_0	$\{A_0\}$	$\{A_0, A_1\}$
	A_1	$\{A_2\}$	$\{A_2\}$
	A_2	$\{A_3\}$	$\{A_3\}$
	$\vdots$	$\vdots$	$\vdots$
	A_{n-1}	$\{A_n\}$	$\{A_n\}$
	A_n	$\emptyset$	$\emptyset$

Table 11.2 A tabular representation of $M^{(n)}$.

(b) We use the Myhill–Nerode theorem (Theorem 4.1) to show that any DFA M_n that accepts $L^{(n)}$ must have at least 2^n states. To this end, let x and y be two binary strings that differ in at least one position within their last n symbols. In other words,

$$x = \xi \alpha_n \alpha_{n-1} \cdots \alpha_1 \quad \text{and} \quad y = \eta \beta_n \beta_{n-1} \cdots \beta_1,$$

where

- $\xi, \eta \in \{0,1\}^*$ are binary strings of possibly different lengths;
- each $\alpha_i \in \{0,1\}$ and each $\beta_j \in \{0,1\}$;
- for at least one $k \in \{1, 2, \ldots, n\}$, $\alpha_k \neq \beta_k$.

We claim that $x \not\equiv_{L^{(n)}} y$, so that (by Theorem 4.1) x and y must lead M_n's initial state to distinct states. Once we establish this, we shall be done, because there are clearly 2^n distinct length-n endings for binary strings.

We prove that $x \not\equiv_{L^{(n)}} y$ by supplying a continuation string z such that one of xz and yz belongs to $L^{(n)}$, while the other does not. To this end, say that $\alpha_k = 1$ and $\beta_k = 0$.

Because we use no properties of x and y other than the ones enumerated above, we lose no generality by assuming that $\alpha_k = 1$, while $\beta_k = 0$.

With this assumption, we choose z to be *any* binary string of length $n - k$: $z \in \{0,1\}^{n-k}$. We then have

$$xz = \xi \alpha_n \alpha_{n-1} \cdots \alpha_{k+1} 1 \alpha_{k-1} \cdots \alpha_1 z,$$

while

$$yz = \eta \beta_n \beta_{n-1} \cdots \beta_{k+1} 0 \beta_{k-1} \cdots \beta_1 z.$$

By design, α_k is the nth symbol from the end of string xz, while β_k is the nth symbol from the end of string yz. Therefore, our assumption about the values of α_k and β_k ensures that $xz \in L^{(n)}$, while $yz \notin L^{(n)}$. By definition, this means that $x \not\equiv_{L^{(n)}} y$, as claimed.

While the proof is now complete, I should not leave you in suspense about the status of all short binary strings with regard to M_n. (Of course, the x's and y's of our proof must have length at least n.) You can satisfy yourself that each binary string v of length $n - h$, for $h > 0$, is in the same class of relation $\equiv_{L^{(n)}}$ as is the length-n string $0^h v$. This means that M_n does not need to have more than 2^n states— so we now understand $L^{(n)}$'s memory requirements (in the sense of Section 5.4) exactly. $\square$

11.2 An Application: The Kleene–Myhill Theorem

11.2.1 A Convenient Enhancement of NFAs

Because NFAs accept only regular languages, nondeterminism is indeed only a mathematical convenience for us. Because we are being kind to ourselves by allowing ourselves this convenience, let us continue to be kind by making the NFA model even easier to use within the context of the important Kleene–Myhill theorem. We do this by enhancing NFAs to allow them to have so-called ε-transitions.

An ε-nondeterministic finite automaton (ε-NFA, for short) $M = (Q, \Sigma, \delta, Q_0, F)$ is an NFA whose state-transition function δ is extended to allow so-called ε-transitions, under which M changes state spontaneously, without polling its input port. Formally, M's state-transition function now has an expanded domain:

$$\delta : Q \times (\Sigma \cup \{\varepsilon\}) \to \mathscr{P}(Q).$$

By allowing M to change state on "input" ε—which is really the absence of an input—we are able to simplify some of our constructions in Section 11.2.2's proof of the Kleene–Myhill theorem. First, though, we show that such transitions do not augment the power of NFAs. This demonstration is crucial to our allegation that the Kleene–Myhill theorem characterizes regular sets, rather than some variant thereof.

Theorem 11.2. *Every language accepted by an ε-NFA $M = (Q, \Sigma, \delta, Q_0, F)$ is regular.*

Proof. For each state $q \in Q$, we define q's ε-reachability set $E(q)$ as follows:

$$E(q) \stackrel{\text{def}}{=} \{p \in Q \mid [p = q] \text{ or } (\exists p_1, p_2, \ldots, p_n \in Q)[q \stackrel{\varepsilon}{\to} p_1 \stackrel{\varepsilon}{\to} p_2 \stackrel{\varepsilon}{\to} \cdots \stackrel{\varepsilon}{\to} p_n \stackrel{\varepsilon}{\to} p]\}.$$

Essentially, $E(q)$ is the set of states that state q can reach spontaneously, i.e., via ε-transitions. (One sometimes sees $E(q)$ called something like the "ε-closure" of state q.)

Using this construct, we now present an NFA $M'' = (Q, \Sigma, \delta'', Q_0'', F'')$ that has no ε-transitions and is equivalent to M in the sense of accepting the same language. Here are the formal specifications of M'''s defining components, δ'', Q_0'', and F''. (Components Q and Σ are inherited from M.)

- $Q_0'' = \bigcup_{q \in Q_0} E(q);$
- $(\forall \sigma \in \Sigma, \, Q' \subseteq Q) \left[\delta''(Q', \sigma) = \bigcup_{p \in \delta(Q', \sigma)} E(p) \right];$
- $F'' = F \cup \{q \in Q_0 \mid F \cap E(q) \neq \emptyset\}.$

M'' thus uses the sets $E(q)$, for $q \in Q$, to systematically "trace through" and "collapse" all of M's ε-transitions. Specifically:

- by definition of Q_0'', M'' starts out in all states that M does, either directly or via ε-transitions;
- by definition of δ'', M'' makes all state transitions that M does, either directly or via ε-transitions;
- by definition of F'', M'' accepts any string that M does, either directly or via ε-transitions.

We leave the formal verification that $L(M) = L(M'')$ as an exercise. $\square$

With the results of this section in our toolkit, we can henceforth wander freely within the world of DFAs, NFAs, and ε-NFAs when studying regular languages. We make extensive use of this freedom in the next section.

11.2.2 The Kleene–Myhill Theorem

The Myhill–Nerode theorem (Theorem 4.1) characterizes the regular languages by exploiting (limitations on) DFAs' abilities to make discriminations among input strings. We shall see imminently that one can also characterize the regular languages via three operations that suffice to build such languages up from the letters of the input alphabet Σ. This characterization culminates in this section's Kleene–Myhill theorem (Theorem 11.3), which occupies a central role in both the theory and applications of the finite automaton model. Theorem 11.3 also gives rise to a useful notation, called *regular expressions*, for assigning operational names to each regular language.

> **Notes.** (*a*) See Section 4.3.2 for a brief discussion of names that are *operational*.
> (*b*) Our use of the plural "names" in our assertion that regular expressions assign operational names to each regular language is no typo. We shall see as we develop the elements of regular expressions that one weakness of this naming scheme for regular languages is that it can take exponential computational resources to determine whether two given expressions denote the same language.

The theorem appeared first in [51] and, in slightly modified form, in [19].

Three basic operations on languages. The Kleene–Myhill theorem describes a sense in which three operations on languages explain the inherent nature of regular languages over a given alphabet Σ. We review the definitions of the operations from Sections 2.1, 2.4.1, and (especially) 6.2.

- The *union* of languages L_1 and L_2: the set-theoretic union $L_1 \cup L_2$.
- The *concatenation* of languages L_1 and L_2:

$$L_1 \cdot L_2 \overset{\text{def}}{=} \{xy \in \Sigma^\star \mid [x \in L_1] \text{ and } [y \in L_2]\}.$$

The following observation about the concatenation of languages is particularly relevant in the context of this section. Because languages need not be "prefix-free"—i.e., because all of the strings $u_1, u_1 u_2, \ldots, u_1 u_2 \cdots u_k$ may belong to L_1—recognizing concatenations is a "very nondeterministic" operation: each encountered prefix $u_1 u_2 \cdots u_i$ in L_1 could be the x that one is looking for (i.e., the first factor of the concatenation), or it could be just a prefix of that x.

- For any language L and integer $k \geq 0$,

$$L^0 \overset{\text{def}}{=} \{\varepsilon\} \quad \text{and, inductively,} \quad L^{k+1} \overset{\text{def}}{=} L \cdot L^k.$$

Clearly, $L^1 = L$. We call each language L^k the kth *power* of language L—using the word "power" as with exponentiation.

Because every power L^k for $k > 2$ involves at least two concatenations of languages, recognizing such powers is "even more nondeterministic" a task than is recognizing a single concatenation.

• The *star-closure* of a language L, denoted by L^*, is the set

$$L^* \stackrel{\text{def}}{=} \bigcup_{i=0}^{\infty} L^i = \{\varepsilon\} \cup L \cup L^2 \cup L^3 \cup \cdots.$$

Note the somewhat unintuitive fact that $\emptyset^* = \{\varepsilon\}$. (In fact, $\emptyset^*$ is the only finite star-closure language!)

Of course, the fact that the star-closure L^* of a language L involves *iterated* concatenation makes it "even more nondeterministic" than any single fixed power L^k of L.

We turn now to the reason for our interest in the operations of union, concatenation, and star-closure.

Regular expressions. We now define a powerful notational mechanism that affords us operational names for every regular language. We call this mechanism *regular expressions*. As you read on, always keep in mind that each regular expression $\mathscr{R}$ is just a (finite) string! Expression R denotes—i.e., is the name of—a (possibly infinite) language L, *but expression $\mathscr{R}$ is not itself a language!*

Table 11.3 presents the inductive definition of a regular expression $\mathscr{R}$ over a (finite) alphabet Σ, accompanied by the "interpretation" of the expression, in terms of the language $\mathscr{L}(\mathscr{R})$ that it denotes.

	Atomic Regular Expressions	
	Regular Expression $\mathscr{R}$	Associated Language $\mathscr{L}(\mathscr{R})$
	$\emptyset$	$\emptyset$
	ε	$\{\varepsilon\}$
For $\sigma \in \Sigma$:	σ	$\{\sigma\}$
	Composite Regular Expressions	
For RE's $\mathscr{R}_1, \mathscr{R}_2$:	$(\mathscr{R}_1 + \mathscr{R}_2)$	$\mathscr{L}(\mathscr{R}_1) \cup \mathscr{L}(\mathscr{R}_2)$
For RE's $\mathscr{R}_1, \mathscr{R}_2$:	$(\mathscr{R}_1 \cdot \mathscr{R}_2)$	$\mathscr{L}(\mathscr{R}_1) \cdot \mathscr{L}(\mathscr{R}_2)$
For any RE $\mathscr{R}$:	$(\mathscr{R}^*)$	$(\mathscr{L}(\mathscr{R}))^*$

Table 11.3 Inductive definition of regular expressions and the languages that they denote.

We sometimes try to enhance legibility by violating our formal rules and omitting parentheses and dots, as when we write $a^* b^*$ for $((a)^*) \cdot ((b)^*)$. We shall always be careful to avoid ambiguities when employing such abbreviations—and when we are being formal, we shall never employ such abbreviations!

The Kleene–Myhill theorem. We now finally can develop the result that exposes a sense in which regular expressions tell the entire story of regular languages. We state the theorem in two quite distinct ways in order to give you two sources of intuition as we embark on the proof.

Theorem 11.3. (The Kleene–Myhill theorem) *A language is regular if and only if it is definable by a regular expression.*
In other words, the family of regular languages over an alphabet Σ is the smallest family of subsets of Σ^* that contains all finite languages over Σ (including $\emptyset$ and $\{\varepsilon\}$) and that is closed under a finite number of applications of the operations of union, concatenation, and star-closure.

We prove Theorem 11.3 via two lemmas. The first lemma shows that every language that is denoted by a regular expression is a regular language.

Lemma 11.1. *If the language $L \subseteq \Sigma^*$ is denoted by a regular expression $\mathscr{R}$, then L is a regular language.*
In other words, the family of regular languages contains all finite languages, and it is is closed under the operations of union, concatenation, and star-closure.

Proof. We showed in Lemma 4.2 that every finite language is regular. Therefore, we need focus only on the closure properties of the regular languages. We present schematic intuitive arguments for these closure properties, that can easily be turned into inductive proofs, and we augment these arguments with a simple running example.

We present in Figure 11.4 the two schematic NFAs, M_1 and M_2, that we use to present our schematic intuitive arguments. In Figure 11.5, we instantiate the

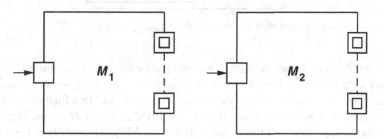

Fig. 11.4 A schematic depiction of NFAs M_1 and M_2. Small squares denote states; squares with inscribed squares are accepting states. The dashed lines indicate that we make no assumptions about the number of accepting states. The arrows point to the start states.

schematic NFAs of Figure 11.4 with two simple explicit NFAs, $\widehat{M}_1$ and $\widehat{M}_2$, that we use to illustrate our schematic constructions. Both $\widehat{M}_1$ and $\widehat{M}_2$ have the input alphabet $\Sigma = \{1\}$, and both accept single-string languages: $L(\widehat{M}_1) = \{11\}$, and $L(\widehat{M}_2) = \{111\}$.

Union. We build an NFA M_{1+2} that accepts $L(M_1) \cup L(M_2)$, as follows. We take M_1 and M_2 and "defrock" their start states: these states still exist, but they are no

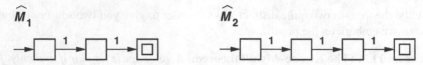

Fig. 11.5 The NFAs $\widehat{M}_1$ and $\widehat{M}_2$. Notational conventions follow Figure 11.4.

longer start states. We then endow M_{1+2} with a (single) new, nonaccepting, start state whose only transitions are ε-transitions to the "defrocked" start states of M_1 and M_2; we illustrate this construction schematically in Figure 11.6. Clearly, the only paths in

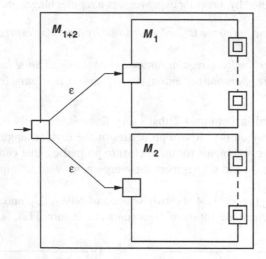

Fig. 11.6 A schematic depiction of the "union" NFA M_{1+2}.

M_{1+2} from the (new) start state to an accepting state consist of the ε-transition from the start state to the "defrocked" start state of either M_1 or M_2—say, without loss of generality, M_1—followed by a path from M_1's start state to a final state of M_1. It follows that M_{1+2} accepts a string if and only if either M_1 or M_2 does. Figure 11.7 illustrates the application of this construction to the NFAs $\widehat{M}_1$ and $\widehat{M}_2$ of Figure 11.5, to produce the "union" NFA $\widehat{M}_{1+2}$; clearly, $L(\widehat{M}_{1+2}) = \{11, 111\}$.

Concatenation. We build an NFA $M_{1.2}$ that accepts $L(M_1) \cdot L(M_2)$ as follows. We take M_1 and M_2 and "defrock" M_2's start state: it still exists, but it is no longer a start state. The start state of M_1 becomes $M_{1.2}$'s start state. Next, we "defrock" M_1's accepting states: these states still exist, but they are no longer accepting states. Finally, we add ε-transitions from M_1's "defrocked" accepting states, to M_2's "defrocked" start state; see Figure 11.8. The net effect of this construction is that whenever $M_{1.2}$ has read a string x that would lead M_1 to one of its accepting states—so that x belongs to $L(M_1)$—$M_{1.2}$ continues to process any continuation of x within M_1, but it also passes that continuation through M_2. It follows that if there is any way to parse the input to M into the form xy, where $x \in L(M_1)$ and $y \in L(M_2)$,

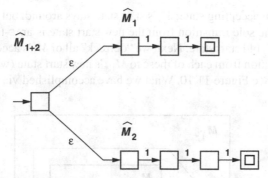

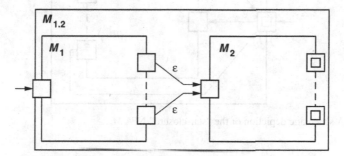

Fig. 11.7 Depicting the "union" NFA $\widehat{M}_{1+2}$.

Fig. 11.8 A schematic depiction of the "concatenation" NFA $M_{1.2}$.

then $M_{1.2}$ will find it—via the inserted ε-transition. Conversely, if that ε-transition leads $M_{1.2}$ to an accepting state, then the successful input must admit the desired decomposition. Figure 11.7 illustrates the application of this construction to two copies of the union NFA $\widehat{M}_{1+2}$ of Figure 11.7, producing the "concatenation" NFA $\widehat{\widehat{M}}_{1.2}$; clearly, $L(\widehat{\widehat{M}}_{1.2}) = \{1111, 11111, 111111\} = \{1^4, 1^5, 1^6\}$.

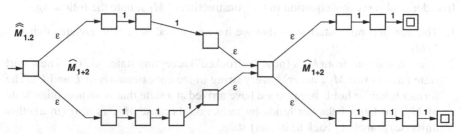

Fig. 11.9 Depicting the "concatenation" NFA $\widehat{\widehat{M}}_{1.2}$.

Star-Closure. We build an NFA $M_{1,*}$ that accepts the language $(L(M_1))^\star$. The only delicate issue here is that we must take care that $M_{1,*}$ accepts ε, as well as all positive powers of $L(M_1)$. Taking care of this delicacy first, we give $M_{1,*}$ a new start state that is also an accepting state: M_1's start state stays around, but it is "defrocked" as a start state. The sole transition from the new start state is an ε-transition to M_1's (now "defrocked") old start state. Next, we "defrock" all of M_1's accepting states, and we add an ε-transition from each of these to $M_{1,*}$'s new start state (which, recall, is an accepting state). See Figure 11.10. What we have accomplished via this construction

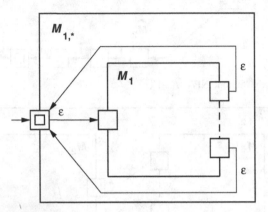

Fig. 11.10 A schematic depiction of the "star-closure" NFA M_*.

is the following. If $M_{1,*}$ reads an input x that would be accepted by M_1, then it hedges its bets. On the one hand, $M_{1,*}$ keeps reading x, thereby seeking continuations of x that also belong to $L(M_1)$; additionally, though, $M_{1,*}$ assumes that the continuation of x is an independent string in $(L(M_1))^*$, so $M_{1,*}$ spawns a universe in which it starts over, in the start state of M_1.

Explained in another way, the NFA $M_{1,*}$ implements the following identity, which holds for arbitrary languages L:

$$L^\star = \{\varepsilon\} \cup L \cdot L^\star.$$

In order to observe this equation in the construction of $M_{1,*}$, note the following.

1. The new accepting start state that we have endowed $M_{1,*}$ with ensures that $\varepsilon \in L(M_{1,*})$.
2. The ε-transitions from M_1's (now "defrocked") accepting states to $M_{1,*}$'s new start state ensures that $M_{1,*}$ accepts every string in the concatenation of L and $L^\star$. The former behavior holds because we have arrived at a state that is an accepting state of M_1; the latter behavior holds by induction, because $M_{1,*}$ is also (in another universe) branching back to its start state.

We thus see that $L(M_{1,*}) = (L(M_1))^\star$, as claimed. Figure 11.11 illustrates the application of this construction to the "union" NFA $\widehat{M}_{1+2}$ of Figure 11.7, thereby producing the "star-closure" NFA $\widehat{\widehat{M}}_{(1+2),*}$.

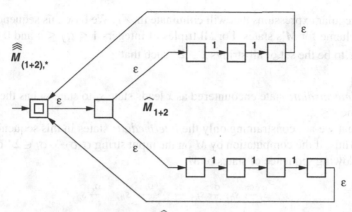

Fig. 11.11 Depicting the "star-closure" NFA $\widehat{\widehat{M}}_{(1+2),*}$.

You should verify that if M_1's start state is an accepting state, then one can easily modify the preceding construction so that one need not add a new start state to $M_{1,*}$.

The preceding constructions and explanations provide the intuition underlying a formal proof of this lemma that is based on an induction on the length of the input string. We leave the task of formalizing the induction as an exercise. □

The second lemma shows that every regular language is denoted by some regular expression.

Lemma 11.2. *If the language $L \subseteq \Sigma^*$ is regular, then L is denoted by a regular expression $\mathscr{R}_L$.*

Proof. We prove the lemma via a famous dynamic programming algorithm for constructing $\mathscr{R}_L$ that originated in [63]. The reader should compare this algorithm with the closely related Floyd–Warshall algorithm for computing transitive closures and related path-oriented problems in graphs; this algorithm originated in [25, 105], and it appears in major texts on algorithms such as [20].

Focus on an arbitrary DFA $M = (Q, \Sigma, \delta, q_0, F)$, and let $L = L(M)$. Let us (re)name M's states (which constitute the set Q) as an indexed *sequence*, $s_1, s_2, \ldots, s_n$, with $s_1 = q_0$. We use the indices of these state names to orchestrate the dynamic program that we use to obtain a regular expression $\mathscr{R}_L$ that denotes $L(M)$. (Using our earlier notation, as in Table 11.3, we could write $\mathscr{L}(\mathscr{R}_L) = L(M)$.) Note that the *only* aspect of the way we (re)name M's states that is relevant to our algorithm is our assigning q_0 the name "s_1." We do need to know q_0's name, because it tells us where—i.e., in which state—computations by M start. Other than that one piece of information, we just need an indexing of M's states that allows us to refer unambiguously to state #1 (which is "s_1" in our proposed naming scheme), state #2 (which is "s_2" in the scheme), and so on.

As the next step in developing our dynamic program, we need to derive a systematic sequence of sublanguages of $L(M)$ that will allow us to build up increasingly complex regular expressions that will culminate in $\mathscr{R}_L$. We base this sequence on our naming scheme for M's states. For all triples of integers $1 \leq i, j \leq n$ and $0 \leq k \leq n$, define $L_{ij}^{(k)}$ to be the set of all strings $x \in \Sigma^*$ such that

1. $\delta(s_i, x) = s_j$;
2. every *intermediate* state encountered as x leads state s_i to state s_j has the name s_ℓ for some $\ell \leq k$.

   Note that we are constraining only the *intermediate* states in this sequence, not s_i or s_j. Thus, if the computation by M on the input string $\sigma_1 \sigma_2 \cdots \sigma_\ell \in \Sigma^*$ describes the following sequence of states of M,

$$s_i \xrightarrow{\sigma_0} s_{h_1} \xrightarrow{\sigma_1} s_{h_2} \xrightarrow{\sigma_2} \cdots \xrightarrow{\sigma_{\ell-1}} s_{h_{\ell-1}} \xrightarrow{\sigma_\ell} s_j, \tag{11.1}$$

then we are asserting that $h_m \leq k$ for every $m \in [1, \ell - 1]$, but we are making no restriction on either i or j.

Let us consider the computational implications of the preceding conditions.

When $k = 0$, there is no intermediate state in the sequence (11.1), so that

$$L_{ij}^{(0)} = \begin{cases} \{\sigma \mid \delta(s_i, \sigma) = s_j\} & \text{if } i \neq j, \\[2mm] \{\varepsilon\} \cup \{\sigma \mid \delta(s_i, \sigma) = s_j\} & \text{if } i = j. \end{cases}$$

Note that the first line of this definition implicitly specifies $L_{ij}^{(0)}$ to be the empty set $\emptyset$ when $i \neq j$ and there is no σ such that $\delta(s_i, \sigma) = s_j$.

When $k > 0$, we can derive an exact expression for the set $L_{ij}^{(k)}$ in terms of sets $L_{ab}^{(kc)}$ whose index c is strictly smaller than k, via the following intuition. The set $L_{ij}^{(k)}$ consists of all strings that lead state s_i to state s_j via a sequence of intermediate states, each having an index *no larger than* k; $L_{ij}^{(k)}$ therefore consists of:

- the set $L_{ij}^{(k-1)}$ of all strings that lead state s_i to state s_j via intermediate states whose indices are strictly smaller than k, *unioned with* ...

  - the set $L_{ik}^{(k-1)}$ of all strings that lead state s_i to state s_k via intermediate states whose indices are strictly smaller than k, *concatenated*[1] *with* ...
  - the set $\left(L_{kk}^{(k-1)}\right)^\star$ of all strings that lead state s_k back to itself via intermediate states whose indices are strictly smaller than k, *repeated as many times as you want, concatenated with* ...
  - the set $L_{kj}^{(k-1)}$ of all strings that lead state s_k to state s_j via intermediate states whose indices are strictly smaller than k.

Representing this recipe symbolically, we have

---

[1] Because languages and sets of strings are the same things within computation theory, this operation of *concatenation* has been defined in Section 6.2.

$$L_{ij}^{(k)} = L_{ij}^{(k-1)} \cup L_{ik}^{(k-1)} \cdot \left(L_{kk}^{(k-1)} \right)^* \cdot L_{kj}^{(k-1)}.$$

Because M has n states in all, the set $L_{ij}^{(n)}$ comprises *all* strings that lead M from state s_i to state s_j. Finally, L is the union of the sublanguages that lead M from its initial state s_1 to some accepting state; i.e.,

$$L = \bigcup_{s_\ell \in F} L_{1\ell}^{(n)}.$$

This completes the proof, because we have now determined how to construct L from (perforce, finite) subsets of Σ, via a finite number of applications of the operations of union, concatenation, and star-closure. One can translate the recipe for so constructing L directly into a regular expression $\mathscr{R}_L$, for one can view a regular expression as precisely such a recipe.

Of course, the regular expression $\mathscr{R}_L$ is finite, because there are only $n^3 + n^2$ sets $L_{ij}^{(k)}$.

In order to concretize the message of Lemma 11.2, we present a sample invocation of the dynamic programming algorithm that constitutes its proof, using the simple DFA M_1 of Figure 3.2. Table 11.4 presents the tableau produced by the dynamic program, with regular expressions simplified to enhance readability. As the table (and visual inspection because M_1 is so simple) indicates,

$$L(M_1) = L_{11}^{(3)} = (a^3)^*,$$

using a "pidgin" regular expression whose meaning should be clear.

The observant reader will recognize that no feature of the regular-expression-producing algorithm of Lemma 11.2 demands that we start with a DFA; the algorithm works also if one starts with an NFA. We illustrate this by starting with the simple NFA M_1' of Figure 11.1. Table 11.5 presents the tableau produced by the dynamic program, with regular expressions simplified to enhance readability. As the table (and visual inspection because M_1' is so simple) indicates,

$$L(M_1') = L_{13}^{(3)} = (0+1)^* 1 (0+1)(0+1)^*,$$

using a "pidgin" regular expression whose meaning should be clear.

k	A regular expression for $L_{ij}^{(k)}$		
0	$L_{11}^{(0)} = \varepsilon$	$L_{22}^{(0)} = \varepsilon$	$L_{33}^{(0)} = \varepsilon$
	$L_{12}^{(0)} = a$	$L_{23}^{(0)} = a$	$L_{31}^{(0)} = a$
	$L_{13}^{(0)} = \emptyset$	$L_{21}^{(0)} = \emptyset$	$L_{32}^{(0)} = \emptyset$
1	$L_{11}^{(1)} = \varepsilon$	$L_{22}^{(1)} = \varepsilon$	$L_{33}^{(1)} = \varepsilon$
	$L_{12}^{(1)} = a$	$L_{23}^{(1)} = a$	$L_{31}^{(1)} = a$
	$L_{13}^{(1)} = \emptyset$	$L_{21}^{(1)} = \emptyset$	$L_{32}^{(1)} = a^2$
2	$L_{11}^{(2)} = \varepsilon$	$L_{22}^{(2)} = \varepsilon$	$L_{33}^{(2)} = \varepsilon + a^3$
	$L_{12}^{(2)} = a$	$L_{23}^{(2)} = a$	$L_{31}^{(2)} = a$
	$L_{13}^{(2)} = a^2$	$L_{21}^{(2)} = \emptyset$	$L_{32}^{(2)} = aa$
3	$L_{11}^{(3)} = \varepsilon + a^2(\varepsilon+a^3)^\star a$ $= (a^3)^\star$	$L_{22}^{(3)} = \varepsilon + a(\varepsilon+a^3)^\star a^2$ $= (a^3)^\star$	$L_{33}^{(3)} = \varepsilon + (\varepsilon+a^3)^\star a^3$ $= (a^3)^\star$
	$L_{12}^{(3)} = a^2(\varepsilon+a^3)^\star a^2$ $= a(a^3)^\star$	$L_{23}^{(3)} = a(\varepsilon+a^3)^\star$ $= a(a^3)^\star$	$L_{31}^{(3)} = (\varepsilon+a^3)^\star a$ $= a(a^3)^\star$
	$L_{13}^{(3)} = a^2(\varepsilon+a^3)^\star$ $= a^2(a^3)^\star$	$L_{21}^{(3)} = a(\varepsilon+a^3)^\star a$ $= a^2(a^3)^\star$	$L_{32}^{(3)} = (\varepsilon+a^3)^\star a^2$ $= a^2(a^3)^\star$

Table 11.4 Executing the dynamic programming algorithm of Lemma 11.2 on the DFA M_1 of Figure 3.2.

k	A regular expression for $L_{ij}^{(k)}$		
0	$L_{11}^{(0)} = (\varepsilon+0+1)$	$L_{22}^{(0)} = \varepsilon$	$L_{33}^{(0)} = (\varepsilon+0+1)$
	$L_{12}^{(0)} = 1$	$L_{23}^{(0)} = (0+1)$	$L_{31}^{(0)} = \emptyset$
	$L_{13}^{(0)} = \emptyset$	$L_{21}^{(0)} = \emptyset$	$L_{32}^{(0)} = \emptyset$
1	$L_{11}^{(1)} = (0+1)^\star$	$L_{22}^{(1)} = \varepsilon$	$L_{33}^{(1)} = (\varepsilon+0+1)$
	$L_{12}^{(1)} = (0+1)^\star 1$	$L_{23}^{(1)} = (0+1)$	$L_{31}^{(1)} = \emptyset$
	$L_{13}^{(1)} = \emptyset$	$L_{21}^{(1)} = \emptyset$	$L_{32}^{(1)} = \emptyset$
2	$L_{11}^{(2)} = (0+1)^\star$	$L_{22}^{(2)} = \varepsilon$	$L_{33}^{(2)} = (\varepsilon+0+1)$
	$L_{12}^{(2)} = (0+1)^\star 1$	$L_{23}^{(2)} = (0+1)$	$L_{31}^{(2)} = \emptyset$
	$L_{13}^{(2)} = (0+1)^\star 1(0+1)$	$L_{21}^{(2)} = \emptyset$	$L_{32}^{(2)} = \emptyset$
3	$L_{11}^{(3)} = (0+1)^\star$	$L_{22}^{(3)} = \varepsilon$	$L_{33}^{(3)} = (0+1)^\star$
	$L_{12}^{(3)} = (0+1)^\star 1$	$L_{23}^{(3)} = (0+1)(0+1)^\star$	$L_{31}^{(3)} = \emptyset$
	$L_{13}^{(3)} = (0+1)^\star 1(0+1)(0+1)^\star$	$L_{21}^{(3)} = \emptyset$	$L_{32}^{(3)} = \emptyset$

Table 11.5 Executing the dynamic programming algorithm of Lemma 11.2 on the NFA M_1' of Figure 11.1.

Chapter 12
Nondeterminism in Computability Theory

12.1 Introduction

We begin our study of nondeterminism in computability theory by crafting a non-deterministic version of the online Turing machine (OTM) of Section 3.3; we call this enhanced model an *NTM*, for "nondeterministic Turing machine." The NTM model provides us a "bookend" to match our study of nondeterministic finite automata (NFAs) in Chapter 11. Whereas NFAs arise from adding nondeterminism to the computationally weakest model in our study, namely FAs, NTMs arise from adding nondeterminism to the computationally most powerful model in our study—indeed, the computationally most powerful possible model, period, according to the Church–Turing thesis.

The major lesson of this chapter is that just as nondeterminism does not enhance the computational power of FAs (Theorem 11.1), it also does not enhance the computational power of TMs. The reasons for these conclusions are quite distinct, though. We found in Theorem 11.1 that the FA model is *so weak* computationally that its nondeterministic version can exploit only a bounded degree of nondeterminism—and one can always use a bigger deterministic FA to simulate this degree of nondeterminism without actually resorting to nondeterminism. (Quantifying the adjective "bigger" in the preceding statement: we were always able to simulate the degree of nondeterminism accessible to an n-state NFA M using a 2^n-state DFA to simulate M.) In contrast, we find in this chapter that the TM model is *so powerful* computationally that it can simulate unbounded amounts of nondeterminism.

A word of warning is in order. In order to establish the ability of deterministic TMs—or any of their computationally equivalent kin—to simulate nondeterministic TMs, we need to modify the "online" feature of our *online* TM model, for technical reasons that we discuss in the next section. (In anticipation: online computation is inherently a deterministic phenomenon: the formalism via which a nondeterministic computational model accepts input strings is incompatible with the way online computational models accept input strings.)

A.L. Rosenberg, *The Pillars of Computation Theory*, Universitext,
DOI 10.1007/978-0-387-09639-1_12, © Springer Science+Business Media, LLC 2010

12.2 Nondeterministic Turing Machines

12.2.1 The NTM Model

We describe the formal extension of OTMs to NTMs only sketchily, because so much of the development parallels our extensions of OAs to NOAs and of FAs to NFAs.

Let $M = (Q_{poll}, Q_{aut}, \Sigma, \Gamma, \delta, q_0, F)$ be a (deterministic) OTM, so that its state-transition function δ associates a single element of the set $Q \times \Gamma \times \{N, L, R\}$ with each element of the set $((Q_{poll} \times \Sigma) \cup Q_{aut}) \times \Gamma$. We render M *nondeterministic*, i.e., make it an *NTM M'*, by:

- letting M' initiate its computations in a *set Q_0* of initial states, rather than in a single state q_0;
- replacing δ by an extended state-transition function δ' that maps each element of the domain

$$((Q_{poll} \times \Sigma) \cup Q_{aut}) \times \Gamma$$

to a (possibly empty) *subset* of

$$Q \times \Gamma \times \{N, L, R\},$$

rather than to a single element of that set.

As with general NOAs, the extension of the state-transition function δ to the extended version, δ', affords us a formal mechanism for allowing M' to spawn alternative universes (or equivalently, to make nondeterministic guesses) as M' processes an input string. This mechanism means that a "computation" by M' can be viewed as a *tree of moves*, analogous to an NOA's computational tree of moves, as depicted in (10.1). It is essential to stress the important differences between the *trees of moves* that we encounter here with NTMs and the simpler *trees of moves*, as in (10.1), that we encountered with NOAs; cf. Section 10.1.

1. The "state" associated with each node of the NTM M''s *tree of moves* is one of M''s *total states* (or configurations), as defined in Section 3.3. As indicated in that section, each such node has the form

$$C = \langle w, \gamma_1 \cdots \gamma_m q \gamma_{m+1} \cdots \gamma_n \rangle, \tag{12.1}$$

indicating that as a result of the (nondeterministic) branches M' has taken leading to this node:

   a. M' has read the string $w \in \Sigma^\star$ at its input port;
   b. M' is in (internal) state q;
   c. M''s read/write head is positioned on symbol $\gamma_{m+1} \in \Gamma$;
   d. M's tape is entirely blank, except possibly for the region delimited by the string $\gamma_1 \cdots \gamma_m \gamma_{m+1} \cdots \gamma_n \in \Gamma^+$.

2. OTMs and NTMs have both polling and autonomous internal states; FAs and OAs have only polling states. This fact has two consequences.

   a. Whereas all nodes at each level ℓ of an NFA's *tree of moves* represent situations in which the NFA has read $\ell - 1$ input symbols, there is no such synchrony or uniformity at the levels of an NTM's *tree of moves*.
   A practical consequence of this difference is that the *total states* at the nodes of an NTM's *tree of moves* must record the portion of the input string that the NTM has read to that point. As one can see in (10.1), this is not necessary with an NFA's or an NOA's *tree of moves*, because all states at each tree level have read the same portion of the input string as they progress from the root of the tree to the current node.

   b. Some or all branches of an NTM's *tree of moves* may be infinite, representing branches along which the NTM never halts.
   Of course, this is related to the absence of level-by-level synchrony in NTMs' *trees of moves*. NOAs' and NFAs' *trees of moves* process one input symbol per tree level, so the trees are always finite.

As with our other nondeterministic models, we say that the NTM M' *accepts* a string $x \in \Sigma^*$ if it accepts x in at least one of the universes it spawns while processing x. This means that some node in M''s *tree of moves* contains a configuration of the form (12.1) where the input string w is x, and where the internal state q is a polling, accepting state. As usual, $L(M')$, the *language accepted by* M', is the set of all input strings that M' accepts.

OAs vs. TMs. The reader may be wondering why we bother with troublesome models such as TMs and NTMs, given that OAs and NOAs seem to behave so much more smoothly (as in our discussion of *trees of moves*). The reason, simply, is that OAs and NOAs lack the structure to tell us *how* to achieve the desired behavior on a real computer. Because of this lack of a mechanism for specifying how any state transition is actually performed, either in hardware or software, there is nothing that an OA cannot compute—but there is also nothing that an OA *can* compute. What we observe from our development of computability theory (and, in the next chapter, complexity theory) is that the world gets messy when you start worrying about *how* to achieve specific behaviors. The work of Gödel and Turing tells us that this messiness is inherent: we cannot avoid it just by fiddling with our (logical or computational) models.

In view of the preceding paragraph, why do we bother with the OA/NOA model at all? Precisely to understand what features of computational systems depend only on the fact that they are state-transition systems! We thus get conceptually important—albeit too-abstract-to-implement—versions of the Myhill–Nerode theorem (Theorem 3.1) and the NOA-OA subset construction (Theorem 10.1) from our study of the OA/NOA model. But we cannot use this abstract model to replace a computational model (such as TMs) that is based on real, implementable computation.

12.2.2 Deterministic Simulation of Nondeterminism: NTMs and OTMs

This section is devoted to the algorithmic details of our claim that nondeterminism does not enhance the computational power of OTMs and their equivalently powerful kin (cf. the Church–Turing thesis). We begin with the preparatory work needed to determine carefully what such a result actually states!

As suggested in Section 12.1, the problem that we must confront is that *nondeterministic* computation is, in some senses, incompatible with (deterministic) *online* computation. The incompatibility arises from a certain "sequentiality" that is inherent in the notion of *online* computation. In detail, a central property of (deterministic) online computation by some computing device M is that for all n, M must announce its acceptance/rejection decision about the string comprising the first n letters of the current input word *before* it reads the $(n+1)$th input letter. One discerns this regimen with all of the online models we have discussed: OAs, FAs, and OTMs. For all of these models, all of the accepting/rejecting states are *polling* states. (In fact, all states of OAs and FAs are polling states.) It is impossible to enforce this type of "sequentiality" on a general computation by a *nondeterministic* TM M, because M may poll its input port at different rates along different branches of its nondeterministic computation. Specifically, M will generally have both polling and autonomous states, and a state transition that splits universes can send M into a polling state in one universe and into an autonomous state in another universe. When this occurs, distinct nodes at the same level of M's *tree of moves* may have read different amounts of the current input string. As a consequence, M may accept a string $\sigma_1 \sigma_2 \cdots \sigma_k \sigma_{k+1}$ at level ℓ along one branch of its *tree of moves*—which means after $\ell - 1$ nondeterministic steps—while it accepts string $\sigma_1 \sigma_2 \cdots \sigma_k$ only at some level $\ell' \gg \ell$ along some other branch—which means after $\ell' \gg \ell$ nondeterministic steps. The resulting asynchrony in the *trees of moves* of an NTM thus robs the model of the "sequentiality" that is inherent to online computation. So what can we do to compare these "apples and oranges"? We mention a few options.

1. *Banish autonomous states from our computational models.*
   In fact, two of the models we have been studying, OAs and FAs, have only polling states. For both of these, the subset construction demonstrates that nondeterminism does not enhance the model's computing power; cf. Theorems 10.1 and 11.1.
2. *Bound the asynchrony in the* tree of moves *of a nondeterministic computation.*
   The intention here is not to let different branches of a nondeterministic computation get "too far" out of synchrony. In the presence of such a bound, a deterministic simulator can be sure that if the NTM has not accepted an input string by such and such a time, then it will never accept the string. This allows a kind of online behavior with a bounded built-in delay.
   This is done in our study of time-restricted computation in Chapter 13. We study there OTMs and NTMs M that are embellished with *timing functions*. Each OTM (resp., NTM) M has an associated timing function f_M such that if M accepts an input string x, then it does so within $f_M(\ell(x))$ steps (resp., within $f_M(\ell(x))$

nondeterministic steps, i.e., at some node that is at level $\leq f_M(\ell(x))$ in M's "tree of moves"). We present in Chapter 13 an analogue of Theorems 10.1 and 11.1 for time-restricted TMs, as part of our buildup to the **P**-vs.-**NP** problem.

3. *"Disable" the "sequentiality" inherent in online computation.*

Let us consider again why an OTM's need for "sequentiality" can be incompatible with nondeterministic computation. Say that an OTM M is (perforce, deterministically) simulating a computation by an NTM M', and it discovers a node in M''s *tree of moves* in which M' accepts a string x—but M has not yet discovered a node in M''s *tree of moves* in which M' accepts some prefix x' of x. Then M cannot accept string x, because it does not yet know how to make an accept/reject decision about string x'. If there were some way to infer the proper decision about x' from the decision about x, then M's dilemma would disappear!

We are going now to introduce a natural mechanism that "disables" the "sequentiality" inherent in online computation in a way that allows us to compare deterministic and nondeterministic computing devices. We then prove that in the presence of this mechanism, every NTM can be simulated by an OTM—which means that nondeterminism does not enhance an OTM's computing power.

Note that had we chosen to develop computation theory around Turing's original Turing machine model [104], then we would never have run up against the "sequentiality" issue at all—for his original model is not an online one. We have chosen to base our development on *online* TMs in order to stress the essential unity of the models as one progresses from FAs through OTMs to OAs.

A mechanism for "disabling" "sequentiality." Let Σ be a finite alphabet that contains a designated symbol $\bullet$ that we call a *point*. Say that a language $L \subseteq \Sigma^*$ is *pointed* if every word in L contains precisely one occurrence of $\bullet$, which occurs at the end of the word. One can phrase this condition symbolically by asserting that each word in L has the form $x\bullet$ for some $x \in (\Sigma \setminus \{\bullet\})^*$, or equivalently:

$$L \subseteq \left(\Sigma \setminus \{\bullet\}\right)^* \cdot \{\bullet\} = \left\{x\bullet \mid x \in \left(\Sigma \setminus \{\bullet\}\right)^*\right\}.$$

Thus, the symbol $\bullet$ functions essentially as a period does in many natural languages (such as English).

Theorem 12.1. *For every nondeterministic Turing machine M that accepts a pointed language, we can construct a (deterministic) online Turing machine M^* such that $L(M^*) = L(M)$.*

Moreover, there exists a constant $c_M > 1$ such that if M accepts a length-n word $x \in L(M)$ within t_x nondeterministic steps—ie., via a path of length $\leq t_x$ in its tree of moves—then M^* accepts x via a computation that has $\leq c^{t_x}$ (deterministic) steps.

Proof. Say that the NTM $M = (Q_{\text{poll}}, Q_{\text{aut}}, \Sigma, \Gamma, \delta, Q_0, F)$ accepts the pointed language $L(M) \subseteq \left(\Sigma \setminus \{\bullet\}\right)^* \cdot \{\bullet\}$. We describe the algorithmic and representational issues that allow a deterministic OTM M^* to simulate M.

As discussed earlier, the nondeterministic computation by M in response to an input string $x \in \Sigma^*$ can be viewed as M's generating a *tree of moves*, each of whose nodes represents a configuration of M, of the form (12.1). A simulation of M by the deterministic TM M^* thus has two levels. At the higher level resides the question of how M^* represents each of M's trees of moves and how M^* uses this representation to orchestrate its threading of a tree. At a lower level resides the detailed strategy for how M^* simulates each nondeterministic step by M; this step involves using each node of M's tree of moves to generate the successors of that node in the tree. We discuss these two levels in turn.

Informally, our strategy is to have M^* process an input string x by expanding—*in a breadth-first manner*—the tree of moves that M generates while processing x. The reader will note that this procedure is quite analogous to the way a computer plays a game such as chess: each of M's trees of moves (one for each input string) is a direct analogue of a game tree.

The reader should begin to ponder why we insist that the simulating OTM M^* expand M's trees of moves in a *breadth-first* manner. As a hint, what would happen if M^* unfortunately started following a branch along which M never halted? How does a breadth-first expansion of a tree of moves prevent this?

How M^* simulates each tree of moves. As M branches nondeterministically to generate its tree of moves, M^* explicitly generates the tree, in a breadth-first fashion. M^* orchestrates this generation process by using a data structure that processes M's configurations in a first-in-first-out (FIFO) *queue*-like order.[1] In what follows, we represent a FIFO queue into which the elements $a, b, \ldots$ have been loaded, in that order—so that a will be the first element to come out, b the second, etc.—as follows:

$$> \ldots, b, a >$$

(Of course, in our case, these elements are configurations of M.) M^*'s simulation proceeds as follows.

1. M^* begins the simulation by inserting M's initial configuration,

$$C_0 = \langle \varepsilon, \boxed{\text{B}} \, q_0 \boxed{\text{B}} \, \boxed{\text{B}} \, \rangle,$$

into the initially empty queue. Pictorially, the queue now appears as

$$> C_0 >.$$

2. Inductively, a step of M^*'s simulation begins with the queue containing some sequence of configurations of M:

$$> C_m, C_{m-1}, \ldots, C_{i+1}, C_i >.$$

To aid exposition, we have indexed the configurations in the order in which they were inserted into the queue—which (by definition of the queue data structure) is also the order in which they will be extracted from the queue.

---

[1] Our brief discussion of queues in Section 9.8.1.C should give the reader enough background for the current discussion. Readers seeking more information should consult a text on algorithms, such as [20].

In the following, M^* will perform the indicated tasks while in autonomous states—except for the moves in which M^* is explicitly polling the input port.

a. M^* extracts the oldest configuration, C_i in our example, from the queue. Say that

$$C_i = \langle w, \gamma_1 \cdots \gamma_m q \gamma_{m+1} \cdots \gamma_n \rangle. \tag{12.2}$$

Recall that this configuration indicates that at the nondeterministic step being simulated, M is in internal state q and is scanning symbol $\gamma_{m+1} \in \Gamma$ on its worktape.

b. i. If q is an autonomous state of M, then M^* consults M's program to determine $\delta(q, \gamma_{m+1})$.

   ii. If q is a polling state of M, then

   A. If the input string w that M has already read is pointed, i.e., if $w = u\bullet$ for some $u \in (\Sigma \setminus \{\bullet\})^*$, and if q is an accepting state, then M^* enters an accepting polling state.

   B. If either of the preceding conditions does not hold, then M^* enters a nonaccepting polling state.

   In either case, M^* determines the next input symbol σ that M would see—we detail later how M^* does this. M^* then consults M's program to determine $\delta(q, \sigma, \gamma_{m+1})$.

c. Having determined $\delta(q, \gamma_{m+1})$ when q is an autonomous state or $\delta(q, \sigma, \gamma_{m+1})$ when q is a polling state, M^* now knows the set of $k \geq 0$ new configurations

$$\{C_{i1}, C_{i2}, \ldots, C_{ik}\}$$

that M will spawn at this nondeterministic step. M^* inserts these k new configurations—in arbitrary order—into the FIFO queue, with the appropriate time-indices:

$$> C_{m+k}, C_{m+k-1}, \ldots, C_{m+1}, C_m, C_{m-1}, \ldots, C_{i+1} >.$$

The order in which the k new configurations are inserted into the queue is immaterial, because all we care is that they are all guaranteed to get processed.

3. M^* then repeats the cycle of simulating the next nondeterminsitic step of M.

This completes the overview of how M^* simulates one of M's trees of moves. We turn now to the details of how M^* processes each node of the tree.

How M^* processes nodes in M's tree of moves. We endow M^* with nine worktapes, one of which is two-dimensional, in order to describe the detailed simulation perspicuously. Careful bookkeeping can certainly reduce this number. The reader seeking a *single*-worktape version of M^* can apply the techniques from Section 9.8.2 to convert our realization of M^* to one that uses only a single linear worktape.

M^*'s worktapes play the following roles.

1. M^* uses one *two-dimensional* worktape—the PROGRAM TAPE—to record M's δ function in tabular form. Each row of this table can be read as a *case statement* in a program: For each *polling* state of M, this entry has the form

    Case: STATE = q; INPUT = σ; WORK-SYMBOL = γ::
    >   NEW-STATE = q';
    >   NEW-WORK-SYMBOL = γ';
    >   NEW-HEAD-DIRECTION = $D \in \{N, L, R\}$

    The "new" entries come from the equation $\delta(q, \sigma, \gamma) = \langle q', \gamma', D \rangle$.
    For each *autonomous* state of M, this entry has the form

    Case: STATE = q; WORK-SYMBOL = γ::
    >   NEW-STATE = q';
    >   NEW-WORK-SYMBOL = γ';
    >   NEW-HEAD-DIRECTION = $D \in \{N, L, R\}$

    The "new" entries come from the equation $\delta(q, \gamma) = \langle q', \gamma', D \rangle$.

2. M^* uses one worktape—the INPUT tape—to record the input string $w \in \Sigma^*$ that it has read thus far. Note that w may be quite a bit longer than the input strings that M has read in the configurations at many of the nodes of M's *tree of moves*. Specifically, M's autonomous states may cause it to "lag" in reading the input along certain branches.

3. M^* uses one worktape—the QUEUE—to implement the *FIFO queue* that will control M^*'s threading of M's *tree of moves*. The use of the QUEUE was described in the high-level portion of our description of M^*'s simulation of M.

4. M^* uses one worktape as the ASSEMBLY TABLE on which it assembles M's k new configurations, $C_{i1}, C_{i2}, \ldots, C_{ik}$, from M's current configuration, C_i, and M's current nondeterministic move, $\delta(q, \gamma_{m+1})$ or $\delta(q, \sigma, \gamma_{m+1})$.

5. One worktape will serve as the SCRATCH TAPE on which M^* will extract from the configuration C_i of M that is currently being processed the arguments needed to determine M's next nondeterministic move. These arguments are M's internal state q, the worktape symbol γ_{m+1} that M's read/write head is currently scanning, and—when q is a polling state—the new input symbol σ.

- M^* transfers the oldest queue entry—which we have been calling C_i; cf. (12.2)— from the queue onto the CONFIG-SCRATCH tape. M^* then transfers the string w from C_i to the INPUT-SCRATCH tape (which is distinct from the INPUT tape, as we shall see momentarily), and it transfers both q and γ_{m+1} to the WORK-SCRATCH tape.

- If q is an autonomous state, then M^* uses the pair $\langle q, \gamma_{m+1} \rangle$ from the WORK-SCRATCH tape as an index into the PROGRAM tape, to determine $\delta(q, \gamma_{m+1})$, whose value it then records on the STATE-CHANGE scratch tape.

- If q is a polling state, then M^* compares the contents w of the INPUT-SCRATCH tape with the contents x of the INPUT tape. Either the two strings are identical or w is a prefix of x; the latter occurs when autonomous states have caused M to lag in reading the input on this branch of its tree of moves.

  If w is a prefix of x, then M^* determines the next letter that M would read at this point; this is a letter $\sigma \in \Sigma$ such that $w\sigma$ is a (not necessarily proper) prefix of x.

M^* replaces the field "w" in C_i with "$w\sigma$," to indicate that after the step, M will have read input $w\sigma$ to this point.

If $w = x$, so that M is "up to date" on this branch, then M^* enters a polling state and awaits the next input symbol at the input port. If $w = x$ is *pointed*, so that $w = y\bullet$ for some $y \in (\Sigma \setminus \{\bullet\})^*$, then if q is an accepting state of M, then M^* enters an *accepting* polling state (so that it accepts w); if q is *not* an accepting state of M, then M^* enters a *nonaccepting* polling state (so that it does not accept w).

In either case, M^* is now in a polling state, awaiting a new symbol at its input port. If this symbol never comes, then the computation stalls—which is how online computations generally halt. If a new symbol comes—call it σ—then M^* replaces the field "w" in C_i with "$w\sigma$," to indicate that after the step, M will have read input $w\sigma$ to this point.

In either of the preceding cases, M^* now uses the triple $\langle q, \sigma, \gamma_{m+1} \rangle$ from the WORK-SCRATCH tape as an index into the PROGRAM tape, to determine $\delta(q, \sigma, \gamma_{m+1})$, whose value it then records on the STATE-CHANGE scratch tape.

- M^* uses the contents of both the STATE-CHANGE and SCRATCH tapes to assemble the new configurations that M has spawned during this nondeterministic step. Each entry on the STATE-CHANGE tape tells how to transform C_i into one of the new configurations—by changing the internal state, rewriting the current worktape symbol, and shifting the read/write head. (Note that if there had been a need to extend the current input string w, then this would have been done in the preceding step of the simulation.

The preceding simulation algorithm seems to be rather complicated, but that impression is due to our striving for a level of detail that will make it clear how a Turing machine—rather than a real computer—can keep track of all necessary details and perform all necessary manipulations. The reader might well be able to understand the simulation algorithm better from just the high-level portion of the description.

A final note regarding simulation time. Note that all of M's trees of moves have node-degrees bounded above by $c'_M = 3|Q||\Sigma||\Gamma|$. (To see why, look carefully at the state-transition function δ.) It follows that M^*'s breadth-first searches through these trees takes time that is exponential in the length of M's shortest path to an accepting node. The base of this exponential is a simple function of c'_M that accounts for the time that M^* needs to manage and manipulate each of M's configurations. Details are left to an exercise. □

12.3 Nondeterminism as Unbounded Search, 2

This section complements Section 10.2's algorithmic characterization of nondeterministic "computation" with a characterization that is *logical*, in the sense that it incorporates the search into the *specification* (in logical notation) of the NOA's computation.

A logical view of nondeterministic search. Let us focus on an arbitrary semidecidable set/language $A \subseteq \Sigma^\star$, for some fixed finite alphabet Σ. By definition, A's semidecidability resides in the fact that its semicharacteristic function[2] κ'_A is semicomputable. Another way to look at this is that A's semidecidability resides in the existence of a program P_A such that, for all $x \in \Sigma^\star$:

$$P_A \begin{cases} \text{halts on every input } x \text{ that belongs to } L, \\ \text{loops forever on every input } x \text{ that does not belong to } L. \end{cases}$$

Any program that (semi)computes κ'_A operates in the prescribed manner.

The existence of programs such as P_A means that we can actually define the set/language A in terms of the behavior of such a program; namely,

$$A = \{x \in \Sigma^\star \mid P_A \text{ halts on input } x\}. \tag{12.3}$$

The advantage of this manner of defining A is conceptual, not computational. To see the latter point first, note that we know from our study of the halting problem (Section 9.3) that there is no algorithm that will decide, for all programs P_A and inputs $x \in \Sigma^\star$, whether P_A halts on input x.

> If you need some refreshing on this topic in order to appreciate the preceding sentence, consider the case that $A = \text{DHP}$, and x is the program P_{DHP} that computes the semicharacteristic function κ'_{DHP} of the diagonal halting problem DHP. We showed in the proof of Theorem 9.1 that there is no algorithm that will decide whether P_{DHP} halts when run with a copy of itself as input.

Thus, defining A via (12.3) cannot help us to decide membership in A. However, we do derive a conceptual benefit, which we describe now.

Let x, y, and z be finite strings over some alphabet Σ. Recall that via encodings, each such string may be viewed as an uninterpreted string, or as a program, or (cf. Section 3.3) as a computation by by some given program. Compare the predicate

$$P_1(x,y) \equiv \text{program } x \text{ halts on input } y$$

with the predicate

$$P_2(x,y;z) \equiv \text{string } z \text{ is the computation by program } x \text{ on input } y.$$

We begin the comparison with the simple observation that $P_2(x,y;z)$ implies $P_1(x,y)$. This is because the existence of a finite computation by program x on input y means that program x halts on input y. A bit subtler observation—but certainly no less important—is the following fundamental computational difference between the two predicates. Whereas predicate $P_1(x,y)$ will be semidecidable for any "reasonable" computational model,[3] it will not generally be decidable, for the halting-problem-related reasons given a few sentences ago. In contrast, for any "reasonable"

---

[2] Recall that for all $x \in \Sigma^\star$, $\kappa'_A(x) = 1$ when $x \in A$ and is undefined otherwise.

[3] See our discussion of "reasonable" models in Section 1.1.B.

computational model, predicate $P_2(x,y;z)$ is *always decidable!* (One can always check that the alleged computation is an actual one.)

Because we have assumed nothing about the set/language A that seeded our discussion, other than A's being semidecidable, we can infer from the preceding discussion the following very important observation.

Theorem 12.2. *For every semidecidable predicate $P_{s\text{-}d}(x)$, there exists a decidable predicate $P_{dec}(x,y)$ such that for all x, $P_{s\text{-}d}(x) \equiv (\exists y)P_{dec}(x,y)$.*

Proof. Let's go through the chain of reasoning that proves the theorem, while referring back to the discussion that preceded the theorem.

We start with the semidecidable predicate $P_{s\text{-}d}(x)$. We derive from it the semidecidable set/language

$$A_{P_{s\text{-}d}} = \{x \in \Sigma^\star \mid P_{s\text{-}d}(x)\},$$

and from this set/language the program $P_{A_{P_{s\text{-}d}}}$ that computes $A_{P_{s\text{-}d}}$'s semicharacteristic function. We thereby derive the following alternative definition of $A_{P_{s\text{-}d}}$:

$$A_{P_{s\text{-}d}} = \{x \in \Sigma^\star \mid \text{ program } P_{A_{P_{s\text{-}d}}} \text{ halts on input } x\}.$$

Because of these two ways of defining the single set $A_{P_{s\text{-}d}}$, we know that the two defining predicates are logically equivalent:

$$(\forall x \in \Sigma^\star) \left[P_{s\text{-}d}(x) \equiv [\text{program } P_{A_{P_{s\text{-}d}}} \text{ halts on input } x] \right].$$

We now define predicate $\widetilde{P}$ via

$$(\forall x \in \Sigma^\star) \left[\widetilde{P}(x;y) \equiv [\text{string } y \text{ is the computation of program } P_{A_{P_{s\text{-}d}}} \text{ on input } x] \right].$$

As discussed earlier, predicate $\widetilde{P}$ is decidable. Moreover,

$$(\forall x \in \Sigma^\star) \left[[\text{program } P_{A_{P_{s\text{-}d}}} \text{ halts on input } x] \equiv (\exists y)\widetilde{P}(x;y) \right].$$

We conclude from the preceding reasoning that

$$(\forall x \in \Sigma^\star) \left[P_{s\text{-}d}(x) \equiv (\exists y)\widetilde{P}(x;y) \right].$$

We have thus expressed the arbitrary semidecidable predicate $P_{s\text{-}d}$ as an existential quantification of the decidable predicate $\widetilde{P}$. $\quad\square$

What does Theorem 12.2 say that is relevant to our study of nondeterminism? It tells us that the essence of nondeterminism is the ability to make guesses (or to "split universes"). This ability is equivalent computationally to the ability to perform unbounded searches that seek an appropriate value for a "hidden" variable. Theorem 12.2 tells us that this searching ability is precisely what is needed to translate the world of the solvable/decidable/computable to the world of semisolvable/semidecidable/partially computable. This insight will have extremely important consequences as we turn to complexity theory, in Chapter 13.

Chapter 13
Complexity Theory

13.1 Introduction

Perhaps the major—certainly the most dramatic—strength of computability theory is its robustness across widely varying models, as expressed in the Church-Turing thesis. Probably the major weaknesses of computability theory are:

- its accepting as "computable" many functions that we have no idea how to compute—such as the run-of-7's function discussed in Section 9.7,
- its accepting as "computable" many functions that although computable in an idealized world of limitless resources, cannot be computed within the known universe: our sun will burn out, and all known matter will be exhausted, before the computation is completed.

In the light of these weaknesses, we must view parts of computability theory as mathematical abstractions that can never be realized physically.

The preceding assessment is not intended to disparage or undervalue computability theory. The theory provides invaluable insights into the nature of computation, and it provides a conceptual framework for reasoning about computation that has been rigorously tested and challenged for more than six decades. The assessment suggests, though, that computability theory deals best with "big issues" and that it needs to be refined if one wants to bring issues of practicality, or even feasibility, into the discussion. We now embark on a study of the major underpinnings of *complexity theory*, a field that can be viewed as refining the conceptual tools that computability theory uses to expose the inherent nature of computation—just so issues of practicality or feasibility can be discussed. Complexity theory achieves its refinements by embellishing many of the core notions of computability theory with efficiency-exposing parameters. These parameters help to make you aware of the quantitative consequences of your algorithmic decisions, so that you are less likely to unintentionally dedicate to a single computation every atom in the universe or every moment left to our sun.

A.L. Rosenberg, *The Pillars of Computation Theory*, Universitext,
DOI 10.1007/978-0-387-09639-1_13, © Springer Science+Business Media, LLC 2010

Note the qualifier "unintentionally": no mathematical theory will prevent you from ignoring the quantitative consequences of your allocation of computational resources. The most that one can expect is that the theory will enable you to estimate these consequences.

Let us expand on the preceding discussion. If one had to select just one "big" issue with which to justify the existence of computability theory, one would do well to choose the issue of encoding, as embodied in the kindred notions of (mapping) reduction and completeness. As we embark on our study of complexity theory, we shall observe the fundamental role that parameterized versions of these notions play in exposing and explaining the computational characteristics of myriad problems that are as important to the practitioner as to the theorist. Let us prepare the ground for these obervations via the following anecdotes.

By the early 1960s—not long after the development of "real" digital computers—both theoreticians and practitioners involved in computational fields such as combinatorial optimization began looking for ways to enlist the new field of digital computing in their quest for solutions to their computational problems. One of the most influential practitioners in this regard was the renowned expert in combinatorial optimization Jack Edmonds. Edmonds noticed that there were myriad computationally—and economically—significant problems for which the only known algorithms took time *exponential in the size of the problem*. We briefly describe just two of these "computable but practically intractable" problems, in order to hint at how varied such problems can be and to indicate informally how the indicated time bound manifests itself in concrete situations. The list of such problems could easily be expanded into the many thousands, and beyond.

1. **The traveling salesman problem.** One is given a set of n cities that the eponymous "salesman" must visit, along with a matrix C of intercity travel "costs." Each matrix entry $C(i,j)$ is the "cost" of traveling from city i to city j. The "cost" could be based on expenditures such as airfare or mileage or tolls or . . . . We put the word "cost" in quotes because the measure used need not obey any specific "reasonable" laws—such as the triangle inequality for distance-related costs.
   To solve the problem: One must find a minimum-"cost" tour of the cities, beginning and ending in the same city and visiting each intermediate city precisely once. The *"exponential size"* of the known computations that solved the problem meant that these computations would take $2^{\Omega(n)}$ steps in order to solve an n-city instance of the problem.

2. **The Boolean minimization problem.** One is given a logical (or, Boolean) expression E that specifies a logic function F. To clarify terms here:

   - E is a mathematical expression whose variables range over the set of logical, or, Boolean, values $\{0,1\}$, whose constants come from this set, and whose operations manipulate elements of this set.
   - F is a function that maps $\{0,1\}^k \to \{0,1\}$, where k is the number of variables in the expression E.

   To solve the problem: One must find a shortest expression $\widehat{E}$ that is logically equivalent to E, in the sense that it specifies the same logic function F.

The *"exponential size"* of the known computations that solved the problem meant that these computations would take $2^{\Omega(n)}$ steps in order to solve the problem for an expression that had n variables.

The reader seeking immediate gratification, in the form of a *long* list of problems to add to the two just presented, should consult the early encyclopedic compilation [28] by Michael R. Garey and David S. Johnson.

From a quite different point of departure, at around the same time, a number of mathematical logicians—whom we shall call *computational logicians* for reasons that will be clear imminently—were making use of the new tool, the digital computer, to extend what was known computationally about the *propositional calculus*.

The *propositional calculus* is the logical calculus that reasons about logical, or Boolean, expressions that are built using just the Boolean operations/connectives—*with no quantifiers* (such as $\exists, \forall$). See Section 13.3.5 for a detailed definition.

Specifically, these logicians wanted to enlist this tool in the venture of determining, given a logical expression E as input, whether E was a theorem of the propositional calculus. The computation that made this determination would provide a proof of E if it were a theorem and a refutation of E if it were not. It was known that one could, in fact, make these determinations, because the set of theorems of the propositional calculus is decidable. In fact, it had been known since roughly 1920—see [107]— that one could make these determinations *via a conceptually simple computation*, because (in contrast to the *quantified* logic of the *predicate calculus*) the propositional calculus is *semantically complete*. In plain language, semantic completeness means the following. Let E be a logical expression that contains k Boolean variables. Say that you instantiate the variables in E with the logical constants $0, 1$ in every possible way. (Of course, there are 2^k distinct instantiations, each of which triggers an evaluation of E with its variables appropriately instantiated.) The semantic completeness of the propositional calculus means that E is a theorem of the propositional calculus if and only if E evaluates to 1, or TRUE under all 2^k instantiations.

Expressions that evaluate to TRUE under all instantiations of logical values for their variables are called *tautologies*. Thus, another way to define semantic completeness is to say that the set of theorems of the propositional calculus is coextensive with, or equal to, the set of tautologies.

As we discussed in Section 9.1, Gödel's seminal work [30] showed that any (quantified) logical system that can express even rather primitive facts about positive integers is *incomplete*. This means, intuitively, that, in contrast to the propositional calculus, there are "true" statements that cannot be proved.

Early discussions about the completeness of the propositional calculus were in the "spirit" of computability theory, in that no promises were made about how hard it would be computationally to prove or refute a candidate theorem. In fact, however, it was easy to get an upper bound on the complexity of these computations, because the proofs of completeness were constructive: If the k-variable expression E was a theorem, then one could derive a proof of E from the computations that effected the 2^k instantiations of logical values into E, and if E was not a theorem, then one could derive a refutation of E from these computations. That was the good news. The bad news was that the constructions derived from the proof were *exponential in the size of the potential theorem*—i.e., all known ways of performing the required instantiation-plus-evaluation computations on a k-variable expression took $2^{\Omega(k)}$ steps on some expressions. And the news got even worse, because all of the (many!) proof/refutation generators that were developed for the propositional calculus could be shown to suffer from exponential worst-case time-complexity. Because this compexity severely limited the size of candidate theorems that early digital computers could prove or refute, the computational logicians kept trying to find *efficient*—meaning subexponential time—algorithms for proving or refuting candidate theorems in the propositional calculus. To make a long story short, all of the computational logicians' attempts were running into the same problem as Edmonds was with his optimization algorithms: every algorithm they crafted took, in the worst case, time exponential in the size of the input.

So, we had two groups of unhappy (human) computers. Each was confronting a class of computational problems that digital computers should, theoretically, help with, yet both were frustrated by their inability to use computers to solve all but unsatisfyingly small instances of their problems. Was there an inherent reason for the apparent intransigence of the problems that these groups wanted to solve, or was it just a lack of appropriate algorithmic insights on their parts?

Around 1970, Stephen A. Cook [18] took a giant step in explaining—albeit to date not alleviating—the frustration of both the combinatorial optimizers and the computational logicians. In that source, Cook proposed a theoretical approach to studying the complexity of a large range of computational problems that was based formally on computability theory but that refined that theory by incorporating measures of the time needed to compute a function (or, equivalently, solve a computational problem or decide [membership in] a language). Essentially the same theory was discovered independently, and roughly contemporaneously, by Leonid Levin [56], using an algorithmic, rather than computability-theoretic, framework. Remarkably, Cook and Levin, followed within a year by Richard M. Karp [48], and thereafter by a host of other computation theorists, used the *(computational) complexity theory* that resulted from Cook's formalization to show that all of the problems mentioned in this section—and myriad others—are, in a sense that we shall formalize, *encodings of one another*.

Step #1 in the conceptual development of complexity theory can be viewed as the following insight, by Cook, Levin, Karp, and their followers, about encodings.

If   one restricts the computability-theoretic notion of reducibility by demanding that the reducing function be *efficiently* computable, in terms of its time requirements,

and if   one "paints with a sufficiently broad brush," by measuring time coarsely, focusing on *polynomial time*, rather than, say, on linear time or quadratic time,

then   one finds that the combinatorial optimizers and the computational logicians had, in fact, been working on encodings of the same computational problems!

The time-restricted encodings that resulted from this insight thus allowed one to relate computational problems that on the surface, had nothing to do with one another.

As we begin, in the following sections, to study reductions that relate specific problems, you should ask yourself how the coarse measurement of time residing in the phrase "polynomial time" helps.

Step #2 in the conceptual development of complexity theory is manifest in an insight that is orthogonal to the preceding one and that may have emerged from Cook's initial focus on theorem provers for the propositional calculus. The challenge of determining whether a k-variable Boolean expression E is a theorem (or equivalently, a tautology) requires one to deal with a universal quantifier that ranges over a set of cardinality 2^k: E must evaluate to TRUE under *every* instantiation of its variables with truth values. However, the calculation needed to process each individual instantiation is simple, requiring, in fact, time that is *linear* in the size of expression E.

You should verify the preceding assertion about the complexity of evaluating E under a single instantiation of its variables with truth values.

Perhaps, then, it would be computationally less onerous to try to answer the question
   Is expression E a theorem?
indirectly, by focusing instead on the complementary question
   Is expression $\overline{E}$ refutable?
(Expression E is *refutable* if there exists an instantiation of E's variables with truth values under which E evaluates to FALSE.) This indirection would convert the universal quantifier that characterizes tautology—TRUE under *all* instantiations—with an existential quantifier—FALSE under *some* instantiation. What is exciting about this change of focus is how it is impacted by the fact—see Section 12.3—that existential quantifiers in a computational setting betoken *searches* that can be represented by allowing one's computational platforms to be *nondeterministic*. In fact, it is easy to verify that when viewed as a language—which is how we view all computational problems; cf. Section 2.4.1—the set of refutable logic expressions can be accepted by a *nondeterministic* Turing machine that operates in *(nondeterministic) linear time*: The TM uses its nondeterminism to "guess" an instantiation that refutes E, and it then deterministically evaluates E under that instantiation, to see whether its "guess" was correct.

Combining Steps #1 and #2, subsequent work by Levin, Karp, and their followers showed that an incredible variety of computational problems that had resisted all quests for subexponential-time (deterministic) solution—including many of Edmonds's optimization-oriented problems—shared the following property with propositional theorem-proving. A language-theoretic formulation of some version of the problem could be solved efficiently by a *nondeterministic* Turing machine. One might need to focus on "a version" of the problem—as with Cook's shift of focus from theorems to refutable expressions—and one might need to settle for "efficient" nondeterministic computation—not necessarily linear time—but the broad-brush commonality was there to observe. Moreover, if one substituted the phrase "time polynomial in the size of the problem" for "efficient," then the preceding observation could be sharpened: The language-theoretic formulation of each problem could be solved by a nondeterministic Turing machine that operated in (perforce, nondeterministic) polynomial time.

Thus arose the problem that many count as the premier unresolved problem in computation theory, the now-famous **P**-*vs.*-**NP** *problem*. This is the problem of determining whether the family **NP** of *all* languages that can be recognized in polynomial time by *nondeterministic* "algorithms" is coextensive with the corresponding family **P**, which is defined in terms of *deterministic* algorithms.

> If you prefer non-language-theoretic terminology, then the **P**-*vs.*-**NP** problem can be specified via the following question. Can every *nondeterministic* "algorithm" that operates in polynomial time be simulated by a *deterministic* algorithm that operates in polynomial time?

The **P**-*vs.*-**NP** problem dominates this chapter, in deference to its status as *the* preeminent open problem in today's world of computation theory—perhaps even of today's world of computer science.

A final note is needed to round out this introduction. The phrase "albeit to date not alleviating" used in introducing Cook's seminal work on computational complexity betokens the fact that computational researchers have yet to resolve the original fundamental question: Does there exist a (deterministic) polynomial-time algorithm that solves any of the problems—such as the traveling salesman problem or the boolean minimization problem or the problem of deciding theoremhood in the propositional calculus—whose exponential-time algorithms motivated the invention of complexity theory as we know it. What the researchers have accomplished, though, is of immense conceptual importance! We know now that if a polynomial-time algorithm can be found for any of these apparently intractable problems—or if one could prove that no such algorithm exists—then this algorithm, or this proof, could be adapted to yield a similar result for *every one* of these problems.

Once the conceptual benefits of the Cook–Levin–Karp formulation of complexity theory was recognized (which was almost instantaneous!), it took no time (read: mere months) for computational researchers to adapt the formulation to encompass a number of measures of computational complexity other than time; over the ensuing decades, today's rich theory of computational complexity evolved. Because of the (then-)new, improved conceptual settings, people were able to systematically study a broad repertoire of complexity measures for a broad range of problems, instead of being restricted to ad hoc studies of limited scope, such as we reviewed in Sections 5.4 and 5.5. (Notably, the preceding two studies predated Cook's work by only roughly five years.)

> As one can guess from the last sentences of the preceding paragraph, the work of Cook, Levin, and Karp did not arise in a vacuum. People had been trying for several years to craft a theory of computational complexity that would be embraced by both the computational theorists (because it gave access to sophisticated theoretical results) and the computational practitioners (because it either explained or—even better—alleviated some of the computational intransigence of significant "real" problems). Decidedly nontrivial progress had been made in crafting complexity theories that captured the interest of many computational theorists, but none of these theories had achieved overwhelming success in the theoretical community or anything beyond casual interest in the practical community. The Cook–Levin–Karp was recognized almost instantaneously by both theorists and practitioners.

Our introductory discussion reveals two biases that are shared by most "producers" and "consumers" of complexity theory.

1. Among the various resources that one expends while computing, *time* is at (or near) the top of the list in importance.
   Any such general pronouncement must be examined in the light of real-world factors that either strengthen or weaken the thesis. In a dynamic field such as computing, such factors can appear and disappear quickly, due to technological changes. Here are a few moderating factors relating to the importance of *time complexity*.

   - *Strengthening* the pronouncement is the decreasing cost of many hardware resources expended in computing—especially *memory* and *storage*.
   - *Strengthening* the pronouncement is the fact that adding (cheap) extra memory to a computing system inevitably slows down a processor's access to the memory.
   - *Weakening* the pronouncement is the growing importance of controlling *power* consumption—as mobile computing becomes more prevalent.
   - *Weakening* the pronouncement is the growing importance of overall work completion *amounts*, rather than *rates*—as computational modalities such as Internet-based computing allow one access to massive amounts of computing power—cf. [54]—whose efficiency is of less importance than its effectiveness (better late than never ...).

2. As one focuses on the theoretical aspects of *time* complexity, no question comes closer to dominating the agenda than the question of how efficiently deterministic computing devices—which are physically realizable as specified—can simulate nondeterministic computing devices—which are idealized abstractions. As we have seen in Sections 10.2 and 12.3, nondeterministic "algorithms"[1] can often be used as a convenient abstraction and/or shorthand for specifying deterministic algorithms that are preceded by some kind of search. The key question is how much computational resource the searches take, as compared to the deterministic algorithms that they introduce. The **P**-vs.-**NP** problem will be our primary vehicle for studying the interaction between deterministic and nondeterministic time-restricted computation. But the reader should be aware that the principles underlying our study are relevant in a broad range of situations that go far beyond the combinatorial and computation-logic problems that were the original motivation for studying how efficiently deterministic computations can simulate nondeterministic ones.

This chapter is devoted to developing the underpinnings of complexity theory. *Time* complexity—and, more specifically, the important **P**-vs.-**NP** problem—dominates our discussion. We do, however, briefly discuss also space (or memory) as a complexity measure, both because of its intrinsic conceptual interest and to illustrate by example how differently distinct complexity measures can behave. This

---

[1] As we note in several places, these nondeterministic pseudoalgorithms are not really algorithms because they cannot be realized physically as specified.

approach allows us to introduce the principles underlying the most important issues in complexity theory, while giving the reader a broader perspective that should afford her access to the large, dynamic literature on the theory.

13.2 Time and Space Complexity

There are many ways to measure the complexity of computation. Some of these ways are quite model-specific, such as measuring the consumption of power in battery-powered computing devices. Other ways transcend the specific characteristics of the device doing the computing: it is hard to imagine a computational medium that does not have some accompanying notion of *time* and *space*. Because of our focus on pervasive underlying principles, we restrict attention to *time* and *space* as measures of computational complexity. These ubiquitous measures allow us to illustrate almost all of the concepts and tools that one encounters in studying any measure.

Even having decided to focus solely on the time and space complexity measures, we must spend some time discussing *how* to measure the consumption of these resources, because, as we see in this section, model-specific details influence the formal development.

The decisions that we must make regarding how to formalize our complexity measures become much more difficult when the computing devices of interest are *nondeterministic*. Our first decision is to base our study of complexity on the NTM model introduced in Section 12.2. (This is a variant of the dominant model in the complexity theory literature, hence positions the student well for accessing that literature.) One can easily build on the robustness results for OTMs, in Section 9.8, to conjecture that a version of complexity theory based on the NTM model will be robust against a large variety of changes to the model. However, as one contemplates formulating a theory of the complexity of NTM computations, one finds immediately that the model's nondeterminism—which is an essential feature if we are to study the **P**–vs.–**NP** problem—leads to two behavioral characteristics that make it hard to quantify how much resource (time and space in our study) an NTM has expended during a computation. You can follow the details of the model in Section 12.2 as we discuss two problems that we henceforth refer to as the *nondeterministic-complexity puzzle*.

1. First, a rather minor problem. NTMs accept words but never reject them. This fact introduces an asymmetry into the way one measures time or space consumption that is largely absent in the deterministic case. Of course, the asymmetry is not totally absent even with *deterministic* OTMs, which by never halting, can fail to accept a word without rejecting it.
2. Second, a major problem. An NTM M can accept an input word x along many branches of the computation tree $\mathcal{T}_M(x)$ that it generates while processing input x. (Recall that the branching in $\mathcal{T}_M(x)$ corresponds to M's spawning distinct "universes" while processing x.) By definition, M accepts input x precisely when

x leads M along *at least one* branch in $\mathcal{T}_M(x)$ from the tree's root node—which is M's initial total state—to an accepting node.

The problem arises because of the highlighted phrase "at least one." M may accept x along many branches, and distinct accepting branches may differ in lengths—which (intuitively) measures time—and amount of memory used—which (intuitively) measures space. To phrase the problem differently, M may use different amounts of time and space in different "universes." The problem we must address is, which branch of M's computation is the most appropriate one for measuring M's time consumption while processing x? Which branch is most appropriate for measuring M's space consumption?

To this point we have often invited the reader to think about the simple cases of NFAs or NOAs when trying to garner intuition about nondeterministic computation. This place is different! NFAs and NOAs give no intuition regarding the preceding two problems, because all states of an NFA or NOA are polling states, so that all of the "universes" of an NFA or NOA process inputs in lockstep.

Without further ado, we turn to our resolution of the several formulational challenges we have been discussing.

13.2.1 On Measuring Time Complexity

We are able to introduce the topic of deterministic time complexity at a quite general level, by discussing OAs rather than more finely structured computational models. It is only when we extend the topic to nondeterministic models that we must migrate to the more structured NTM model.

A. The Basic Measure of Time Complexity

Let M be a (deterministic) OA that processes words over the input alphabet Σ, and let $t : \mathbb{N} \to \mathbb{N}$ be a nondecreasing total function that we employ as a *time-bounding* function. We say that M *operates within time* $t(n)$ if the following holds for all words $x \in \Sigma^\star$. If M halts on input x—thereby either accepting or rejecting the input—then its computation on input x consists of $\leq t(\ell(x))$ steps. See Sections 3.1 and 9.8.

We turn now to the *time*-related version of the *nondeterministic-complexity puzzle*. We resolve the puzzle's two issues in what can be viewed as an "optimistic" way. We measure an NTM M's time consumption while processing an input word x only at the *accepting* nodes in M's computation tree $\mathcal{T}_M(x)$. And, among competing accepting nodes, we focus on a *shallowest* accepting node in the tree, i.e., an accepting node that M has reached via the shortest (nondeterministic) computation, or, as fast as possible. This leads us to the following definition.

Let M be an NTM that processes words over the input alphabet Σ; let $t : \mathbb{N} \to \mathbb{N}$ be a total nondecreasing function. We say that M *operates within (nondeterministic) time* $t(n)$ if the following holds for all words $x \in \Sigma^\star$. If M accepts input x, then in at least one of its accepting computations on input x, M reaches an accepting state within $t(n)$ (nondeterministic) steps, i.e., along a tree brach of length $\leq t(n)$.

B. The Classes P and NP

The importance of the **P**-vs.-**NP** problem stands on two features of the problem.

Identifying "polynomial-time" as "efficient." Computational practitioners such as Edmonds and the earlier-referenced computational logicians were suffering with seemingly unavoidable *exponential-time* (*deterministic*) algorithms for the problems they wanted to compute. They proposed using *polynomial-time* complexity as the desirable, "efficient," counterpoint to the undesirable, "inefficient," exponential time. Of course, the now universally accepted focus on *polynomial-time* computation as the exemplar of "efficient" computation is an imperfect one. As several researchers have demonstrated, an *exponential-time* algorithm that operates within time $t(n) = (1 + 10^{-13})^n$ is almost always to be preferred to a *polynomial-time* algorithm that operates within time $t(n) = 10^{13} \cdot n^{100}$. (We hyperbolize to make our point.) However, as a first-order approximation to the truth, identifying "polynomial-time" as "efficient" is a reasonable abstraction—as long as one recognizes that it is *only* a first-order approximation to the truth.

As we have suggested earlier, the seminal work in [18], [56], [48], and myriad subsequent studies allowed the practitioners referred to in the preceding paragraph to refine the notion of "inefficient" that characterized their computational problems to a much smaller class of problems than "exponential-time" ones. Specifically, these sources showed that appropriate variants of their problems could be solved by *nondeterministic* "algorithms" that operate in (nondeterministic) polynomial time. Specifically, language-theoretic versions of their problems belonged to the family **NP** of languages that can be recognized in (nondeterministic) polynomial time. Importantly, as we note in the next paragraph, the family **NP** is *robust*, in that membership in the family is unaffected by a wide variety of structural changes to one's nondeterministic computational model.

Robustness and model independence. The study of time complexity is quite model dependent, unless one paints with a "broad brush" by dealing with rather coarse time classifications. To illustrate the point, if we were to study the family of languages that can be recognized in time *linear* in the length of the input string, then we would be studying quite distinct language families under the following computational models:

- OTMs with 1 one-dimensional read/write worktape;
- OTMs with 2 one-dimensional read/write worktapes;
- OTMs with 1 two-dimensional read/write worktape;
- OTMs with 2 two-dimensional read/write worktapes;
- *Off-line* TMs with 1 one-dimensional tape. This is, essentially, Turing's original model [104]. It starts each computation with the input string as the sole contents of its tape; it uses the tape both as a record of the input and as scratch memory.

Adding yet more worktapes and/or changing the dimensionalities of the worktapes, or allowing, say, tree-structured worktapes would lead to yet other families. The same type of model dependence would be observed with other "narrow" complexity classes, such as *quadratic* time or *cubic* time. Indeed, the results developed in Section 5.5 can be used to expose a variety of instances of this model's sensitivity.

On the other hand, if we paint with a rather broad metaphorical brush, by allowing substantial variation in computation times, then we can regain quite a bit of the model independence that is a compelling feature of computability theory (although nowhere near the robustness that led to the Church–Turing thesis). Specifically, if we allow *polynomial variation* in our timing functions, then all of the variants of the TM mentioned in the preceding paragraph lead to the same time-restricted language families. (Instances of this robustness/invariance can be observed in our analyses of the constructions of Section 9.8.) Thus, we find complexity theory populated with "polynomial time" instead of, e.g., "linear time"; "exponential time" instead of, e.g., "time 2^n"; and so on.

Recall once again computation theory's long history of converting general computational problems to language-theoretic problems that retain the salient structure of the original problems. The level of model independence that one achieves by allowing *polynomial variation* in a timing function has led to a strong identification of the family **P** of languages that can be recognized within time *polynomial* in the length of the input (say, on an OTM) as the family of languages that can be recognized *efficiently*.

> We have come full circle, in a sense. The original focus on polynomial-time computation was an *intuitive* view of "polynomial-time" as a counterpoint to the inefficiency of exponential-time computation. In light of the work by Cook, Levin, Karp, and their successors, we now have *technical* reasons for identifying polynomial-time computation as "efficient" computation—because of the model independence that this choice affords us.

For the total range of features of the language families **P** and **NP** that we have discussed in this introduction, the computational community—practitioners and theorists alike—now identify the formal problem known as "**P**-vs.-**NP**" with the informal quest for efficient algorithms for the enormous collection of significant computational problems that originated with Edmonds's optimization problems.

Before leaving this topic, we should clarify that the identification of exponential time-complexity as the enemy of efficiency has been refined, but not abandoned, by our focus on the class **NP** as our computational target. As the following pair of corollaries of Theorem 12.1 indicate, **NP** is a subfamily of the family of languages that can be recognized in (deterministic) exponential time. (We leave the straightforward proofs of these corollaries to the reader.)

The first corollary of Theorem 12.1 quantifies the dramatic expansion of time requirements if one deterministically simulates a nondeterministic TM M via the strategy of breadth-first threadings of M's computation trees.

Corollary 13.1 *For every NTM M there is a constant $c_M > 1$ for which the following holds. If M accepts $L(M)$ within (nondeterministic) time $t(n)$, then there is a (deterministic) OTM that accepts $L(M)$ within (deterministic) time $c_M^{t(n)}$.*

The preceding result specializes to the following significant observation.

Corollary 13.2 *The class **NP** is contained in the class of languages that are recognized by OTMs that operate in exponential time.*

13.2.2 On Measuring Space Complexity

Contrasting with the model sensitivity that accompanies (plagues?) measures of time complexity, space complexity is a rather model-robust measure. Indeed, it is this robustness that lends significance to the OA-based results of Section 5.4. Almost any way of instantiating the abstract OA model with a concrete *online* computational model, such as a variant of OTM, will change the results of Section 5.4 by only a constant factor. As the preceding sentence suggests, the one major way of instantiating the abstract OA model that *does* have a profound impact on space complexity is to drop the *online* feature of OAs. This impact, which we observe in the proof of Lemma 13.2, gives rise to the technical (rather than definitional) development in this section.

One's immediate reaction when faced with the task of defining the space complexity of an OTM M is to count how many squares of worktape M uses when processing an input string x, as a function of $\ell(x)$. One quickly finds, though, that this approach may not appropriately measure the *complexity* of deciding the language $L(M)$, because it conflates two distinct uses of the worktape:

- as "passive" memory that records (portions of) the input for later reference,
- as memory that participates in the actual processing of an input string.

Instead of discussing these two roles of memory abstractly, we focus our discussion around an illustrative specific example language.

A. The Pointed Palindromes as a Driving Example

We focus on a variant of an old linguistic friend (from Chapters 4, 5, and 6), the language of *palindromes*. This language provides a dramatic example of the two roles played by a TM's worktapes and thereby helps us home in on a desirable formal measure of space complexity. We focus here on a *pointed* version of the palindromes, which will allow us to isolate more easily the "passive" use of memory, which is less interesting, from the "active" use of memory, which is what we want our measure of space complexity to capture.

> The reader may recall the notion of a "pointed" language from Section 12.2.2. As in that section, this notion allows us here to avoid certain technical inconveniences inherent in the implicit demands of *online* computation.

Consider the language L_{ppal} of *pointed palindromes* over the alphabet $\{0,1\}$. Each word in L_{ppal} has the form $x\bullet$, where

1. $x \in \{0,1\}^\star$,
2. $\bullet \notin \{0,1\}$,
3. x reads the same forward and backward, i.e., is a palindrome.

We know from Lemma 5.6 that any OA M that accepts the (unpointed) palindromes must employ $2^{\Omega(n)}$ distinct states when processing input words of length n; the proof of that lemma extends with only clerical changes to the *pointed* palindromes L_{ppal}.

These changes are a valuable exercise to make sure that you understand the proof technique of Lemma 5.6.

If M is, in fact, an OTM, rather than a generic OA, then this lower bound on the number of states of M when M is viewed as an OA translates into a similar-size lower bound—i.e., also of order $2^{\Omega(n)}$—on the number of *total states* of M when M is viewed as an OTM. (See Section 3.3 for a discussion of total states in OTMs.) This latter lower bound, in turn, yields the lower bound $\Omega(n)$ for the number of squares of worktape that M must employ when processing input words of length n.

Let us expand on the last of these lower bounds, since it relates intimately to how we decide to measure space complexity. M's total state includes both its internal state and the contents of its worktape. (Let us use, as always, Q for M's set of internal states and Γ for M's worktape alphabet.) If at some step of a computation, M's worktape has length l, then M can go through $|Q| \cdot l \cdot |\Gamma|^l$ distinct total states before it must change this length. (The word recorded on the tape can be any of the $|\Gamma|^l$ length-l words over the alphabet Γ, and M's read/write head can reside on any of the tape's l squares.) This reckoning illustrates why the lower bound on M's use of tape squares during a computation is exponentially smaller than the lower bound on the number of total states that M employs during the computation.

Now, M's worktape is its only *flexible* memory: M can always use more squares of worktape, if needed, but its state-set Q has fixed size. Therefore, we infer from Lemma 5.6 the following bound on the space complexity of the language L_{ppal}.

Lemma 13.1. *Any OTM that recognizes the language L_{ppal} must use $\Omega(n)$ squares of worktape when processing input words of length n.*

The unsatisfying aspect of the preceding conclusion is that almost all of the $\Omega(n)$ required tape squares are used "passively," just to record the portion of the input word that has been read thus far at M's input port. While this observation about M's tape usage is obvious at an intuitive level, we can actually verify a form of the observation formally, as follows. Let us depart from the *online* TM model in a way that presents a TM with a read-only record of the entire input word *at no cost in space complexity*. This departure gives us a Turing machine model that we dub the *Input-Recording TM* (*IRTM*, for short). An instance of the model is depicted in Figure 13.1, wherein the IRTM has already read seven input symbols, which form the string $\sigma_1 \sigma_2 \sigma_3 \sigma_4 \sigma_5 \sigma_6 \sigma_7$.

The significance of the IRTM model is that it allows us to focus on the memory that is used to *process* input strings—what we have called "active" memory—rather than on the memory that is used to record the strings—what we have called "passive" memory. The insights that one can gain from the discriminatory power afforded by IRTMs is illustrated rather dramatically by the following counterpoint to Lemma 13.1.

Lemma 13.2. *There exists an IRTM M_{ppal} that recognizes the language L_{ppal} while using $O(\log n)$ squares of its read/write worktape when processing inputs of length n.*

Proof. One can design the IRTM M_{ppal} promised by the lemma so that it implements a computational analogue of two "fingers" that point to successive pairs of symbols

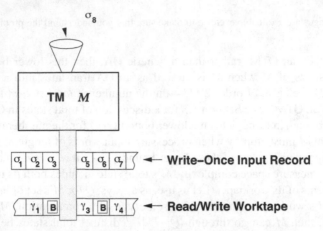

Fig. 13.1 The IRTM, with a write-once input-recording tape and a conventional read/write worktape.

of x that must be identical if x is a palindrome. When processing inputs of length n, the "fingers" must be able to point to one pair of symbols that are distance $n-1$ apart in the input, one pair that are distance $n-2$ apart, and so on, down to distance 1 or 2 (depending on the parity of $\ell(x)$). Specifying either the absolute positions where the "fingers" point or the relative positions determined by "interfinger" distances requires $\Theta(\log n)$ bits, whence M_{ppal}'s space requirements. Let us now flesh out this strategy.

We design M_{ppal} while viewing TM tapes as linear (i.e., one-dimensional) arrays that are accessed via pointers (the read/write heads). We can simplify our design of M_{ppal} by viewing the IRTM as a real machine with real tapes, rather than as a program with data structures. We simplify the description of how M_{ppal} processes an input word by describing first an IRTM M that has *two* read/write worktapes, on each of which M uses $\lceil \log_2(n+1) \rceil$ tape squares when processing inputs of length n. We digress to establish the following proposition, which tells us how to convert this IRTM M into the sought IRTM M_{ppal}.

Proposition 13.1 *Consider an IRTM M that has c read/write worktapes. Say that M uses $\leq s$ squares on each tape while processing some input word x. One can replace M by an IRTM M' that has a single read/write worktape and uses $O(s)$ bits on its worktape while processing x.*

(Note that the IRTM M may have a large worktape alphabet, but we insist that M' have a 2-letter worktape alphabet; hence M uses "$\leq s$ *squares* on each tape" while M' uses "$O(s)$ *bits*" on its tape.)

Proof (Sketch). We start with an IRTM M'' whose single read/write worktape has c *tracks*. As in Section 9.8.2, M'' achieves the effect of having c tracks on its tape by using a worktape alphabet

$$\Gamma_{M''} = (\Gamma_M \times \{\boxed{\text{B}}, \blacktriangle\}) \times \cdots \times (\Gamma_M \times \{\boxed{\text{B}}, \blacktriangle\}),$$

where:

- Each track is encoded by one instance of the set product $(\Gamma_M \times \{\boxed{B}, \blacktriangle\})$. The elements of Γ_M provide the contents of the worktape squares, and the special symbol $\blacktriangle$ provides a pointer to the square currently under scan. (Given the role of the symbol $\blacktriangle$ as a pointer, there will be just a single instance of $\blacktriangle$ for each track.)
- There are c instances of the 2-set subproduct $(\Gamma_M \times \{\boxed{B}, \blacktriangle\})$ in the indicated $2c$-set product.

We design the IRTM M' to emulate M'' step by step, using an encoding of each letter of $\Gamma_{M''}$ as a bit-string of length

$$(1 + \lceil \log_2 |\Gamma_M| \rceil) \, c.$$

It should be clear that if M uses $\leq s$ squares of each of its c read/write worktapes, then M' uses $\leq (1 + \lceil \log_2 |\Gamma_M| \rceil) cs$ bits on its single read/write worktape—because c and $|\Gamma_M|$ are fixed constants. $\square$

We return to the lemma. We now set out to design the auxiliary IRTM M that decides L_{ppal} while using *two* read/write worktapes, call them T_1 and T_2. We then rely on Proposition 13.1 to convert M into the desired IRTM M_{ppal}.

M begins by recording its entire input $x\bullet$ on the worktape that serves as its write-once input record. It accomplishes this by remaining in polling states (therefore, reading the input) until it encounters the "point" $\bullet$, which indicates that the input is complete. (If there is any continuation of this string, then M enters a "dead" polling state, because no continuation of $x\bullet$ can belong to L_{ppal}.)

As M reads and records $x\bullet$ (on its input record), it uses one of its read/write worktapes, say T_1, to count in binary from 0 to $\ell(x)$. Of course, this uses $O(\log \ell(x))$ squares of T_1.

> You probably know already that $\lceil \log_2(n+1) \rceil$ bits are necessary and sufficient for representing the integer n in binary. As a valuable exercise, you should prove this useful "everyday" fact, by induction on n.

M then checks whether x is a palindrome by checking, in turn, that the following pairs of symbols of x match: the first and last symbols, the second and second-to-last symbols, the third and third-to-last symbols, and so on. M orchestrates this sequence of checks by noting that the first two symbols to be compared are $\ell(x) - 1$ squares apart on the input record, the next two symbols to be compared are $\ell(x) - 3$ squares apart, and so on, until the final two symbols to be compared are either adjacent, i.e., 1 apart (if $\ell(x)$ is even) or 2 apart (if $\ell(x)$ is odd). (When $\ell(x)$ is odd, one obviously does not have to waste time checking that the central symbol matches itself.) These simply decreasing distances are easily—and compactly—computed using the binary numeral that M has stored on tape T_1, as follows.

Details of M's check for palindromes. M decrements the numeral on read/write worktape T_1, so that it now contains $\ell(x) - 1$; it ensures that T_2 contains 0; and it executes the following process as long as T_1 contains (the numeral of) a positive integer. M initially performs an *odd-numbered phase* of the following process.

- Odd-Numbered Phase

    1. If T_1 contains 0, then M halts and announces that the word x *is* a palindrome, so that the input word $x\beta$ *does* belong to L_{ppal}. Otherwise, M continues.
    2. M stores the symbol σ that it finds in its current square on the input record in its internal memory, i.e., in its internal state.
    3. M moves rightward on the input record the number of squares specified by the numeral stored on T_1. Specifically, for each square of the input record that M encounters in its rightward trajectory, M decrements the numeral on T_1, simultaneously incrementing the numeral on T_2 (in preparation for the next phase). When T_1 contains 0, M knows that it has completed its rightward journey on the input record.

       M now compares the symbol σ' that it finds on the current square of the input record with the symbol σ that it has stored in its internal memory.
    4. If $\sigma' \neq \sigma$, then M halts and announces that the word x is *not* a palindrome, so that the input word $x\beta$ *does not* belong to L_{ppal}.

       If $\sigma' = \sigma$, then x *could be* a palindrome, so M prepares for the next phase of its computation, by:
       a. subtracting 2 from the numeral currently on T_2,
       b. shifting one square to the left on the input record,
       c. exchanging the contents of tapes T_1 and T_2; after this exchange, T_2 will contain 0, and T_1 will contain (the numeral of) some integer of the form $\ell(x) - 2k - 1$.
    5. M now executes an *even* phase of this process.

- Even-Numbered Phase
  An *even* phase is identical to an *odd* phase, with the following two changes:

    – The roles of the read/write worktapes T_1 and T_2 are interchanged.
    – The roles of "left" and "right" as directions for M's movement on the input record are interchanged.

We illustrate M's operation by applying it to one even-length input word, $10001\bullet$, and one odd-length input word, $1001\bullet$. (We illustrate the process on two pointed palindromes because accommodating nonpalindromes is so simple.) The computations by M on these two inputs are virtually identical. In both cases, M begins by reading the entire input and recording all but the $\bullet$ on its input record. M simultaneously initializes read/write worktape T_1 to the appropriate binary numeral: 101 in the case of input word $10001\bullet$ (because $101_2 = 5$) and 100 in the case of input word $1001\bullet$ (because $100_2 = 4$). In either case, it immediately decrements the numeral on T_1, and it initializes read/write worktape T_2 to the binary numeral 0. We now describe in detail the processing of input $10001\bullet$, leaving to the reader the almost identical processing of input $1001\bullet$.

Phase 1. 1. M stores the initial symbol, 1, from (square 1 of) the input record in its internal state.
      2.   Guided by T_1's numeral 100, M moves 4 ($= 100_2$) squares to the right.

3. *M* checks that the terminal symbol from the input record, 1, is identical to the
    internally stored symbol 1.

    At this point, T_1 contains 0, and T_2 contains the base-2 numeral 100.

    M prepares for the next phase by decrementing T_2 to 10 and moving one square
    left on the input record.

Phase 2. 1. *M* stores the penultimate symbol, 0, from (square 4 of) the input
    record in its internal state.

2. Guided by T_2's value 10, *M* moves 2 (= 10_2) squares to the left.

3. *M* checks that the second-from-left symbol from the input record, 0, is identical
    to the internally stored symbol 0.

    At this point, T_1 contains the base-2 numeral 10, and T_2 contains 0.

    M prepares for the next phase by decrementing T_1 to 0 and moving one square
    right on the input record.

Phase 3. Having determined that T_1 contains 0, *M* accepts the input word.

> If *M* did not shortcut the processing of the input by checking whether the "guiding" work-
> tape T_i contains 0, *M* would have to perform a useless subcomputation on the input string.
> Specifically, on input 10001•, the 0 on T_1 would have *M* perform the useless comparison of
> the central 0 from 10001 with itself. On input 1001•, the 0 on T_2 would have *M* perform a
> virtual "negative" rightward move on the input record.

The IRTM *M* thus decides whether a length-*n* input string belongs L_{ppal} using
$O(\log n)$ squares of read/write worktape. Using Proposition 13.1, we can now convert
M to the desired IRTM M_{ppal}. □

The space that language L_{ppal} demands for *processing* input strings—the "active"
memory—is thus *exponentially smaller* than the total space—"active" memory plus
"passive" memory—that it demands from an OTM.

> Lemma 13.2 allows us to reinterpret the bounds of Lemmas 5.6 and 13.1 in the following
> way. The latter two lemmas tell us that any OA or OTM that decides L_{ppal} must be able to
> *review* input that it has read earlier. The *online regimen* that OAs and OTMs observe forces
> them to explicitly commit memory resources in order to be able to review previously read
> input.

We close this subsection by noting that the palindromes are but one language
among many that have quite different requirements in terms of the amounts of "active" and "passive" memory that they demand. To cite just two languages that we
have visited earlier in the book, there are analogues of Lemmas 13.2 and 13.1 for:

- the language of "squares"

$$L = \left\{ xx \mid x \in \{0,1\}^\star \right\},$$

which we encountered as L_4 in Application 4 of Section 5.1;

- the database language of Section 5.5:

$$L_{DB} = \left\{ \xi_1 : \xi_2 : \cdots : \xi_m :: \eta_1 : \eta_2 : \cdots : \eta_n \right\}, \text{ where, for some positive integer } k:$$

- every ξ_i and η_j is a length-k binary string: $\xi_i, \eta_j \in \{0,1\}^k$;
- the symbols ":" and "::" do *not* belong to $\{0,1\}$;
- $m = 2^k$;
- $\eta_n \in \{\xi_1, \xi_2, \ldots, \xi_m\}$.

The fact that many such languages exist lends weight to the message of Lemmas 13.2 and 13.1 concerning the varied ways in which a computation can use memory.

B. Space Complexity: Deterministic and Nondeterministic

We have seen in subsection A that computational problems can demand space/memory for two quite distinct reasons:

- to allow the computing device to *remember* the input string: what we are calling "passive" memory,
- to assist the computing device in *processing* the input string: what we are calling "active" memory.

Although both uses of memory can legitimately enter into one's assessment of the space complexity of a problem, most complexity theorists concentrate solely on the "active" memory demanded by a problem. We shall follow that practice. Accordingly, we employ *IRTMs* as the primary computational model underlying the study of space complexity.

Let M be a (deterministic) IRTM with input alphabet Σ; let M have a single read/write worktape, in addition to its input record. We say that M *operates within space $s(n)$* for a nondecreasing total function $s : \mathbb{N} \to \mathbb{N}$ if the following holds for all words $x \in \Sigma^\star$. If M halts on input x—thereby either accepting or rejecting the input— then its computation on input x uses no more than $s(\ell(x))$ squares of its read/write worktape. When the function $s(n)$ is used in this way, we call it a *space-bounding function*.

> As we saw in subsection A, our focus on IRTMs that have a *single* read/write worktape is— within constant factors—just for convenience. Proposition 13.1 tells us that we increase the space-bounding function by only a constant factor if we replace an IRTM that has multiple read/write worktapes by one that has a single worktape.

As in our deliberations on time complexity, we resolve the *nondeterministic-complexity puzzle* for space complexity "optimistically." We measure a nondeterministic IRTM M's space consumption while processing input word x only at the *accepting* nodes in M's computation tree $\mathscr{T}_M(x)$. Among competing accepting nodes in $\mathscr{T}_M(x)$, we focus on a *most compact* one, i.e., an accepting node at which M has used the smallest number of squares of its read/write worktape. This leads us to the following definition.

Let M be a nondeterministic IRTM with input alphabet Σ; let M have a single read/write worktape, in addition to its input record. We say that M *operates within (nondeterministic) space $s(n)$*, for a nondecreasing total function $s : \mathbb{N} \to \mathbb{N}$, if the following holds for all words $x \in \Sigma^\star$. If M accepts input x, then in at least one of its accepting computations on input x, M uses no more than $s(\ell(x))$ squares of its read/write worktape. In other words, there is a branch of $\mathscr{T}_M(x)$ that ends at an accepting total state, along which M uses no more than $s(\ell(x))$ squares of read/write worktape.

13.3 Reducibility, Hardness, and Completeness in Complexity Theory

The three notions that occupy us in this section need no introduction, since each is a quantified version of the amply discussed analogous computability-theoretic concept from Section 9.4 (reducibility) or Section 9.6 (Hardness and Completeness). As we assert in Section 13.1, the importance of these notions to complexity theory is at least as great as their importance to computability theory. The role of the pillar "ENCODING" in our understanding of the power and limits of digital computation cannot be overstated!

In the next subsection, we lay the groundwork needed to craft quantified versions of the three notions highlighted in this section.

13.3.1 A General Look at Resource-Bounded Computation

Let $r : \mathbb{N} \to \mathbb{N}$ be a nondecreasing total *resource-bounding* function. As we have discussed earlier, although modern computing technology has given rise to a lengthy list of computational resources that are both interesting and significant, the only resources that we shall bound in our introductory study of complexity theory are *time* and *space*. For the former of these, $r(n)$ is a *time-bounding* function that we usually label $t(n)$; for the latter, $r(n)$ is a *space-bounding* function that we usually label $s(n)$. Extrapolating from our discussion in Section 13.2, to which the reader should refer now:

- Let M be a *deterministic* OA. We say that M *operates within the resource bound* $r(n)$ if the following holds for each input word $x \in \Sigma^\star$. If M halts on input x—thereby either accepting or rejecting the input—then its computation consumes $\leq r(\ell(x))$ units of the function $r(n)$'s associated resource.
- Let M be a *nondeterministic* OA. We say that M *operates within the resource bound* $r(n)$ if the following holds for each input word $x \in \Sigma^\star$. If M accepts input x—i.e., if there is an accepting total state in $\mathscr{T}_M(x)$—then there is a branch of $\mathscr{T}_M(x)$ from M's initial total state to an accepting total state along which M consumes $\leq r(\ell(x))$ units of the function $r(n)$'s associated resource.

For some computational resources, such as space, one will likely want to specialize the preceding definitions to computational models such as the OTM or the IRTM that have more detailed structure than do OAs.

We are now ready to present quantified versions of the three notions highlighted in this section.

.3.2 Efficient *Mapping Reducibility*

A. Defining efficient mapping reducibility. Focus on a set R of resource-bounding functions $\{r : \mathbb{N} \to \mathbb{N}\}$ and on a specific computational model that is appropriate for the functions in R, in the sense of the comments at the end of the preceding subsection. (For instance, we may use IRTMs as the underlying model when R consists of *space*-bounding functions.) We call R a *resource-bounding class* (or sometimes, a *resource bound*, for short) if it is *compositionally comprehensive*,[2] in the following sense. Say that the function $f_1 : \Sigma^* \to \Sigma^*$ is computable within the resource bound $r_1(n) \in R$, and the function $f_2 : \Sigma^* \to \Sigma^*$ is computable within the resource bound $r_2(n) \in R$. Then R is *compositionally comprehensive* if there exists a resource-bounding function $r_3(n) \in R$ such that the function $f_1 \circ f_2 : \Sigma^* \to \Sigma^*$, which is the composition of functions f_1 and f_2, is computable within the resource bound $r_3(n)$.

Language $A \subseteq \Sigma^*$ is (mapping-)reducible to language $B \subseteq \Sigma^*$ within resource-bound R, written $A \leq_R B$, if and only if
    there exists a total function $f : \Sigma^* \to \Sigma^*$ *that is computable*
        within resource bound $r(n)$ for some $r(n) \in R$
    such that   for all $x \in \Sigma^*$, $[x \in A]$ if and only if $[f(x) \in B]$.
As with unquantified mapping reducibility, we call the encoding function f a *reduction function*.

Given our focus on the **P**-vs.-**NP** problem, the following specialization of the preceding definition is central to our development of complexity theory.

Language A is polynomial-time (mapping-)reducible to language B, written $A \leq_{\text{poly}} B$, if and only if there exists a total function $f : \Sigma^* \to \Sigma^*$ *that is computable within time $t(n)$ for some polynomial $t(n)$* such that for all $x \in \Sigma^*$, $[x \in A]$ if and only if $[f(x) \in B]$.
We often abbreviate the phrase "polynomial-time" by "*poly-time*."

Of course, the preceding definition makes sense within our study only because the set F of all polynomials that map $\mathbb{N}$ to $\mathbb{N}$ is closed under functional composition. This closure means that when the functions in F are used as time-bounding functions, the set F is indeed a resource bound.

The reader has surely noted our reluctance to remove the qualifier "mapping" from all terms related to reductions and reducibility—although we do deemphasize the qualifier by parenthesizing it. The reason for this is that as with computability theory, there are genres of reducibility, other than mapping reducibility, that are used in studying complexity theory. The only alternative to m-reducibility that has received considerable attention in complexity theory is *Turing reducibility*. Indeed, within complexity theory, mapping reducibility is often called *Karp reducibility* because of its use in [48], one of the original sources of complexity theory as we study it; and Turing reducibility is often called *Cook reducibility* because of its use in [18], one of the other original sources.

---

[2] The term "compositionally comprehensive" is not common. We introduce it here for the purpose of unifying results that usually appear separately in the literature.

In analogy with Lemma 9.3, we have the following exceedingly important technical lemmas whose proofs are left as exercises.

Lemma 13.3. *The relation "is mapping-reducible to within resource-bound R," which we denote by $\leq_R$, is transitive. In other words, for any three languages $A, B, C \subseteq \Sigma^*$, if $A \leq_R B$ and $B \leq_R C$, then $A \leq_R C$.*

By specializing the abstract resource bound R to poly-time computation, we obtain the following corollary of Lemma 13.3.

Lemma 13.4. *The relation "is polytime mapping-reducible to," which we denote by $\leq_{\text{poly}}$, is transitive. In other words, for any three languages $A, B, C \subseteq \Sigma^*$, if $A \leq_{\text{poly}} B$ and $B \leq_{\text{poly}} C$, then $A \leq_{\text{poly}} C$.*

As a final commentary on our definition of quantified mapping reducibility, we present the following analogue of Lemma 9.2, which plays the same role for complexity theory as Lemma 9.2 plays for computability theory: Both lemmas show that the appropriate version of m-reducibility does capture essential components of the intuition that the ability to decide language B effectively (for computability theory) or efficiently (for complexity theory) "helps" one decide language A effectively or efficiently. One can easily adapt the following lemma to many resource bounds other than poly-time computation.

Lemma 13.5. *Let A and B be languages over the alphabet Σ, and say that $A \leq_{\text{poly}} B$.*

(a) *If $B \in$ **NP** (resp., $B \in$ **P**), then $A \in$ **NP** (resp., $A \in$ **P**).*
(b) *Contrapositively, if $A \notin$ **NP** (resp., $A \notin$ **P**), then $B \notin$ **NP** (resp., $B \notin$ **P**).*

B. Exemplifying efficient mapping reducibility. We present a simple but illustrative quantified m-reduction that shows the concept "in action." We need a few definitions.

The (CNF) satisfiability problem, SAT. We focus on the set of logical expressions that are in *conjunctive normal form* (*CNF*, for short). As the following three examples suggest, these are logical expressions that have the form of a logical product (conjunction) of logical sums (disjunctions) of *logical variables*, i.e., variables that range over the set $\{0, 1\}$ (which are used here as *truth values*):

$$E_1 = P \wedge Q \wedge R,$$
$$E_2 = (A \vee B \vee \overline{C}) \wedge D \wedge (\overline{A} \vee \overline{B}),$$
$$E_3 = (X \vee Y) \wedge (\overline{X} \vee \overline{Y}).$$

Note that each expression is constructed using (logical analogues of the) Boolean operations to interconnect *literals*. Each literal is an occurrence of a logical variable— say, for illustration, the variable X that appears in expression E_3—in either its *true form*, i.e., the form X, or its *complemented form*, i.e., the negated form $\overline{X}$. (Note that variable X appears in both its true form and its complemented form in E_3.)

A *satisfying assignment* for a logical CNF expression E is an instantiation of truth values for the *variables* in E—and thereby by inheritance, to the literals that appear in E—under which E evaluates to 1, or TRUE. Let us illustrate these concepts with expression E_3. Consider the assignment

$$\text{Assignment to } variables\text{: } X \leftarrow 1; Y \leftarrow 0$$

to the two variables that appear in E_3. This variable assignment induces the following assignment to the four literals that appear in E_3:

$$\text{Assignment to } literals\text{: } X \leftarrow 1; Y \leftarrow 0; \overline{X} \leftarrow 0; \overline{Y} \leftarrow 1$$

Under this assignment, expression E_3 evaluates to the constant expression

$$(1 \vee 0) \wedge (0 \vee 1) \equiv 1,$$

so that this is indeed a satisfying assignment for E_3. The reader should verify that the assignment

$$\text{Assignment to } variables\text{: } X \leftarrow 0; Y \leftarrow 1$$

is also a satisfying assignment for E_3, but that the other two possible assignments, under which both variables get the same truth value, are *not* satisfying assignments for E_3.

> For reasons suggested by the nature of the satisfying and unsatisfying assignments for E_3, the Boolean function specified by the expression is called *exclusive or* and is usually denoted by xor. The reader can verify that this function is just (mod 2) addition of the variables X and Y, when they are viewed as ranging over the base-2 digits.

We say that an expression E is *satisfiable* if it admits a satisfying assignment; otherwise, E is *unsatisfiable*. We have just shown that expression E_3 is satisfiable.
The *(CNF) satisfiability problem*, which is generally referred to by the nickname SAT, is the following set of logical expressions in conjunctive normal form

$$\text{SAT} \overset{\text{def}}{=} \{x \mid x \text{ is a CNF formula that admits a satisfying assignment}\}.$$

The CLIQUE problem. A *clique* in an undirected graph $\mathcal{G}$ is a set of nodes S every two of which are connected by an edge; $|S|$ is the *size* of the clique.
The problem CLIQUE is to decide, of a given graph $\mathcal{G}$ and integer $k \in \mathbb{N}$, whether $\mathcal{G}$ contains a clique of size k.

Lemma 13.6. *SAT* $\leq_{\text{poly}}$ *CLIQUE.*

Proof. Let us be given an arbitrary CNF expression

$$E = D_1 \wedge D_2 \wedge \cdots \wedge D_m,$$

where each D_i is a disjunction of literals

$$D_i = \ell_1^{(i)} \vee \ell_2^{(i)} \vee \cdots \vee \ell_{k_i}^{(i)}.$$

We use the structure of expression E to generate a graph $\mathcal{G}_E$ with the following two properties.

1. One can produce $\mathcal{G}_E$ from E in time polynomial in the size of E, as measured by the number of bits required to write E as a binary string.
2. Expression E is satisfiable *if and only if* graph $\mathcal{G}_E$ has a clique of size m. (Note that m is the number of disjuncts—or logical sums—that make up E.)

We begin by specifying $\mathcal{G}_E$'s structure, i.e., its nodes and edges.

- We give $\mathcal{G}_E$ a node $v_j^{(i)}$ for each literal $\ell_j^{(i)}$ that appears in E. $\mathcal{G}_E$ thus has $k_1 + k_2 + \cdots + k_m$ nodes.
- We given $\mathcal{G}_E$ an edge $(v_b^{(a)}, v_d^{(c)})$ for every pair of literals $\ell_b^{(a)}, \ell_d^{(c)}$ that appear in E *such that*

  1. $a \neq c$
  This means that literals $\ell_b^{(a)}$ and $\ell_d^{(c)}$ appear in different disjuncts of E.

  2. $\ell_b^{(a)} \neq \overline{\ell_d^{(c)}}$
  This means that one can simultaneously satisfy $\ell_b^{(a)}$ and $\ell_d^{(c)}$; i.e., there is an instantiation of truth values for the variables in E under which both $\ell_b^{(a)}$ and $\ell_d^{(c)}$ map to 1.

A standard technique for representing a graph such as $\mathcal{G}_E$ is via its *adjacency matrix*. This is a $(k_1 + k_2 + \cdots + k_m) \times (k_1 + k_2 + \cdots + k_m)$ matrix of 0's and 1's that is built as follows. We first (re)name the nodes of $\mathcal{G}_E$ by the set of positive integers $\{1, 2, \ldots, (k_1 + k_2 + \cdots + k_m)\}$. We then populate the matrix with 0's and 1's by placing 1 in matrix-entry (i, j) just when there is an edge in $\mathcal{G}_E$ between nodes i and j; we place a 0 in this entry if the edge does not exist. The reader should be able to verify that given expression E, we can construct an adjacency matrix for $\mathcal{G}_E$ in time polynomial in the size of E. *The procedure that accomplishes this construction is the reduction function for the poly-time reduction we are now providing.* We have thus achieved the first of our two goals.

We now correlate the satisfiability of E with the presence of cliques in $\mathcal{G}_E$.

Say first that expression E is satisfiable. E's satisfiability means that there exists a subset of the literals that appear in E, one from each of E's m disjuncts, that can all be mapped to 1 simultaneously by some single instantiation of the *variables* that appear in E. (Perhaps several such subsets exist, but at least one does.) Let $S_{\text{lit}} = \{\ell_{i_1}^{(1)}, \ell_{i_2}^{(2)}, \ldots, \ell_{i_m}^{(m)}\}$ be such a subset. By the way we have specified the edge-set of $\mathcal{G}_E$, there must be an edge between every pair of nodes that correspond to the literals in S_{lit}. (Distinct literals in S_{lit} come from distinct disjuncts of E, and every pair of literals in the set can simultaneously be satisfied.) *The set of edges of $\mathcal{G}_E$ that we have just described is a clique of size m within $\mathcal{G}_E$.*

Say next that graph $\mathcal{G}_E$ has a clique of size m. This means that there exists a set of nodes $S_{\text{node}} = \{v_{i_1}^{(1)}, v_{i_2}^{(2)}, \ldots, v_{i_m}^{(m)}\}$ of $\mathcal{G}_E$ for which there is an edge between every two nodes in the set. By the way we have constructed $\mathcal{G}_E$, this means that the set of literals that correspond to the nodes in S_{node} betokens the existence of a set of literals in E, with one literal per disjunct, that are simultaneously satisfiable. *The set of literals in E that we have just described admits a satisfying assignment for E—which means that E is satisfiable.*

The reduction is now complete. $\square$

13.3.3 Hard Problems and Complete Problems

A quantified notion of mapping reducibility that is based on a well-conceived resource bound R can sometimes expose valuable information about a significant family $\mathbf{F}$ of languages. We have already observed one instance of this assertion in our study of computability theory. In Chapter 9, we saw how the kindred notions of *m-reducibility, Hard languages, and Complete languages* immeasurably enhance our understanding of decidability and semidecidability. The kindred *quantified* versions of these three notions play at least as important a role within complexity theory.

> One could argue that the complexity-theoretic versions of these notions are *more* important than their computability-theoretic cousins because of the implications of the former for a broad range of *practical* computational problems (such as the combinatorial optimization problems that we discussed in Section 13.1).

Let $\mathbf{F}$ be a family of languages, and let R be a resource bound. A language/problem A is:

- $\mathbf{F}$-*Hard* (with respect to resource bound R) if every language $B \in \mathbf{F}$ is mapping-reducible to A within resource bound R; symbolically, $B \leq_R A$.
- $\mathbf{F}$-*Complete* (with respect to resource bound R) if

  – A is $\mathbf{F}$-Hard with respect to resource bound R,
    and
  – $A \in \mathbf{F}$.

Thus, one can view every $\mathbf{F}$-Complete language as being (one of) the computationally hardest languages within the family $\mathbf{F}$, at least with respect to the resource bound R. Specifically, for every language $B \in \mathbf{F}$, there is an encoding of instances of B as instances of A via an encoding function that is "efficient" with respect to resource bound R.

The concrete instantiation of the preceding notions that likely has the longest reach into the world of practical computing is the one in which:

- $\mathbf{F}$ is the family **NP** of languages that are accepted by nondeterministic Turing machines that operate within time that is polynomial in the size of the input;
- the resource bound R comprises the set of polynomial timing functions, so that the associated mapping reducibility is $\leq_{\text{poly}}$.

Specialized to this situation, Hardness and Completeness are defined as follows. A language $A \subseteq \Sigma^\star$ is:

- **NP-**_Hard_ if every language $B \in$ **NP** is poly-time reducible to A; symbolically, $B \leq_{poly} A$.
- **NP-**_Complete_ if

  - A is **NP-**_Hard_

  and

  - $A \in$ **NP**.

The informal assertion that the **NP**-Complete languages are the "hardest" ones in the class **NP** is justified, via Lemma 13.5, by the fact that:

If any **NP**-Complete language A were in **P** (i.e., were poly-time decidable),
then every language in **NP** would be in **P** (i.e., would be poly-time decidable).

13.3.4 An NP-Complete Version of the Halting Problem

Just as it was not clear a priori (in Section 9.6) that mapping-Complete languages existed, it is not clear a priori that there exist **NP**-Complete languages. In fact, though, a time-restricted version of the halting problem HP turns out to be a "hardest" language in the class **NP**.

Before proceeding, a word of caution is in order. We must proceed carefully when discussing the classes **P** and **NP** because the constitution of these classes depends on the computational model being used. As discussed in several places earlier, we cannot be as free in thinking about an arbitrary "reasonable" abstract computational model when developing complexity theory as we could when developing computability theory. Specifically, polynomial variation in timing functions is not a "broad enough brush" to allow us to develop a theory of computational complexity that admits a full-blown analogue of the Church-Turing thesis. To suggest what the problem is, consider two rather different abstract computational models that we discussed in Section 9.8: the OTM, which performs all of its calculations by manipulating strings symbol by symbol, and the register machine (RM), which performs all of its calculations by performing arithmetic on (arbitrary-size) integers. As we noted in Section 9.8, OTMs and RMs are equivalent in computing power: they compute the same class of functions; they decide the same class of languages; they semidecide the same class of languages. However, if we assess the time that OTMs and RMs take to perform their computations using measures that are natural for each model's native data type—strings for OTMs and integers for RMs—then we are unintentionally favoring RMs by an _exponential_ variation. This is because an RM can "magically" act on enormous integers in a single step, while an OTM—or any equivalent string-processing computational model—must operate digit-by-digit on _numerals_ that represent these integers. The problem is highlighted most easily via an example.

Recall the _subset-sum_ problem that we discussed briefly in Section 1.1.2.

The subset-sum problem.
Generic instance: n positive integers, $m_1, m_2, \ldots, m_n$, plus a *target integer t*
To decide: Is there a subset of the integers m_i that sum to t?

Say that we discover an algorithm $\mathscr{A}$ for the subset-sum problem that makes the required decision on an arbitrary instance of the problem, $\langle m_1, m_2, \ldots, m_n; t \rangle$, within $O(nt)$ steps. (One finds such an algorithm in Section VI.6.1 of [65].) What is the time complexity of this algorithm? It depends on one's computational model!

If one implements algorithm $\mathscr{A}$ on an RM, then a natural measure of the size of the problem instance $\langle m_1, m_2, \ldots, m_n; t \rangle$ would be $\Theta(n \cdot \max\{m_1, m_2, \ldots, m_n, t\})$. The reasoning is that integers are the natural data type for the RM, and this instance can be specified by n integers, each of size $\leq \max\{m_1, m_2, \ldots, m_n, t\}$. When size is measured in this way, the time-complexity of algorithm $\mathscr{A}$ is *quadratic* in the size of the problem instance.

If one implements algorithm $\mathscr{A}$ on an OTM, then a natural measure of the size of the problem instance $\langle m_1, m_2, \ldots, m_n; t \rangle$ would be $\Theta(n \cdot \log(\max\{m_1, m_2, \ldots, m_n, t\}))$. The reasoning is that strings, or numerals, are the natural data type for the OTM, and this instance can be specified by n numerals, each of size $\leq \log(\max\{m_1, m_2, \ldots, m_n, t\})$. When size is measured in this way, the time-complexity of algorithm $\mathscr{A}$ is *exponential* in the size of the problem instance.

The problem illustrated by the preceding story arises because the natural notion of size for an integer N is N's magnitude, while the natural notion of size for a string x is $\ell(x)$, x's length. Thus, if one measures time or space for a given model in terms that are natural for that model's natural data type, then polynomial variation in resource-bounding functions is not adequate to "level the playing field" for models as different as OTMs and RMs. This fact would force us to study the complexity of computations by the two models using different (but parallel) theories.

How then do we formulate a single theory of computational complexity that will handle all computational models? The convention that mainstream complexity theory has used since its invention in 1971 [18] is to mandate that

We measure time and space complexity as a function of the size of the input problem instance to our computational model, as measured by the number of bits *needed to write down the input.*

This measure deprives RMs of the advantage that accrues when they operate on large integers in single steps.

We noted in Section 9.8 that even the preceding, apparently narrow, convention does allow a fair amount of flexibility; e.g., we can endow a TM with any fixed number of tapes of any fixed dimensionality without altering the theory we are developing.

On to our first **NP**-Complete language/problem!

The *poly-time halting problem*, $\mathrm{HP}^{(\mathrm{poly})}$, is the set of ordered triples of strings of the form[3]

$$\langle x, y, 0^t \rangle \tag{13.1}$$

such that

- both x and y are binary strings: $x, y \in \{0, 1\}^\star$;

---

[3] Note that "0^t" in this context denotes a string of t instances of 0; exponentiation thus denotes *iterated concatenation* here, *not* iterated multiplication.

- $t \in \mathbb{N}$; i.e., t is a nonnegative integer;[4]
- the NTM (encoded by) x—note that x is *nondeterministic*—accepts input y in $\leq t$ *nondeterministic* steps.

As usual, we use a computationally efficient pairing function to turn $\mathrm{HP}^{(\mathrm{poly})}$ into a set of strings, rather than a set of triples of strings—so that $HP^{(\mathrm{poly})}$ becomes a legitimate language.

We could accomplish this conversion of triples to strings in a variety of ways. The simplest would probably be to encode $\mathrm{HP}^{(\mathrm{poly})}$'s constituent triples (13.1) as binary strings, via a simple string-oriented encoding such as the following. We could encode

  0 as 000
  1 as 010
  , as 0110
  ⟨ as 01110
  ⟩ as 011110

and thereby encode each triple of the form (13.1) as a uniquely—and easily—decipherable binary string. Under this encoding, the triple

$$\langle 0101,\ 00111,\ 0000000 \rangle$$

would be encoded as the binary string (with spaces inserted to enhance legibility)

01110 000 010 000 010 0110 000 000 010 010 010 0110 000 000 000 000 000 000 000 011110

We now verify our claim that $\mathrm{HP}^{(\mathrm{poly})}$ is **NP**-Complete.

Theorem 13.1. *The poly-time halting problem* $\mathrm{HP}^{(\mathrm{poly})}$ *is* **NP**-*Complete.*

Proof. We consider in turn the two conditions needed for **NP**-Completeness.

$\mathrm{HP}^{(\mathrm{poly})}$ **is NP-hard.** Somewhat surprisingly, this part of the proof is quite straightforward, especially after seeing Theorem 9.6. Let $A \subseteq \{0,1\}^*$ be an arbitrary language in **NP**. There exist, by definition, an NTM x and a (timing) polynomial p such that for all $y \in \{0,1\}^*$,

$$[y \in A] \iff [x \text{ accepts input } y \text{ in } \leq p(\ell(x)) \text{ nondeterministic steps}]. \tag{13.2}$$

By definition of $\mathrm{HP}^{(\mathrm{poly})}$, the righthand assertion in (13.2) is equivalent to the assertion that

$$\langle x,\ y,\ 0^{p(\ell(x))} \rangle \in \mathrm{HP}^{(\mathrm{poly})}.$$

If the language A is specified and presented via its accepting NTM x, then the transformation that converts x and y into the string $\langle x,\ y,\ 0^{p(\ell(x))} \rangle$ is a total poly-time computable function. We conclude, therefore, that $A \leq_{(\mathrm{poly})} \mathrm{HP}^{(\mathrm{poly})}$. Since A was an arbitrary language in **NP**, this half of the proof is done.

$\mathrm{HP}^{(\mathrm{poly})}$ **belongs to NP.** We begin this more-complicated half of the proof by reviewing what we need to show. The language $\mathrm{HP}^{(\mathrm{poly})}$ belongs to **NP** if and only if there

---

[4] An encoding of integer t by a string of t letters, such as 0^t, is called a *tally encoding* or a *unary representation* of integers.

is an NTM M that accepts the language. If this NTM M existed, then in response to an input triple of strings $\langle x, y, 0^t \rangle$, M would proceed *nondeterministically*—don't forget that M is itself nondeterministic—to an accepting state if and only if when processing the input y, NTM x proceeds nondeterministically to an accepting state within t nondeterministic steps. We thus have two nondeterministic processes going on here:

1. the time-restricted process via which NTM x accepts its input string y,
2. the process via which NTM M simulates x in order to decide whether to accept its input triple $\langle x, y, 0^t \rangle$.

In order to demonstrate the existence of NTM M in a perspicuous manner, we are going to invoke some of the poly-time robustness of (N)TMs that we demonstrated in Section 9.8. Specifically, we showed there that:

> *Every (N)TM that uses several worktapes of any fixed dimensionalities can be simulated in poly-time by an (N)TM that uses a single worktape that is one-ended and one-dimensional (i.e., linear).*

We exploit this flexibility in the (N)TM model by:

- allowing M to have two worktapes, one two-dimensional and one linear, and
- insisting that the x component of every triple $\langle x, y, 0^t \rangle$ that is a candidate for membership in $\mathrm{HP}^{(\mathrm{poly})}$ specify an NTM that uses a one-ended one-dimensional worktape.

The computation of M on an input triple $\langle x, y, 0^t \rangle$ proceeds as follows. Say that x specifies the NTM $x = (Q^{(x)}, \Sigma^{(x)}, \Gamma^{(x)}, \delta^{(x)}, Q_0^{(x)}, \{q_{\mathrm{acc}}^{(x)}, q_{\mathrm{rej}}^{(x)}\})$.

A. M begins by creating a $(t+1) \times (t+3)$ array that we call a *(computation) tableau* (plural: *tableaux*). M constructs the tableau by nondeterministically specifying, in each of the $t+1$ rows of the tableau, a string that has the form of a length-$(t+1)$ total state of NTM x, flanked on both ends by the special end-denoting delimiter symbol #, which does *not* belong to $\Gamma^{(x)}$. In more detail, M constructs the tableau by

- "guessing" (by splitting universes) one symbol at a time,
- using the "0^t" component of its input to regulate both the number of rows in the tableau and the number of symbols in each row,
- using its finite-state control to ensure that each row of the tableau belongs to the set $\#\Gamma^k Q \Gamma^{t-k}\#$ for some integer $k \in \mathbb{N}$, i.e., that each row

  - begins and ends with an occurrence of the special symbol #,
  - contains exactly one state-symbol $q \in Q^{(x)}$.

Note that the uniform length of the rows in the tableau—which greatly simplifies M's computation—may force M to violate the formal syntax of "total state" mandated in Section 3.3, by padding some rows of a tableau with occurrences of the blank symbol $\boxed{\mathrm{B}}$, in order to achieve the desired uniform row-length of $t+1$ symbols. A "typical" row of a tableau thus has the form

$$\boxed{\#}\,\boxed{\gamma_1}\cdots\boxed{\gamma_i}\,\boxed{q}\,\boxed{\gamma_{i+1}}\cdots\boxed{\gamma_k}\,\boxed{\mathrm{B}}\cdots\boxed{\mathrm{B}}\,\boxed{\#} \qquad (13.3)$$

where each symbol γ_i belongs to x's worktape alphabet $\Gamma^{(x)}$, and q belongs to x's state-set $Q^{(x)}$.

> Note: it is possible that a given row of a tableau never appears in any actual computation by x, despite the fact that the row belongs to the set $\#\Gamma^k Q \Gamma^{t-k}\#$ for some integer $k \geq 0$, hence has the form required of a total state. M checks for this possibility as it processes its way through the tableau.

M records the tableau it has been generating nondeterministically on its two-dimensional tape; it uses its one-dimensional tape to record the NTM-program x. Note that M has been splitting universes at every step throughout the dual process—namely, copying x onto its linear tape and creating the tableau on its two-dimensional tape—so that the process ends with each universe having its own tableau and each tableau occurring in some universe.

> **(a)** Throughout, we describe the tableau as though the states of NTM x and the symbols in both its input alphabet $\Sigma^{(x)}$ and worktape alphabet $\Gamma^{(x)}$ were *atomic symbols*. In fact, because NTM M must be "universal," in the sense of being able to simulate *any* NTM x on any of its valid input strings y, we leave unspecified—but always within our awareness—some standard encoding that M uses to specify the programs that we are calling (N)TMs. There is nothing sophisticated going on here; it is just another instance of our encoding everything as binary strings. One could imagine, for instance, that the states of an (N)TM are encoded by strings in the set 110^*11, while the letters of an (N)TM's worktape alphabet are encoded by strings in the set 0^*1. Such an encoding will allow the strings that represent total states of every (N)TM to be parsed easily into their semantically meaningful constituents.
>
> **(b)** The fact that M ends up with exponentially many universes does not jeopardize our argument, because M has proceeded for only $O(t^2)$ nondeterministic steps, and it has created a bounded number of new universes at each step. Specifically: M *never creates more than* $|\Gamma^{(x)}| + 1$ *new universes at a step*.

B. Focus on an individual universe.

> This request is assuring you that the computation we describe here happens *independently* in each universe—as mandated by the notion of nondeterminism.

M wants to check whether the successive rows of this universe's populated tableau represent a sequence of total states $C_0, C_1, \ldots, C_t$ of NTM x on input y that forms an accepting computation. What does this mean? By definition of "accepting computation," the following conditions must hold simultaneously.

1. C_0 is a valid initial total state of NTM x.
   This means that C_0 is of the form $\#q\boxed{B}\boxed{B}\cdots\boxed{B}\#$, where $q \in Q_0^{(x)}$ is one of x's initial states, and there are t occurrences of $\boxed{B}$. C_0 must have this form because x's worktape is blank at the beginning of every computation.
2. C_t is a valid accepting total state of NTM x.
   This means that C_t has the form $\#\xi q_{\text{acc}}^{(x)} \eta\#$, where (i) $\xi, \eta \in (\Gamma^{(x)})^*$; (ii) $\ell(\xi) + \ell(\eta) = t$; (iii) $q_{\text{acc}}^{(x)}$ is the halt-and-accept state of NTM x.
3. Each total state C_{i+1}, where $i \in [0,t]$, is a valid successor of C_i in a (nondeterministic) computation of input string y by NTM x.

The exact meanings of the relevant notions, such as "accept state" and "successor total state," can be gleaned from the definition of a computation by a TM, in Section 3.3, and of a computation by an NTM, in Section 12.2.

Let us flesh out the process of checking condition #3. Recall that each row has precisely one state symbol (from x's state-set $Q^{(x)}$) and precisely t worktape symbols (from x's worktape alphabet $\Gamma^{(x)}$). Say that in the universe we are focusing on, M is simulating NTM x on input y. Consider an arbitrary row $r \in [0,t]$ of the tableau that M created within this universe. Focus on the 3-symbol substring of row r "where the action is": the state symbol $q \in Q^{(x)}$, the symbol d immediately to q's left, and the symbol e immediately to q's right. Next, focus on the 3-symbol substring, call it abc, of row $r+1$ that corresponds to the 3-symbol substring dqe of row r. In order to study how these two 3-symbol substrings interact in x's computation, we look carefully at the 2×3 subtableau of M's tableau created by the substrings dqe and abc:

$$
\begin{array}{ll}
\text{row } r+1: & a\,b\,c, \\
\text{row } r: & d\,q\,e.
\end{array}
\tag{13.4}
$$

Keep in mind that:

- $q \in Q^{(x)}$,
- $d, e \in (\Gamma^{(x)} \cup \{\#\})$,
- $a, b, c \in (Q^{(x)} \cup \Gamma^{(x)} \cup \{\#\})$, with precisely one of the three symbols in $Q^{(x)}$.

Under what conditions can this subtableau appear as part of a valid computation by NTM x on input y? We branch on the nature of state q to answer this question.

(1) $q = q_{\text{acc}}^{(x)}$.

In this case, we must have $a = d$, $b = q$, and $c = e$. This convention allows us to meaningfully fill out the tableau, even when some of x's nondeterministic universes terminate before a full t steps.

For the other two categories of states of x, we consult x's state-transition function, $\delta^{(x)}$. Because $\# \notin \Gamma^{(x)}$, M aborts its simulation in this universe if $e = \#$: this value for e cannot correspond to any situation that x would encounter in a computation having t or fewer (nondeterministic) steps. (More technically, the case $e = \#$ is prohibited by the fact that $\delta^{(x)}$ is not defined on the set $Q^{(x)} \times \{\#\}$.) We continue our analysis, therefore, under the assumption that $e \in \Gamma^{(x)}$.

(2) q **is an autonomous state**.

In this case, the possible (nondeterministic) behaviors of x depend only on the contents of the subtableau. Say that for some direction $D \in \{N, L, R\}$,

$$
\langle q', e', D \rangle \in \delta^{(x)}(q, e).
\tag{13.5}
$$

This means that when NTM x is in (the autonomous) state q and is reading symbol $e \in \Gamma^{(x)}$ on its worktape, one of x's possible moves involves transitioning to state q', rewriting symbol e on the worktape by symbol $e' \in \Gamma^{(x)}$, and moving the worktape's read/write head (one square) in direction D. Then the subtableau (13.4)—focus on the row-$(r+1)$ portion—must have a form mandated in the following table:

$$
\begin{array}{|ll|}
\hline
\text{row } r+1: & d\ q'\ e' \quad \text{if } D=N \\
\text{row } r: & d\ q\ e \\
\hline
\text{row } r+1: & q'\ d\ e' \quad \text{if } D=L \\
\text{row } r: & d\ q\ e \\
\hline
\text{row } r+1: & d\ e'\ q' \quad \text{if } D=R \\
\text{row } r: & d\ q\ e \\
\hline
\end{array}
\tag{13.6}
$$

(3) q is a polling state.

The analysis when q is a polling state differs from the (preceding) analysis of the case in which q is an autonomous state only in the determination of x's valid options as it transitions from total state C_r to total state C_{r+1}. Say that the input y to NTM x has the form

$$ y = \sigma_1 \sigma_2 \cdots \sigma_n, $$

where each σ_i belongs to $\Sigma^{(x)}$. Say that p of the $r-1$ internal states of x that appear in the sequence $C_0, C_1, \ldots, C_{r-1}$ of total states are polling states. This means that when x is in internal state q within total state C_r, it is reading the $(p+1)$th symbol, σ_{p+1}, of y. This, in turn, means that the potential moves of x in this situation are specified precisely by the set $\delta^{(x)}(q, \sigma_{p+1}, e)$. It follows that the valid row-$(r+1)$ successors of the row-r subtableau d, q, e in (13.4) are delimited by the following condition,

$$ \langle q', e', D \rangle \in \delta^{(x)}(q, \sigma_{p+1}, e), $$

rather than by (13.5). The remainder of the analysis mirrors the case in which q is an autonomous state, hence is left to the reader.

The issue of timing. We turn finally to the critical matter of verifying that M operates within nondeterministic polynomial time. Because M clearly takes $O(t^2)$ nondeterministic steps to construct the tableaux corresponding to input $\langle x, y, 0^t \rangle$, we focus our analysis only on the checking that M must perform on each tableau (within that tableau's universe). We organize our analysis around the three conditions that M must check.

Checking the first two enumerated conditions (the "start" condition and the "accept" condition) takes M (nondeterministic) time $O(t \cdot \ell(x))$, because it requires only a scan of two of the tableau's rows. The constant factor hidden by the big-O is explained thus: the quantity $O(\ell(x))$ includes the time M spends dealing with its encoding of x's states and worktape symbols; the quantity $O(t)$ accounts for the time M spends scanning both rows 0 and t of the tableau. Each of the t checks M performs for the third condition (the "consecutiveness" condition) takes time $O(t^2 \cdot \ell(x))$. To wit, all of the "action" in each $C_i \to C_{i+1}$ check involves at most three consecutive positions in each total state: the state symbol and its successor, and possibly also its predecessor. The valid transformations of each such pair of 3-symbol substrings are delimited by the NTM-program encoded by string x. M's tasks for each of the t necessary consecutiveness checks consist of the following subprocesses.

- M identifies the relevant 3-symbol substrings for both C_i and C_{i+1}, by scanning along the then-current row r; this takes nondeterministic time $O(t \cdot \ell(x))$.

- By scanning NTM-program x on its linear tape, M verifies, in nondeterministic time $O(\ell(x))$, that the 3-symbol substrings represent a valid $C_i \rightarrow C_{i+1}$ transition under NTM-program x.
- M verifies, in nondeterministic time $O(t \cdot \ell(x))$, that the portions of total states C_i and C_{i+1} to the left of the 3-symbol substrings are identical, as are the portions of total states C_i and C_{i+1} to the right of the 3-symbol substrings.

Because the preceding process is executed t times, the total nondeterministic time for M's simulation of NTM x on input y is $O(t^2 \cdot \ell(x))$, which is indeed polynomial in the length, $\ell(x) + \ell(y) + t$, of input $\langle x, y, 0^t \rangle$.

It follows that $\mathrm{HP}^{(\mathrm{poly})} \in \mathbf{NP}$, completing the proof. $\quad\square$

We have thus found an **NP**-Complete problem, but a very abstract one. What led to the explosive growth in the development of and the appreciation for complexity theory was the discovery of **NP**-Complete problems that related to computational problems that were important in the "real world." We turn now to the historically first discovered of these.

13.3.5 The Cook-Levin Theorem: The **NP**-Completeness of SAT

If $\mathrm{HP}^{(\mathrm{poly})}$ were the only known **NP**-Complete problem, it is a fair guess that the theory of **NP**-completeness would not have caused much of a ripple *outside* the confines of the theoretical computer science community.

> **NP**-Completeness theory would certainly have found its place of honor *within* the community even if it did not have implications for "practical" computational problems. Nondeterminism had been known for years to explain important aspects of computational structure in terms of unbounded search (cf. Sections 10.2 and 12.3). The theory of **NP**-Completeness demonstrates that nondeterminism can play a similar role in explaining complexity-related aspects of computational structure in terms of resource-bounded search.

However, in the space of barely six months in 1971–1972, two seminal studies demonstrated via a large number of examples that the importance of **NP**-completeness to our understanding of "real" computation would be hard to overstate. In the first of these pioneering works [18], Stephen A. Cook established the fundamental notions of the theory of **NP**-Completeness and exhibited the first examples of "real" computational problems that are **NP**-Complete. In the second of the works [48], Richard M. Karp augmented Cook's list of "real" **NP**-Complete problems with a broad, varied repertoire of practically significant combinatorial problems. (A third, roughly contemporaneous, pioneering study of **NP**-completeness [56] by Leonid Levin, which is of no less scientific importance than the two others, became known in Europe and North America somewhat later.) These studies showed that the property of being **NP**-Complete is shared by variants of myriad problems of indisputed "real," "practical," importance. These problems belong to a broad range of domains, from scheduling, to logic design, to constraint satisfaction, to resource allocation, to theorem proving, and on and on. One major shared "behavioral" characteristic of these problems is that no one knows—to this day!—how to solve large instances of any of them in time that grows slower than *exponentially* in the sizes of the instances.

We reiterate from earlier sections that the "size" of a problem is the number of bits required to write down (i.e., specify, or, describe) an instance of the problem.

This section is devoted to establishing the **NP**-Completeness of the historically first of these "real" problems, the CNF-satisfiability problem, SAT.

A bit of historical background may lend some perspective to Cook's original formulation of the notion of **NP**-Completeness and of his identification of SAT as being **NP**-Complete. In Section 9.1, we briefly discussed Gödel's incompleteness theorem and its devastating impact on the quest for algorithms that would automatically prove theorems in even simple logical systems: in short, no such algorithms can exist! There is, however, one significant logical system, the *propositional calculus* (a quantifier-free logical system), for which theorem-proving algorithms do exist. The propositional calculus enjoys a property that broadens the class of algorithms that one can use to establish or refute the theoremhood of sentences: The calculus is *semantically complete*, meaning that its theorems are precisely the *tautologies*, i.e., the sentences that are true under all assignments of truth values to variables. This means that the aspects of the propositional calculus that are relevant to this section can be formulated "semantically," by analyzing the truth-oriented behavior of logical sentences, rather than "syntactically," by means of proofs crafted in some deduction-oriented logical system.[5] We now flesh out the relevant details of the "semantic" approach we shall use here. This requires a short digression, to provide definitions.

The propositional calculus. Let $P, Q, R, \ldots$, be *propositional variables*, i.e., abstract variables that range over the set $\{0, 1\}$ of *truth values* (which we represent by 0 and 1). Consider any logical expression E that is formed from variables and truth values, using the three traditional *logical connectives* or (denoted by $\vee$), and (denoted by $\wedge$), and not (denoted by an overline, as in $\overline{X}$); cf. Section 2.1. The completeness of the propositional calculus allows us to define "theorem" in terms of the "semantic" notion of tautology (as defined in Section 13.1).

Lemma 13.7. *A logical expression E of the propositional calculus is a* theorem *if and only if E evaluates to 1 under every instantiation of truth values for the propositional variables in E, i.e., if and only if E is a tautology.*

By historical convention, when discussing theoremhood in the propositional calculus, one usually restricts attention to logical expressions that are written in *Disjunctive Normal Form* (*DNF*, for short), meaning, as a logical sum (or *disjunction*) of logical products (or *conjunctions*). Here are a few examples of DNF expressions.

$$P \vee Q \vee R,$$
$$(A \wedge B \wedge \overline{C}) \vee D \vee (\overline{A} \wedge \overline{B}),$$
$$(X \wedge Y) \vee (\overline{X} \wedge \overline{Y}).$$

---

[5] There are many excellent introductions to mathematical logic that employ the classical deductive ("syntactic") approach to the material. I recommend [15] for a classical, "pure," approach to the topic, and [89] for a very interesting approach that is tailored to a mathematician's way of "thinking logically."

A word about the logical connectives and (conjunction) and or (disjunction) is in order here. In the world of mathematics, we are usually quite pragmatic about the constructs that we use to express our (mathematical) thoughts. We use quantifiers—typically the universal quantifier for all ($\forall$) and the existential quantifier there exists ($\exists$)—without even considering that when the domains over which the quantifiers range are finite, we can replace a universally quantified statement by a quantifier-free conjunction, and an existentially quantified statement by a quantifier-free disjunction. As a trivial but illustrative example, if the propositional variable X always assumes either the value 1 or the value 2 (so that it ranges over the set $\{1, 2\}$), then the quantified expressions

$$(\forall X)P(X) \quad \text{and} \quad (\exists X)Q(X)$$

are, respectively, logically equivalent to the quantifier-free expressions

$$[P(1) \text{ and } P(2)] \quad \text{and} \quad [Q(1) \text{ or } Q(2)].$$

This translation from quantified to logically equivalent quantifier-free expressions is "transparent" to day-to-day mathematics, but it lies at the crux of (the proof of) the main result of this section.

Lemma 13.7 makes it clear that there is a conceptually simple algorithm for deciding whether a given logical expression E is a theorem of the propositional calculus. Let E contain n propositional variables and have length $O(n)$, with the constant factor in the big-O accounting for all literals, logical connectives, and grouping symbols that occur in E. Then one can determine whether E is a theorem by instantiating all 2^n truth values for the variables in E and evaluating each of the resulting variable-free expressions. In accord with the lemma, one can accept (as a theorem) every expression E that *never* evaluates to 0, and one can reject those expressions that ever do. The problem with the preceding algorithm is complexity-oriented rather than conceptual. The algorithm always works, but it takes time $\Omega(2^n)$ to decide the theoremhood (or lack thereof) of expression E, even though E has size $O(n)$. While the reason for the exponential time bound is clear with this naive algorithm—there are exponentially many truth-value assignments—it turns out that every known algorithm for deciding the theoremhood of sentences in the propositional calculus suffers the same inefficiency: all of them take exponential time in dealing with sufficiently long sentences. Is this inevitable? If so, why?

Cook [18] had the following wonderful insight. The universal quantifier in the definition of "theorem" (really of tautology)—"E evaluates to 1 under *every one* of the exponentially many instantiations of truth values for the variables in E"—could conceivably condemn every theorem-proving algorithm to having exponential time complexity just because there are exponentially many instantiations to check. Cook decided, therefore, to focus instead on algorithms that identify the class of *refutable* logical expressions, i.e., the *nontheorems*! One thus changed one's focus to expressions E that satisfy the following.

Lemma 13.8. *A logical expression E is a* nontheorem *of the propositional calculus— i.e., is* refutable—*if and only if its negation $\overline{E}$ evaluates to 1 under some instantiation of truth values to the propositional variables in E.*

We call an instantiation of truth values to the variables that causes $\overline{E}$ to evaluate to 1 a *satisfying assignment* for $\overline{E}$. Thus, the language SAT is the set of CNF sentences that are negations of nontheorems. The "practical" importance of SAT is attested to by the myriad computational problems that have some form of constraint-satisfaction as a significant component of their solutions.

As we shift our focus from logical expressions E in DNF to the negations $\overline{E}$ of these expressions, we:

1. flip the governing quantifier from a universal one to an existential one.
   Instead of needing E to evaluate to 1 under *every* instantiation of truth values, we now need $\overline{E}$ to evaluate to 1 under *some* instantiation of truth values.
2. shift our focus from expressions in DNF, i.e., logical sums (disjunctions) of logical products (conjunctions) to expressions in *CNF* (conjunctive normal form), i.e., logical products of logical sums, as in the following examples:

$$P \wedge Q \wedge R,$$
$$(A \vee B \vee \overline{C}) \wedge D \wedge (\overline{A} \vee B),$$
$$(X \vee Y) \wedge (\overline{X} \vee \overline{Y}).$$

While the second of the preceding changes is just a matter of convenience—De Morgan's laws allow us to translate from expression E in DNF to its negation $\overline{E}$ in CNF in a single linear-time sweep—the first change is exceedingly important conceptually. We have discussed earlier—see Section 12.3—how to translate an existential quantifier in the specification of a computational problem as a nondeterministic search. With respect to the notion "refutable," this translation tells us that the procedure for showing that an expression $\widetilde{E}$ in CNF is *not* a theorem of the propositional calculus can be organized as a two-step process:

1. a search for a refuting assignment for the variables in $\widetilde{E}$.
   This search takes linear *nondeterministic* time, via a sequence of "guessed" truth values for the variables in $\widetilde{E}$; no one knows how fast the search can be done deterministically.
2. a check to verify that the chosen assignment does indeed refute expression $\widetilde{E}$.
   The check takes linear *deterministic* time, via a linear pass over the constant expression obtained from $\widetilde{E}$ by instantiating the truth values mandated by the assignment.

But we are getting ahead of ourselves.

Theorem 13.2 ([18]). *The language SAT is* **NP-***Complete.*

Proof. In contrast to Theorem 13.1, the proof that the target language—namely, SAT here and HP$^{(\text{poly})}$ in that theorem—belongs to **NP** is rather easy here, while the proof that the language is **NP**-Hard has some complication. Let us focus in turn on the two conditions for **NP**-Completeness.

SAT belongs to NP. Focus on a CNF expression E that contains occurrences of n propositional variables, comprising the set $\mathscr{P} = \{P_1, P_2, \ldots, P_n\}$. Earlier, we described a time-$O(2^n)$ algorithm that would decide whether $E \in$ SAT, by (i) generating all possible truth assignments for the variables in $\mathcal{P}$, then (ii) checking each assignment to see whether it is a satisfying assignment for E. Although this naive approach appears to be quite inefficient when executed deterministically (as described), it actually yields the following efficient nondeterministic "algorithm."

1. The "algorithm" assigns a truth value to each propositional variable in turn, first to P_1, then to P_2, and so on. Thus, the "algorithm" spends nondeterministic time $O(1)$ generating a truth assignment for each variable, so it takes nondeterministic time $O(n)$ to generate the entire truth assignment to the variables in $\mathcal{P}$.

   In more detail: starting with the first propositional variable, P_1, the "algorithm" spawns one universe in which P_1 gets the truth value 1 and another universe in which P_1 gets the truth value 0. (There are now two universes in all.) In each of these universes, the "algorithm" turns to the second propositional variable, P_2. It spawns one universe in which P_2 gets the truth value 1 and another universe in which P_2 gets the truth value 0. (There are now four universes in all.) In all of these universes, the "algorithm" turns to the third propositional variable, P_3, then the fourth, and so on. Once it has run through all of the n propositional variables in $\mathcal{P}$, the "algorithm" nondeterministically resides in 2^n universes, each one containing a unique assignment of truth values to the variables in $\mathcal{P}$. This entire process takes $O(2^n)$ nondeterministic steps, because the processing of each variable P_i consists entirely of (nondeterministically) assigning P_i a truth value, hence takes $O(1)$ nondeterministic steps.

2. In each distinct universe, the "algorithm" checks whether that universe's truth assignment is a satisfying assignment for E, i.e., whether E evaluates to 1 under that assignment. Given an assignment, this determination can be done in (deterministic!) time linear in the number of symbols in formula E. The "algorithm" just replaces each literal in E with the truth value mandated by the assignment that is associated with the current universe, and it evaluates the resulting constant expression.

   We leave as an exercise the verification that this process can, in fact, be accomplished in deterministic linear time. We note only that there is a charming algorithm that employs a single stack to do the evaluation.

In summation, the "algorithm" nondeterministically tests whether expression E belongs to SAT in (nondeterministic) time that is linear in the number of symbols in E. (Note that the "algorithm" does not *decide* whether $E \in$ SAT; it looks only for positive outcomes.)

SAT is NP-Hard. We prove that SAT is **NP**-Hard by means of a poly-time reduction from $\mathrm{HP}^{(\mathrm{poly})}$. This strategy works because $\mathrm{HP}^{(\mathrm{poly})}$ is **NP**-Hard (Theorem 13.1) and poly-time reducibility is a transitive relation (Lemma 13.4).

The reduction we present is, essentially, a demonstration that one can craft, for any valid computational tableau as described in the proof of Theorem 13.1, a CNF expression that describes the tableau; moreover, the CNF expression uses roughly the same number of symbols as occur in the tableau. (Of course, the tableau is a two-dimensional structure, whereas the formula is one-dimensional, but the translation from one format to the other will be a simple one.)

Recall that the tableau corresponding to a potential element, $\langle x, y, 0^t \rangle$, of $\mathrm{HP}^{(\mathrm{poly})}$ is a $(t+1) \times (t+3)$ two-dimensional array. The contents of the tableau correspond in the following way to the computation that NTM x has performed on input y within one of its universes. By assumption, x executes t nondeterministic steps as it tests whether it wants to accept input y. Each branch—i.e., root-to-leaf path—of the computation tree $\mathscr{T}_x(y)$ that x generates as it processes input y thus passes through $t+1$ total states of x. Each of these total states is a row of the tableau.

- Using the conventions of Section 3.3, each total state can be written using $\leq t+1$ symbols. One symbol represents x's current state; the others represent the contents of the $\leq t$ squares of its (linear) worktape that x may visit within any individual universe in its nondeterministic computation. The $t+3$ columns of the tableau thus arise (a) from our padding out short total-state strings with instances of the blank symbol, to make all total-state strings have uniform length $t+1$; (b) from our delimiting each total-state string with an initial and a terminal occurrence of the end-symbol #. See (13.3).
- The sequence of $t+1$ total states is the sequence of rows of the tableau.

Given the shape and contents of the tableau, we can describe its contents using $t^2 + 3t$ skeletal variables. For each integer $i \in [0, t]$, each integer $j \in [0, t+2]$, and each symbol $\zeta \in (Q^{(x)} \cup \Gamma^{(x)} \cup \{\#\})$, the variable $X[i, j, \zeta]$ is TRUE (i.e., assumes the truth value 1) precisely when the symbol ζ resides in position (i, j) of the tableau.

The careful reader will note that in the hope of simplifying the exposition a bit, we are being a bit wasteful here, by allowing more variables than we shall actually use. When we do our accounting later in the proof, we shall see that our profligacy does not jeopardize any of the polynomial-size bounds that our poly-time reduction demands.

As our final preparation for describing the reduction of $\mathrm{HP}^{(\mathrm{poly})}$ to SAT, we introduce the following shorthand notation. Given a set of logical variables $\{Y_1, Y_2, \ldots, Y_k\}$:

$$\bigvee_{i=1}^{k} Y_i \text{ is shorthand for } Y_1 \vee Y_2 \vee \cdots \vee Y_k,$$

$$\bigwedge_{j=1}^{k} Y_j \text{ is shorthand for } Y_1 \wedge Y_2 \wedge \cdots \wedge Y_k.$$

Thus, the symbols $\bigvee_{i=1}^{k}$ and $\bigwedge_{j=1}^{k}$ play the same roles for *logical* sum and product, respectively, as the symbols $\Sigma_{i=1}^{k}$ and $\Pi_{j=1}^{k}$ do for *arithmetic* sum and product, respectively.

Now, finally, on to the reduction! Let $x = (Q^{(x)}, \Sigma, \Gamma^{(x)}, \delta^{(x)}, Q_0^{(x)}, q_{\mathrm{acc}}^{(x)})$ be an NTM. We produce a function Φ that transforms each triple $\xi = \langle x, y, 0^t \rangle$, where $x, y \in \{0, 1\}^\star$ and $t \in \mathbb{N}$, into a CNF expression $\Phi(\xi)$ that has the following properties.

1. The size of $\Phi(\xi)$ is polynomial in the size of ξ. (As mandated by our convention, the "size" of ξ is $\ell(x)+\ell(y)+t$.) The degree of the polynomial must be fixed over all possible triples ξ.
2. Expression $\Phi(\xi)$ is satisfiable if and only if NTM x accepts input y in t nondeterministic steps. In other words, ξ and $\Phi(\xi)$ obey the relation

$$[\xi \in \mathrm{HP}^{(\mathrm{poly})}] \quad \text{if and only if} \quad [\Phi(\xi) \in \mathrm{SAT}]$$

Let us focus on a specific but arbitrary triple $\xi = \langle x, y, 0^t \rangle$ and construct its associated CNF expression $\Phi(\xi)$. Informally, $\Phi(\xi)$ asserts that the tableau corresponding to ξ describes a t-step accepting branch of a nondeterministic computation by NTM x on input y.

$\Phi(\xi)$ is the *conjunction* of the following logical expressions.

- The following conditions collectively ensure that the tableau is well formed, in that each row is a potential total state of x.

  $$- \bigwedge_{i=0}^{t} X[i,0,\#] \;\wedge\; \bigwedge_{i=0}^{t} X[i,t+2,\#]$$

  This condition asserts that every row of the tableau is bounded on the left and the right by the delimiter #.

  $$- \bigwedge_{i=0}^{t}\bigwedge_{j=1}^{t+1} \bigvee_{\zeta \in (Q^{(x)} \cup \Gamma^{(x)})} X[i,j,\zeta]$$

  This condition asserts that the delimiter # occurs *only* at the left and right ends of the rows of the tableau.

  $$- \bigwedge_{i=0}^{t}\bigvee_{j=1}^{t+1} \bigvee_{q \in Q^{(x)}} X[i,j,q]$$

  This condition asserts that every row of the tableau contains a state of x.

  $$- \bigwedge_{i=0}^{t}\bigwedge_{j=1}^{t+1}\bigwedge_{k=j+1}^{t+1}\bigwedge_{q \in Q^{(x)}}\bigwedge_{q' \in Q^{(x)}\setminus\{q\}} (\overline{X}[i,j,q] \vee \overline{X}[i,k,q'])$$

  This condition asserts that only one position in each row contains a state symbol.

- The following condition ensures that row 0 of the tableau contains a valid initial total state of NTM x, meaning an initial state followed by all blanks.

  $$- \bigvee_{q \in Q_0^{(x)}} X[0,1,q] \;\wedge\; \bigwedge_{j=2}^{t+1} X[0,j,\boxed{B}]$$

- The following condition ensures that row t of the tableau contains a valid accepting total state of NTM x, meaning that the halt-and-accept state q_{acc} occurs in the row.

$$- \bigvee_{j=1}^{t+1} X[t,j,q_{\text{acc}}]$$

- Our final task is to ensure that the total state at each row $r+1$ of the tableau, where $r \in [0, t-1]$, is a valid successor of the total state at row r of the tableau. As one can gather from the proof of Theorem 13.1, this is substantially more complicated than expressing the other required conditions. We need to say that (a) every 2×3 window in the tableau whose first row contains a state symbol at its center has a second row that is consistent with x's state-transition function $\delta^{(x)}$; (b) every pair of vertically aligned symbols in the tableau that do not reside in one of these windows are identical. We proceed as follows. To enhance legibility, we reverse our earlier approach and present the English explanation before the logical sentence. Let us refer to the just-mentioned 2×3 windows as *consistency windows*. Focus on an arbitrary row $r \in [0, t-1]$.

If none of positions $j-1$, j, $j+1$ of row r contains a state symbol, then position j does not reside in the 2×3 consistency window for rows r and $r+1$. Therefore, the symbol in position j of row $r+1$ must be identical to the symbol in position j of row r. This fact can be stated, albeit somewhat awkwardly, as the following CNF formula.

$$\bigwedge_{\substack{r \in \{0,1,\dots,t-1\} \\ j \in \{1,2,\dots,t+1\}}} \bigvee_{\substack{\sigma_1 \in (\Gamma^{(x)} \cup \{\#,q_{\text{acc}}\}) \\ \sigma_2 \in (\Gamma^{(x)} \cup \{\#,q_{\text{acc}}\}) \\ \sigma_3 \in (\Gamma^{(x)} \cup \{\#,q_{\text{acc}}\})}}$$

$$\left(\overline{X}[r,j-1,\sigma_1] \vee \overline{X}[r,j,\sigma_2] \vee \overline{X}[r,j+1,\sigma_3] \vee X[r+1,j,\sigma_2] \right)$$

The preceding CNF formula can be understood a bit better if one recalls that the logical connective implies, as in

A implies *B*,

is defined as being logically equivalent to the formula $\overline{A} \vee B$.

If position j of row r does contain a state symbol, then position j is the bottom-center position in the 2×3 consistency window for rows r and $r+1$, which is formed from positions $j-1, j, j+1$ of rows r and $r+1$. This window must be consistent with the function $\delta^{(x)}$, in the sense spelled out in Table (13.6) in the proof of Theorem 13.1. This table indicates that the consistency condition has the following form:

$$\left[X[r,j-1,\sigma_1] \wedge X[r,j,\sigma_2] \wedge X[r,j+1,\sigma_3] \right] \text{ implies}$$

$$\left[X[r+1,j-1,\sigma_1'] \wedge X[r+1,j,\sigma_2'] \wedge X[r+1,j+1,\sigma_3'] \right] \vee \cdots$$

$$\vee \left[X[r+1,j-1,\sigma_1''] \wedge X[r+1,j,\sigma_2''] \wedge X[r+1,j+1,\sigma_3''] \right].$$

The number of alternatives after the implication, and their specific identities, depend, of course, on the function $\delta^{(x)}$.

By our earlier comment, we can convert the preceding implication to the following disjunction:

$$\overline{X}[r, j-1, \sigma_1] \vee \overline{X}[r, j, \sigma_2] \vee \overline{X}[r, j+1, \sigma_3] \vee$$

$$\left[X[r+1, j-1, \sigma_1'] \wedge X[r+1, j, \sigma_2'] \wedge X[r+1, j+1, \sigma_3'] \right] \vee \cdots$$

$$\vee \left[X[r+1, j-1, \sigma_1''] \wedge X[r+1, j, \sigma_2''] \wedge X[r+1, j+1, \sigma_3''] \right].$$

We can now transform this expression to CNF by repeatedly invoking the fact that and distributes over or in the propositional calculus.[6] This distribution at worst squares the size of the original expression.

We have completed the reduction. The proof that it is, in fact a reduction, meaning that a triple ξ belongs to $HP^{(poly)}$ if and only if $[\Phi(\xi) \in SAT]$, follows from the same reasoning as we used to prove Theorem 13.1, so we leave this exercise to the reader.

It remains only to prove that this reduction can be computed in time that is polynomial in the size of the input triple ξ. Because we can simply "read off" the expression $\Phi(\xi)$ from our tableau, it will suffice to show that the size of $\Phi(\xi)$ is polynomial in the size of the input triple ξ. To this end, note the following *very* conservative reckoning.

- As noted earlier, the size, in bits, of the tableau that represents a putative t-step accepting computation of NTM x on input y is bounded above by $O(t^2 \cdot \ell(x))$. To wit, the tableau has $O(t^2)$ cells, each of which contains an encoded symbol whose length is patently $O(\ell(x))$, since x's program contains all possible symbols (hence cannot be short).
- The conditions that ensure the *well-formedness of the computational tableau* comprise $O(t)$ conditions—one for each row of the tableau. The conditions within each row have size $O(t^2 \cdot \ell(x))$, the longest ones being those that prohibit two state symbols in one row. The aggregate size of these conditions is thus $O(t^3 \cdot \ell(x))$.
- The condition that ensures a *valid initial total state* has size $O(t + \ell(x))$.
- The condition that ensures a *valid terminating total state* has size $O(t)$.
- Finally, there are the conditions that ensure that *each total state in the tableau is a valid successor of its predecessor*.

  The first of these guarantees the upward persistence of symbols that cannot be changed at a given step. This condition has size $O(t^2 \cdot (\ell(x))^3)$.

  The second of these guarantees that the second row of each 2×3 consistency window is consistent with the first row. There are t such conditions, each of size $O((\ell(x))^2)$, for an aggregate size of $O(t \cdot (\ell(x))^2)$.

---

[6] The propositional calculus obeys the following distributive law: $(A \wedge B) \vee C \vee D = (A \wedge (C \vee D)) \vee (B \wedge (C \vee D))$.

Adding up the contributions of all of the conditions expressed by $\Phi(\xi)$, we obtain a grand total of $O(t^2 \cdot (\ell(x))^3)$, which is indeed polynomial in the size of the input triple ξ.

We thus have a poly-time reduction of $HP^{(poly)}$ to SAT, whence the latter language is **NP**-Hard because the former one is.

We have thus shown both that SAT belongs to **NP** and that SAT is **NP**-Hard. We conclude that SAT is **NP**-Complete. □

13.4 Nondeterminism and Space Complexity

We now embark on a short excursion into the world of *space complexity*—the study of the memory requirements of deterministic and nondeterministic IRTMs. Our excursion's main focus is on illustrating how differently space complexity behaves from time complexity, both quantitatively and qualitatively. We accomplish this by developing a landmark theorem by Walter J. Savitch that relates the families of languages accepted by *deterministic* and *nondeterministic* IRTMs that operate under *space bounds*. As we develop this theorem, in Section 13.4.1, we contrast its message and proof strategy with the current state of knowledge about the analogous issues for families that are defined via *time bounds*. Our contrast has both a *quantitative* aspect and a *qualitative* one, both arising from the challenge of deterministically implementing the search that is inherent in every computation that is truly nondeterministic.

We have noted twice—in Sections 10.2 and 12.3—that the essential difference between deterministic and nondeterministic computation resides in the latter's prescribing an unbounded search, in addition to the "core" computation. The challenge of incorporating this search into a deterministic simulation of a nondeterministic computation resides in two facts.

1. As we assess the nondeterministic computation's consumption of resources—both time and space—we do not charge the computation for the resources required to perform the search. But we do charge any deterministic simulation of the nondeterministic computation for performing the search! The *quantitative* side of our comparison of *space* vs. *time* complexity resides in noting how the cost of performing nondeterminism's search deterministically expands the space complexity of a computation vs. how it expands the time complexity.
2. A nondeterministic computation offers no prescription for actually executing the search: nondeterminism permits searching via "guessing," an extra-algorithmic notion. The question of how to convert "guessing" to an algorithmic procedure underlies the *qualitative* side of our comparison of *space* vs. *time* complexity.

We now develop our comparison of space vs. time complexity a bit more, to prepare the reader for what to focus on as we develop Savitch's theorem.

Our *quantitative* comparison of space vs. time complexity is based on the demonstration in Savitch's theorem (Theorem 13.3) that a deterministic simulation of a nondeterministic IRTM computation need never do worse than *square* the *space* required for the latter computation. This contrasts with our current inability to avoid *exponentiating* the *time* requirements of a nondeterministic "algorithm" as we simulate it deterministically; cf. Corollary 13.1.

Notes. **(a)** The phrase "current inability" in the preceding sentence emphasizes our current ignorance of the inherent cost in time of simulating nondeterministic computation deterministically. **(b)** Savitch's theorem builds on insights from [98], the first systematic study of the space complexity of TM computations.

Our *qualitative* comparison of *space* complexity vs. *time* complexity is based on the *search-via-counting* strategy that underlies the proof of Savitch's Theorem. This ingenious strategy implements a nondeterministic search through a space of potential solutions via the following two steps.

1. The strategy represents the potential solutions to the problem being solved via the nondeterministic computation as strings over an alphabet that is appropriate to the problem.
   Two illustrations. (a) A potential solution to an n-variable instance of SAT could be represented as a length-n binary string whose kth bit is the candidate truth value for the kth variable in the CNF formula. (b) A potential solution to an instance of CLIQUE that has an n-node graph and a target clique size k could be represented as a sequence of k nodes of the graph, which constitute a potential clique. If the graph's nodes are represented as binary strings, then the potential solution could be a length-$(k \log_2 n)$ string constituted from the symbols "0" "1" "," "(" ")". (We write the alphabet without commas to avoid ambiguity.)
2. The strategy searches through the space of potential solutions by "counting"— in the number base formed by imposing a linear order on the alphabet used to represent the solutions—from 0 through the largest relevant representable number, call it N. We say that N is "relevant" if the numerals corresponding to the potential-solution strings all represent numbers $\leq N$. (In other words, the strategy counts only as high as it needs to in order to search the entire space of potential solutions.) (a) In the case of SAT in the preceding item, there is already a natural order on the alphabet $\{0, 1\}$. Accordingly, the simulator would search through the space of potential solutions by counting in base 2 from 0 to $2^{n+1} - 1$. (b) In the case of CLIQUE in the preceding item, there is no natural order on the suggested five-letter alphabet, so the simulator would impose one by somehow associating the letters with the base-4 numerals $\{0, 1, 2, 3, 4\}$. The simulator would search through the space of potential solutions by counting in base 5 from 0 to $5^{(k \log_2 n)+1} - 1$ (which is the largest integer that one can represent with a base-5 numeral of length $k \log_2 n$.

While often wasteful of time—a more directed search could take *many* fewer steps— the described strategy is *very* compact in its consumption of space.

We say that the strategy is *seemingly* wasteful of time because it could be that all deterministic search strategies use exponential time, in which case, Savitch's strategy *consumes* time but does not *waste* it.

13.4.1 Simulating Nondeterminism Space-Efficiently: Savitch's Theorem

This section is devoted to developing the main ideas about space-compact simulations of nondeterministic IRTMs by deterministic *online* IRTMs. The development culminates in the landmark result known as "Savitch's theorem" (Theorem 13.3). We begin our excursion into space complexity by explaining why we cannot just carry over the ideas that we used in discussing *time* complexity.

When faced with the challenge of determining *deterministically* whether a given nondeterministic IRTM $M = (Q, \Sigma, \Gamma, \delta, Q_0, F)$ accepts a given input word x, we resorted, in the proof of Theorem 12.1, to a *breadth-first* traversal of the computation tree $\mathcal{T}_M(x)$ that M generates while processing x. As noted within that proof, this algorithmic strategy is well suited for this task, because in general one has no way to bound from above how deep in the computation tree $\mathcal{T}_M(x)$ one must search in order to find an accepting state of M. Of course, a simulation of M that is based on a breadth-first traversal of $\mathcal{T}_M(x)$ consumes a lot of space. Specifically, as the simulation is processing level l of $\mathcal{T}_M(x)$, the queue that orchestrates the traversal can contain as many as $(3|Q||\Gamma|)^l$ total states of M, because M could branch to as many as $3|Q||\Gamma|$ distinct total states at every (nondeterministic) state transition. And, each of M's total states at level ℓ of $\mathcal{T}_M(x)$ could employ l squares of M's read/write worktape. It follows that even if M operates within space $s(n)$, the breadth-first simulation strategy could use as many as $c_M^{s(\ell(x))}$ bits of storage/memory, for some constant $c_M > 1$, to determine whether M accepts a string x. Savitch found a way to avoid using exponential space by developing a nonobvious way to search for the (possible) presence of an accepting total state in $\mathcal{T}_M(x)$.

Savitch's simulation strategy begins with the following derivation of an upper bound on the depth one needs to probe in $\mathcal{T}_M(x)$ in order to determine whether M accepts x. Note that this bound allows one to *reject* words that M does not accept, as well as to accept words that M accepts. The general simulation of Theorem 12.1 accomplishes only the positive (acceptance) half of this duo.

A nondecreasing function $a : \mathbb{N} \to \mathbb{N}$ is an *acceptance bounding function* (*acceptance bound*, for short) for a nondeterministic IRTM M if the following holds for every word $x \in \Sigma^*$. If M accepts the input word x, then there is an accepting total state-node at depth $\leq a(\ell(x))$ of the computation tree $\mathcal{T}_M(x)$.

Lemma 13.9. *Let M be a nondeterministic IRTM that operates within space $s(n)$. There exists an absolute constant c_M such that M admits the acceptance bound*

$$a(n) = 2^{c_M \cdot s(n)}. \tag{13.7}$$

Proof. Say that we are guaranteed that M accepts a word x via a branch of its (nondeterministic) computation tree $\mathcal{T}_M(x)$ that uses $\leq s(\ell(x))$ squares of read/write worktape. We claim that the length of this short accepting branch cannot exceed

$$c_M \cdot s(\ell(x)) \cdot |\Gamma|^{s(\ell(x))},$$

for some constant c_M that depends only on M, i.e., that is independent of the length of x. Let us verify this bound.

We note first that M cannot repeat a total state along a shortest branch of $\mathcal{T}_M(x)$. This is because—precisely in the manner of *pumping* within the derivation trees associated with context-free grammars; cf. Section 6.1.3—one could remove any repeating segment of the branch to obtain a shorter branch.

We next see how long M can compute along a branch of $\mathcal{T}_M(x)$ before it must increase the effective size of the worktape. We claim that while M is using m squares of worktape, it can proceed for no more than $|Q|m|\Gamma|^m$ steps—or else it will repeat a total state. To see this, note that for each branch, M can do no more than cycle through all of the $|\Gamma|^m$ possible m-symbol configurations of its worktape, and for each total state, cycle through all of its $|Q|$ internal states, and for each (tape configuration)–(internal state) pair, cycle through all m possible positions for its read/write head. If M were to proceed for even one more step, it would have to repeat a total state along that branch.

Summing up, if M never uses more than $s = s(\ell(x))$ squares of worktape along a shortest accepting branch of $\mathcal{T}_M(x)$, then each such shortest branch can be no longer than

$$T = \sum_{m=1}^{s} |Q|m|\Gamma|^m$$

$$= |Q|\left(\sum_{m=1}^{s} m|\Gamma|^m\right)$$

$$= \frac{|Q||\Gamma|}{|\Gamma|-1}\left(s|\Gamma|^s + \frac{|\Gamma|^s - 1}{|\Gamma|-1}\right).$$

There are numerous excellent texts that show in detail how to evaluate geometric sums such as $\sum_{m=1}^{s} m|\Gamma|^m$. I am partial to a beautiful recursive technique described in [53].

Noting that both $|Q|$ and $|\Gamma|$ are absolute constants—i.e., they depend on the structure of M but are independent of the length of x—we conclude that there exists an absolute constant c_M such that M admits the claimed acceptance bound (13.7). $\square$

We now exploit the bound (13.7) of Lemma 13.9 to develop a space-efficient strategy for determining whether an NTM M's computation tree $\mathcal{T}_M(x)$ contains an accepting node. The only extrinsic requirement that our space-efficient strategy imposes on the space-bounding function $s(n)$ for M upon which we develop the simulation strategy is that $s(n)$ be *constructible*. This means that there exists an IRTM $M^\#$ such that if we feed $M^\#$ any input string of length n, then:

- $M^\#$ can "lay out" a string of $s(n)$ consecutive squares on its read/write worktape. By "lay out," we mean somehow mark these squares to distinguish them from all other squares of the worktape.
- $M^\#$ can accomplish the "laying out" process while using no more than $s(n)$ squares of its read/write worktape.

The second, space-bounding, condition is crucial in the coming proof, because it allows the simulating IRTM to "lay out" a work region on its read/write worktape. We use this facility in line 1 (the initialization step) of the upcoming Program [Does-M-accept?].

Keep in mind, as you proceed to Theorem 13.3, that the overriding challenge in deterministically deciding the language $L(M)$ *space efficiently* is that we cannot afford the space that would be needed to generate and lay out the computation tree $\mathscr{T}_M(x)$: that would require space *exponential* in $s(n)$, rather than the space *quadratic* in $s(n)$ that the theorem actually achieves. Roughly speaking, we are able to use space that is only quadratic in $s(n)$ by generating only *small portions* of one branch of $\mathscr{T}_M(x)$ at a time, and reusing space from one portion to the next.

Theorem 13.3 ([92]). *Let M be a nondeterministic IRTM that operates within space $s(n)$ and that accepts a language L. If the space-bounding function $s(n)$ is constructible, then there exists a deterministic, online IRTM M' that decides language L and that operates within space $(s(n))^2$.*

Note that the *nondeterministic* IRTM M only *accepts* the language L: it accepts all words that belong to L but never rejects any input word. In contrast, the *deterministic online* IRTM M' *decides* L: it accepts all strings that belong to L and rejects all strings that do not.

Proof. We begin with an overview of the simulation strategy, then supply algorithmic details, and end with an analysis of the detailed algorithm.

Overview. Focus on a moment when the nondeterministic IRTM M has read some input $x \in \Sigma^*$, and consider the computation tree $\mathscr{T}_M(x)$ that M generates while processing x. We develop a space-compact (specifically, space $O((s(\ell(x)))^2)$) deterministic online IRTM M' for deciding whether M accepts x. Our IRTM searches within $\mathscr{T}_M(x)$ for an *accepting total state C* of M, i.e., a total state whose internal state is an accepting state. Let C_0 denote the initial total state of M, which is the root of $\mathscr{T}_M(x)$. Using the language and notation of Section 2.5, our strategy seeks an accepting total state C such that $C_0 \Rightarrow C$.

Because of M's space bound, all of the total states along the branch from C_0 to C are strings of length $\leq s(\ell(x)) + 1$; the "+1" accounts for the internal-state symbol that occurs (precisely once) in each total state. This fact allows us to enlist the notion of computational tableau that appears in the proof of Theorem 13.1.

To simplify our task, we note, via Lemma 13.9, that if M accepts the word x, then it does so along some branch of $\mathscr{T}_M(x)$ whose length is no greater than $2^{c_M \cdot s(\ell(x))}$, where c_M is the absolute constant guaranteed by the lemma. This means that we can restrict our search for total state C to the prefix of $\mathscr{T}_M(x)$ obtained by truncating all branches at depth $2^{c_M \cdot s(\ell(x))}$; call the resulting prefix $\widehat{\mathscr{T}}_M(x)$. If $\widehat{\mathscr{T}}_M(x)$ contains the sought accepting total state, then, by definition, M accepts x; if $\widehat{\mathscr{T}}_M(x)$ contains no such total state, then Lemma 13.9 assures us that $\mathscr{T}_M(x)$ also contains no such total state, so that M does not accept x.

It is convenient for the development of our space-compact simulation algorithm to refine the tree-related (ancestor–descendant) relation "$\Rightarrow$" from Section 2.5 in the following way. Within the context of a computation tree $\widehat{\mathcal{T}}_M(x)$, the notation $C_i \Rightarrow C_j$ means that node C_j is a descendant of node C_i within $\widehat{\mathcal{T}}_M(x)$. In computational terms: there is a branch of the (nondeterministic) computation by M on input x along which M arrives at total state C_i, and after some additional number of (nondeterministic) steps—perhaps 0 steps—M arrives at total state C_j. The refinement that we adopt now embellishes the notation "$\Rightarrow$" with an integer $d \in \mathbb{N}$, as in

$$C_i \overset{d}{\Rightarrow} C_j, \tag{13.8}$$

to indicate *how many* steps M must execute in order to reach total state C_j from total state C_i. In more detail, the notation (13.8) betokens the existence of a length-d path of total states in $\widehat{\mathcal{T}}_M(x)$,

$$(C_i = C_{k_0}), C_{k_1}, C_{k_2}, \ldots, C_{k_{d-3}}, C_{k_{d-2}}, (C_{k_{d-1}} = C_j).$$

Note that, by definition of "rooted tree," each node C_{k_a} in this path is the parent of its successor, node $C_{k_{a+1}}$.

Summing up thus far, we determine whether M accepts input x by determining whether $\widehat{\mathcal{T}}_M(x)$ contains an accepting total state C^* such that

$$C_0 \overset{d}{\Rightarrow} C^* \quad \text{for some} \quad d \le d^* = 2^{c_M \cdot s(\ell(x))}. \tag{13.9}$$

Detailed algorithmics. We begin with the simple "envelope" Program Master_M that embodies the online portion of our simulation. In conformity to the phrasing of the theorem, Master_M has M's (nondeterministic) program built in—meaning that we craft a distinct deterministic simulator for every nondeterministic IRTM M. One could craft a stronger, but notationally much more complicated, version of the theorem, in which there is a single "universal" simulator that takes M as an argument.

Our detailed simulation is invoked by a call to Program Master_M. We assume that both of the simulator's tapes—its input record and its read/write worktape—are blank at the moment of this invocation.

We now specify how to determine deterministically and space-compactly whether M accepts the input word $x \in \Sigma^*$ that Program Master_M has received thus far at its input port (and has recorded on its input record). The remaining procedures in our suite answer the question,

"Does M accept x?"

by determining whether there exists an accepting total state C^* in $\mathcal{T}_M(x)$ that satisfies (13.9). As you study the suite of procedures, you should pay special attention to the following two algorithmically interesting issues: How we

1. translate the existential quantifier in the preceding question into a search for the the total state C^*,
2. execute the search in a nonobvious way, by *counting* in a nonstandard number base.

Program Master$_M$

Input	obtained via exterior port

/*Simulate M's behavior on the provided input: accept an input word
$x \in \Sigma^*$ if M accepts it; reject x if M does not accept it*/

---

1. Search for an accepting total state of M in $\widehat{\mathscr{T}}_M(\varepsilon)$
   /*$\widehat{\mathscr{T}}_M(\varepsilon)$ is a fixed finite tree, of depth $\leq 2^{c_M \cdot s(0)}$, so we can
   perform this search in *constant* space*/
2. **if**   an accepting total state exists
3. **then** **return** ACCEPT ε
4. **else** **return** REJECT ε
5. **poll** input port for next input symbol
6. **if**   input port $\leftarrow \sigma$
   /*A new input symbol $\sigma \in \Sigma$ is detected at the input port*/
7. **then** Append σ to the current word on the input record
8. Invoke Program [Does-M-accept?] (word on input record)
9. **if**   Program [Does-M-accept?] returns ACCEPT
10. **then** **return** ACCEPT (word on input record)
11. **else** **if** Program [Does-M-accept?] returns REJECT
12.   **then return** REJECT (word on input record)
13. **goto** line 5

The following program "manages" the process of answering our question about input word x.

Program [Does-M-accept?] x

Input	x

/*Determine whether M accepts input x*/

---

1. $C \leftarrow$ a string of $s(\ell(x)) + 1$ occurrences of $\boxed{\text{B}}$
   /*Initialize the process of generating candidate accepting total
   states via which M accepts x. **Note:** *This initialization is
   possible only because $s(n)$ is constructible.*/
2. $C \leftarrow$ Generate-Next(C, ON)
   /*Generate next candidate *accepting* total state*/
3. **if**   no more candidate accepting total states exist
   /*Program **Generate-Next** with the ON option supplies this
   information*/
4. **then** **return** REJECT
5. **else** **forall** $d \leq 2^{c_M \cdot s(\ell(x))}$
   /*For all valid numbers d of (nondeterministic) steps*/
6.   Test $[C_0 \overset{d}{\Rightarrow} C]$
   /*Is accepting total state C accessible from C_0 in d steps?*/
7.   **if** Program Test returns ACCEPT
8.   **then return** ACCEPT
9.   **else goto** line 2

In order to complement Program [Does-M-accept?] x so that it can answer the question in its name, we have to deal with two issues.

1. *How do we determine* deterministically *whether a proposed candidate C satisfies condition (13.9)?*
   We answer this question by specifying the procedure

   Test $[C_0 \overset{d}{\Rightarrow} C]$

   that is invoked in line 6 of Program [Does-M-accept?].

2. *How do we generate the successive candidate accepting total states?*
   We answer this question by specifying the procedure

   Generate-Next(C, ON)

   that is invoked in line 2 of Program [Does-M-accept?]. We thus implement the existential question, "Does total state C^* exist?" via a procedure for generating all possible candidate total states C one at a time and testing each for being the sought total state C^*.

   Our regimen of

   (generate one new candidate)–(test this candidate)

   is critical in our quest to conserve space.

   We actually need two versions of Program Generate-Next. One version, namely Program Generate-Next with the ON option, implements our search for the *accepting* total state C^*. The other version has the OFF option; it satisfies the needs of Program Test, which is invoked in line 6, for intermediate total states that M encounters on the branch (if it exists) from C_0 to C^* that are *not-necessarily-accepting*.

We deal with the preceding two issues in turn.

1. Vetting candidates for C^*. We focus on vetting a single candidate—call it C_0^*—for the sought total accepting state C^*. Using our quantified ancestor–descendant notation "$\overset{d}{\Rightarrow}$" this amounts to determining whether

$$[C_0 \overset{d^*}{\Rightarrow} C_0^*].$$

The fact that we can delimit the search for C_0^* via the numerical parameter d^* suggests that we can employ a *recursive* procedure for searching through $\widehat{\mathscr{T}}_M(x)$. Specifically, the relation $C_0 \overset{d^*}{\Rightarrow} C_0^*$ holds if and only if there exists a (*not necessarily accepting*) total state C_1^* such that both of the following relations hold:

$$[C_0 \overset{\lfloor d^*/2 \rfloor}{\Longrightarrow} C_1^*] \text{ and } [C_1^* \overset{\lceil d^*/2 \rceil}{\Longrightarrow} C_0^*].$$

The intuition that leads to this insight is that if there is a length-d^* branch from C_0 to C^* in $\widehat{\mathscr{T}}_M(x)$, then there must be *some* total state—call it C_1^*—halfway (to within rounding) along the branch from C_0 to C_0^*. How do we generate candidates for C_1^*? We invoke the same (time-profligate but space-conservative) recipe that we used to generate C_0^*—but we do it now for worktape configurations of lengths $\leq \lfloor d^*/2 \rfloor$ and $\leq \lceil d^*/2 \rceil$, the former length for the branch from C_0 to C_1^* and the latter length for the branch from C_1^* to C_0^*. We then recursively invoke the fact that each of the relations governed by parameter $d^*/2$ (to within rounding) holds if and only if two analogous relations that are governed by parameter $d^*/4$ (to within rounding) hold; to be specific for relation

$$[C_0 \overset{\lfloor d^*/2 \rfloor}{\Longrightarrow} C_1^*],$$

for instance—the other relation is treated similarly—the derived pair of relations have the form

$$[C_0 \overset{\lfloor \lfloor d^*/4 \rfloor \rfloor}{\Longrightarrow} C_2^*] \quad \text{and} \quad [C_2^* \overset{\lceil \lfloor d^*/2 \rfloor \rceil}{\Longrightarrow} C_1^*].$$

Without further ado, we jump from the preceding reasoning to the desired recursive vetting procedure, Program Test.

Program Test $[C_1 \overset{d}{\Rightarrow} C_2]$

Input d, C_1, C_2

/*Determine whether the argument relation holds with the parameter $d \in \mathbb{N}$ and the total states C_1 and C_2*/

1. **if** $d = 1$
2. **then**   **If** C_2 is a valid successor of C_1 in NTM M
3.   **then return** "YES: THE BRANCH EXISTS"
4.   **else return** "NO SUCH BRANCH EXISTS"
   /*Test directly for parent–child relation $C_1 \to C_2$*/
5. **else** $C \leftarrow$ a string of $2^{c_M \cdot s(\ell(x))}$ occurrences of $\boxed{B}$
   /*Initialize the process of generating candidate total states that could occur halfway between C_1, C_2*/
6. $C \leftarrow$ Generate-Next(C, OFF)
   /*Search for a total state halfway between C_1, C_2*/
7.   **if** no more candidate total states exist
   /*Program Generate-Next, with the OFF option, supplies this information*/
8.   **then return** "NO SUCH BRANCH EXISTS"
9.   **else** Test $[C_1 \overset{\lfloor d/2 \rfloor}{\Longrightarrow} C]$; Test $[C \overset{\lceil d/2 \rceil}{\Longrightarrow} C_2]$
   /*Does the candidate intermediate node C work?*/
10.   **if** both calls to Program Test return "YES"
11.   **then return** "YES: THE BRANCH EXISTS"
   /*The candidate intermediate node C works*/
12.   **else goto** line 5

Program Test directly implements the recursive vetting procedure while using Program Generate-Next (with the OFF option) to implement the search for intermediate nodes along the sought branch of $\widehat{\mathscr{T}}(x)$.

2. Searching for candidate total states. Because M operates within space $s(n)$, we can restrict our search for the sought accepting total state C^* to total states that have length $\leq s(\ell(x)) + 1$. Since we do not care how long it takes us to run through all possible candidates, we can accomplish this search compactly via the following "trick."

Recall that every total state of M is a string over the alphabet $\Gamma \cup Q$. Ignore, for simplicity, that total states are very special strings over this alphabet (cf. (3.5)). Incorporating this fact into our algorithms would save us time—which is of no interest here—but would not affect space—which is our only concern. We view the set $\Gamma \cup Q$ as the *set of digits in the number base* $|\Gamma| + |Q|$. It will simplify the detailed algorithmics to use the blank symbol $\boxed{\text{B}} \in \Gamma$ as the symbol for 0, for this allows us always to ignore "leading" blanks. We count in base $|\Gamma| + |Q|$, from the smallest integer that can be represented by a length-$(s(\ell(x)) + 1)$ numeral to the largest integer that can be represented by a length-$(s(\ell(x)) + 1)$ numeral. The former number is 0, which is the value of the numeral $\boxed{\text{B}}\,\boxed{\text{B}}\cdots\boxed{\text{B}}$; the latter number is

$$(|\Gamma| + |Q|)^{s(\ell(x))+2} - 1,$$

which is the value of the numeral $\beta\beta \cdots \beta$, where β denotes the largest digit in the set $\Gamma \cup Q$; both of the indicated numerals contain $s(\ell(x)) + 1$ digits. This counting process automatically enumerates all of the length-$(s(\ell(x)) + 1)$ strings over the alphabet $\Gamma \cup Q$; thereby, the process also automatically enumerates all of the total states of M that could conceivably be the sought total state. Of course, the process also generates a lot of "garbage" strings, i.e., strings that because of their structures, do not represent valid total states of M. However:

- our string-generation procedure can weed out these useless strings before returning any result to the calling procedure—so the calling procedure receives only valid total states;
- we are allowed to waste time *without limit,* as long as we stay within the prescribed number of squares of worktape.

We now flesh out the preceding informally specified procedure to obtain the following program for generating "the next" total state. As noted earlier, we actually need two programs, because we are sometimes called upon to return "the next" total state and sometimes to return "the next" *accepting* total state. We implement this need by means of a *toggle bit* ACC$\in\{$ON, OFF$\}$ that when set to ON, filters out all *nonaccepting* total states; of course, the program always filters out all "garbage" strings.

Analyzing the algorithm. We have amply documented the effectiveness of our suite of simulation procedures, so we concentrate now just on proving that the procedures stay within the advertised space bound of $O((s(n))^2)$. We consider in turn each of the four procedures in the suite.

(1) Program Master$_M$ *uses space* $O(1)$. The program does not directly touch the simulator's read/write worktape. The one "large" item that it processes is the computation tree $\widehat{\mathscr{T}}_M(\varepsilon)$ that M generates when processing the null input. Because that computation tree is fixed, Program Master$_M$ can use its internal memory—i.e., its internal states—to process it.

Program Generate-Next(C, ACC)
Input C (total state); ACC (toggle bit)
/*When ACC is OFF, return the length-($s(\ell(x))+1$) total state that "follows" C numerically. When ACC is ON, return the next *accepting* total state*/

1. $C \leftarrow C + 1$, as a numeral base $|\Gamma| + |Q|$
2. **if**   length(C) exceeds $s(\ell(x)) + 1$
        /*C is viewed here as a numeral, i.e., a string*/
3. **then** **return** "NO MORE CANDIDATE TOTAL STATES EXIST"
4. **else**   **if** ACC is OFF
5.   **then if** C is a valid total state of M
        /*Recall that C is a valid total state precisely when $C \in \Gamma^+ Q \Gamma^+ \boxed{\text{B}}$;
        cf. (3.5)*/
6.   **then return** C
7.   **else goto** line 1
8.   **else if** C is a valid *accepting* total state of M
        /*Recall that a valid total state C is *accepting* if its internal state is*/
9.   **then return** C
10.   **else goto** line 1

(2) Program [Does-*M*-accept?] *uses space* $O(s(\ell(x)))$. The program uses the simulator's read/write worktape in two ways, and always only sparingly.

1. When the program is first invoked, it "constructs" a string of $s(\ell(x)) + 1$ blanks on the worktape. Because the space-bounding function $s(n)$ is constructible, this process utilizes only $O(s(\ell(x)))$ squares of worktape.
2. As the program searches through candidates for an accepting total state that is a descendant of C_0 in $\widehat{\mathscr{F}}_M(x)$, it writes only one total state at a time on the worktape—and it writes this total state over the previous one.

(3) Program Test *uses space* $O((s(\ell(x)))^2)$. This is the program that uses most of the space. In order to appreciate how the space usage evolves, let us briefly illustrate the program in action. Say that one wants to determine whether the following relation holds for some total states $C^{(0)}$ and $C^{(1)}$ and some $d \in \mathbb{N}$:

$$[C^{(0)} \overset{d}{\Rightarrow} C^{(1)}].$$

To simplify notation by avoiding floors and ceilings, let us assume that d is a power of 2. Now, the preceding relation on total states holds if and only if there exists a total state $C^{(2)}$ such that *both* of the following relations hold:

$$[C^{(0)} \overset{d/2}{\Rightarrow} C^{(2)}] \text{ and } [C^{(2)} \overset{d/2}{\Rightarrow} C^{(1)}].$$

The reasoning is that there must be *some* total state—call it $C^{(2)}$—one-half of the way along the branch from $C^{(0)}$ to $C^{(1)}$.

Continuing this reasoning, if the preceding two relations hold between $C^{(0)}$ and $C^{(2)}$ and between $C^{(2)}$ and $C^{(1)}$, then there must be a total state $C^{(3)}$ such that *all three* of the following relations hold:

$$[C^{(0)} \overset{d/4}{\Rightarrow} C^{(3)}] \quad \text{and} \quad [C^{(3)} \overset{d/4}{\Rightarrow} C^{(2)}] \quad \text{and} \quad [C^{(2)} \overset{d/2}{\Rightarrow} C^{(1)}]$$

The reasoning is that there must be *some* total state—call it $C^{(3)}$—one-half of the way along the branch from $C^{(0)}$ to $C^{(2)}$, i.e., one-quarter of the way along the branch from $C^{(0)}$ to $C^{(1)}$.

We'll do just one more step before making our leap, to make sure that the pattern is clear. If the preceding three relations hold between $C^{(0)}$ and $C^{(3)}$ and between $C^{(3)}$ and $C^{(2)}$ and between $C^{(2)}$ and $C^{(1)}$, then there must be a total state $C^{(4)}$ such that *all four* of the following relations hold:

$$[C^{(0)} \overset{d/8}{\Rightarrow} C^{(4)}] \quad \text{and} \quad [C^{(4)} \overset{d/8}{\Rightarrow} C^{(3)}] \quad \text{and} \quad [C^{(3)} \overset{d/4}{\Rightarrow} C^{(2)}] \quad \text{and} \quad [C^{(2)} \overset{d/2}{\Rightarrow} C^{(1)}]$$

The reasoning is that there must be *some* total state—call it $C^{(4)}$—one-half of the way along the branch from $C^{(0)}$ to $C^{(3)}$, i.e., one-eighth of the way along the branch from $C^{(0)}$ to $C^{(1)}$.

Now we make the leap! What is really going on here? The pair of recursive invocations of Program Test at line 9 of Program Test actually *push* instances of relations of the form

$$[C^{(i)} \overset{d/2^k}{\Rightarrow} C^{(j)}]$$

onto a recursion *stack*[7] for later processing. The case $d = 1$ at the beginning of Program Test *pops* one relation instance from the then-current stack. It is this stack that consumes the lion's share of the space on the simulator's worktape. How much space can the stack consume when processing an input word $x \in \Sigma^*$?

- Each entry on the stack is (essentially) a pair of total states of M, plus a numeral (the d-parameter). Because of M's space bound, the two total states consume space $O(s(\ell(x)))$. Because the *maximum* value of the parameter d is $d^* = 2^{O(s(\ell(x)))}$ (cf. (13.9)), the numeral for d can be written in $O(s(\ell(x)))$ bits. It follows that each stack entry consumes space $O(s(\ell(x)))$.
- – Each recursive invocation of Program Test replaces one entry of the stack by two new ones—but these new ones have a d-parameter that is (to within rounding) *one-half* of the parameter of the entry that they replace.
  – When the top stack entry has a d-parameter $d = 1$, then that entry is removed from the stack, so that the stack loses an entry.
  Because of the value of d^*, and because each new top entry in the stack halves the value of the d-parameter of the preceding top entry, the stack never contains more than $\log_2 d^* = O(s(\ell(x)))$ entries.

Thus, the stack contains $O(s(\ell(x)))$ entries, each of size $O(s(\ell(x)))$. It follows that Program Test uses space $O((s(\ell(x)))^2)$.

(4) Program Generate-Next *uses space* $O(s(\ell(x)))$. The program adds 1 to a length-$(s(\ell(x)) + 1)$ numeral, thereby increasing the numeral's length by at most 1. When

---

[7] See Section 9.8.1.B for details about the stack data structure.

such an increase occurs, this branch of the simulation is terminated—so no further increase occurs.

Summing up, we find that our entire simulation uses space $O((s(\ell(x)))^2)$ when processing the input word $x \in \Sigma^*$. The worktape is reinitialized to $O(s(\ell(x)))$ blanks as each new input word x begins to be processed. The space bound of $O((s(n))^2)$ claimed in the theorem is thus validated. $\Box$

We thus see that "determinizing" nondeterministic computations forces us to expand space requirements to a much smaller measure than time requirements—as far as we know.

13.4.2 Beyond Savitch's Theorem

The reader who is intrigued by the search-via-counting strategy will enjoy seeing the strategy exercised vigorously by Neil Immerman and Róbert Szelepcsényi. In [100] and [43], these authors independently, and roughly contemporaneously, proved that every family of languages **F** that is *defined* via a nondeterministic space bound $s(n) \geq \log_2 n$ is closed under complementation. That is, for every language $L \subseteq \Sigma^*$ that belongs to **F**, the complementary language $\overline{L} = \Sigma^* \setminus L$ also belongs to **F**.

> One finds in texts such as [40, 76, 97], which go beyond introductory material on complexity theory, proofs of the Immerman–Szelepcsényi theorem that are "gentler" in exposition than one can expect from research publications such as [100] and [43].

Fleshing out details a bit: The family of languages **F** is *defined* via a nondeterministic space bound $s(n)$ if **F** is the set of all languages that are accepted by a nondeterministic IRTM that operates within space $s(n)$. Two important such space-defined families are:

- **NPSPACE**, the family of languages that are accepted by nondeterministic IRTMs that operate within space $s(n)$ for some *polynomial* $s(n)$;
- the *context-sensitive languages*, the family of languages that are accepted by nondeterministic TMs that operate within space $s(n) = n$.

> Regarding the second example: When $s(n) \geq n$, we no longer need the IRTM model, because a TM has access to enough squares of read/write worktape to record the input it has read thus far.

The result by Immerman and Szelepcsényi, which dates to the late 1980s, settled (in the affirmative) the question whether the family of context-sensitive languages is closed under complementation, a question that had been open since the 1960s.

With respect to our comparison of *time* and *space* complexity, it is notable that no analogue of the Immerman–Szelepcsényi theorem exists for time complexity, even for most individual time-complexity classes. As an important specific instance: no one has yet been able to show whether the family **NP** is closed under complementation— the **NP**-*vs.*-*co***NP** *problem*.

PART V

SAMPLE EXERCISES

Multiplication is vexation,
Division is bad;
The rule of three doth puzzle me,
And practice drives me mad.
                    (Elizabethan MS [1570])

Everyone who teaches this course will bring a unique style and philosophy and perspective to the subject. Each time I taught the course, I added to the list of problems and exercises that I assigned to the students in an attempt to elucidate the material that I was covering and to get the students thinking about computation mathematically. While I often garnered inspiration from the problem sets that I discovered in the many texts that I consulted when teaching the course, I seldom found that the questions formulated by others, no matter how expert in the field, captured exactly the tone that I was trying to set for the course. Not surprisingly, this was particularly true regarding the topics that I always included in the course that did not appear in many— often, in any—other texts. Recognizing that instructors who use this book will react to my favorite problems in much the way that I reacted to those of others, I include here a list of some of my favorites in the spirit of suggestions that I hope will inspire my readers. Please note that while I have attempted to include in this list all of the exercises that I have "left to the reader" in the text, I have undoubtedly missed some.

I hope that you and your students will find that all of the following problems elucidate the material in the text, and that (at least) some of them are interesting and even "inspiring."

Sample Exercises

1. Prove that all three Boolean operations can be expressed using just the single operation set-difference. In other words, find ways of expressing $\bar{S}$, $S \cap T$, and $S \cup T$ using only expressions of the form $S \setminus T$.

2. Consider the operation of composition on binary relations, as defined in Section 2.2.1. Prove that this operation is *associative*. That is, if we denote by $P \circ P'$ the composition of binary relations P and P', then prove that for all binary relations P_1, P_2, P_3,
$$P_1 \circ (P_2 \circ P_3) = (P_1 \circ P_2) \circ P_3.$$

3. Prove in detail the fact alluded to in Section 2.2.2 that an equivalence relation on a set S and a partition of S are just two ways to look at the same notion.

4. Prove that the composition of injections is an injection. Is the composition of surjections necessarily a surjection?

5. Prove that the set Σ^* is infinite whenever Σ is not empty.

6. Establish a one-to-one correspondence between the following two sets. S_1 is the set of finite rooted binary trees in which each nonleaf has two children. S_2 is the set of completely balanced strings of parentheses. The set S_2 is defined inductively as follows.

   - The string () belongs to S_2.
   - For any strings $x_1, x_2 \in S_2$, the string $(x_1 x_2)$ belongs to S_2.
   - No other strings belong to S_2.

7. Evaluate the sum
$$S(n) = \sum_{k=1}^{n} k 2^k.$$

8. Prove that the shortest binary numeral for the integer n has length
$$\lceil \log_2(n+1) \rceil = 1 + \lfloor \log_2 n \rfloor.$$

9. Prove Lemma 2.1: For all alphabets Σ and all languages $L \subseteq \Sigma^*$, the equivalence relation $\equiv_L$ is right-invariant.

A.L. Rosenberg, *The Pillars of Computation Theory*, Universitext,
DOI 10.1007/978-0-387-09639-1, © Springer Science+Business Media, LLC 2010

10. Using first addition and then multiplication as the function in (2.3), present ten strings in the language L_g.

11. Give DFAs that accept the following languages over the alphabet $\{0,1\}$.

   a. The set of all strings that end in 00.
   b. the set of all strings that contain the substring 000 somewhere (not necessarily at the end);
   c. the set of all strings that contain the symbols 011, in that order, but not necessarily consecutively.

12. Prove (3.3). As noted in the text, this can be accomplished via an explicit induction on the length of the input string.

13. Prove that the DOA M' constructed in the proof of Theorem 10.1 accepts the same language as the NOA M from which M' is constructed. The annotations in the construction contain hints for the formal proof.

14. Prove that the following languages are not regular.

   a. $L = \{a^n \mid n$ is a power of $2\}$;
   b. $L = \{a^n b^m \mid n \leq m\}$;
   c. The language L_5 of Lemma 6.2.

15. Use the TRIE (digital search tree) data structure to prove Lemma 4.2: Every finite set of words over any finite alphabet Σ (i.e., every finite subset of $\Sigma^\star$) is a regular language.

   See a text on data structures or algorithms, such as [53] or [20], for information on the simple, but useful, TRIE data structure.

16. Prove that there exist three languages $L_1 \subset L_2 \subset L_3$, all over the alphabet $\Sigma = \{a,b\}$, such that:

   - L_1 and L_3 are regular;
   - L_2 is not regular.

   Message: *The properties of subsets and supersets cannot be used to prove regularity or nonregularity.*

17. Prove, by specifying a finite automaton and arguing that it accepts the desired language, that the following languages are regular.

   a.
   $$L_1 = \{x \in \{0,1\}^\star \mid x \text{ has a 1 as the third symbol from its end}\}$$
   $$= \{y1ab \mid x \in \{0,1\}^\star \text{ and } \{a,b\} \subseteq \{0,1\}\}.$$

   b. L_2 is the set of all strings $\langle x,y \rangle \in (\{0,1\} \times \{0,1\})^\star$ that satisfy the following. If you interpret x and y as binary numerals *in reverse order*—so that the leftmost symbols of x and y are the *low-order* bits of the numerals—then the high-order bit of the *sum* of x and y is 1.

   Just as a notational matter: if $\langle x,y \rangle \in (\{0,1\} \times \{0,1\})^\star$, then x and y are equal in length.

18. Prove—using the (fooling set)–(continuation lemma) argument—that the following languages are *not* regular.

   a. L_3 consists of ordered pairs of equal-length binary numerals, in the same manner as L_2 does. A pair $\langle x, y \rangle$ of length-n numerals belongs to L_3 iff the nth bit of the *product* of x and y is 1.
   b. $L_4 \subseteq \{0, 1, 2\}$ consists of strings of the form $x2y$, where x and y are reverse binary numerals (i.e., the left ends of x and y are the low-order ends). A string $x2y$, where x is of length n, belongs to L_4 just when the nth bit of the *sum* of x and y is 1.
   Thus, whereas the language L_2 of the preceding problem represents the addition operation when numerals are presented *in parallel*, L_4 represents the addition operation when numerals are presented *serially*.

19. Prove that the language $\{a^n b^n c^n \mid n \geq 0\}$ is not regular:

   a. using the "ordinary" form of the pumping lemma (Lemma 6.1);
   b. using the "fooling argument" based on the finite-index lemma.

20. Prove that if an n-state FA M accepts any string—i.e., $L(M) \neq \emptyset$—then it accepts a string of length $< n$.
21. In the $[(3) \Rightarrow (1)]$ portion of the proof of the Myhill–Nerode theorem, we need to show that for all strings $x \in \Sigma^*$, $\delta([\varepsilon], x) = [x]$. Prove that these equations hold.
22. Compute the minimum-state FA that is equivalent to the following one

M	q	$\delta(q, 0)$	$\delta(q, 1)$	$q \in F$?
$\rightarrow$	a	b	a	$\notin F$
	b	a	c	$\notin F$
	c	d	b	$\notin F$
	d	d	a	$\in F$
	e	d	f	$\notin F$
	f	g	e	$\notin F$
	g	f	g	$\notin F$
	h	g	d	$\notin F$

23. Consider the (allegedly) minimal-state FA $\widehat{M}$ constructed in Section 5.2 from the FA M.

   a. Prove that $\widehat{M}$ is well defined, i.e., that it is indeed an FA.
   b. Prove that $\widehat{M}$ accepts the same language as does M, i.e., that $L(\widehat{M}) = L(M)$.
   c. Prove that no FA having fewer states than $\widehat{M}$ accepts $L(M)$.

   (Much of this is just summarizing what we prove in the text.)
24. Prove the following assertions, which point out a technically significant difference between DFAs and NFAs.

   a. There is no integer k such that every regular set is accepted by a DFA having $\leq k$ accepting states.

b. Every nonempty regular set is accepted by some NFA that has just one accepting state, *even when NFAs are* not *allowed to have ε-transitions*.

25. How many distinct regular sets over the alphabet $\{a\}$ are accepted by a DFA having n states and one accepting state? You should restrict attention to languages that are *not* accepted by any DFA having fewer than n states.

26. Prove that the 4-state FA below adds binary integers, in the following sense. We view the FA as emitting the bit 1 whenever it enters an accepting state, and as emitting the bit 0 whenever it enters a rejecting state. We then design an FA that on any input string over the 4-symbol alphabet $\{0,1\} \times \{0,1\}$, say

$$\langle \alpha_0, \beta_0 \rangle \langle \alpha_1, \beta_1 \rangle \cdots \langle \alpha_n, \beta_n \rangle,$$

emits the bit-string

$$\gamma_0 \gamma_1 \cdots \gamma_n$$

just when the bit-string $\gamma_n \gamma_{n-1} \cdots \gamma_0$ comprises the low-order $n+1$ bits of the sum of

$$\alpha_n \alpha_{n-1} \cdots \alpha_0 \text{ and } \beta_n \beta_{n-1} \cdots \beta_0.$$

Note the reversed order of the input and output strings of the FA. The adding FA has four states, $\{q_0, q_1, q_2, q_3\}$; q_0 is the start state; q_1 and q_3 are the accepting states. The state-table for the FA is the following:

δ	$\langle 0,0 \rangle$	$\langle 0,1 \rangle$	$\langle 1,0 \rangle$	$\langle 1,1 \rangle$
q_0	q_0	q_1	q_1	q_2
q_1	q_0	q_1	q_1	q_2
q_2	q_1	q_2	q_2	q_3
q_3	q_1	q_2	q_2	q_3

Hint: Interpret the states via outputs and carries.

27. Using the same formal framework as in the preceding problem, prove that finite automata *cannot* multiply. That is, there is no FA that, on every input string

$$\langle \alpha_0, \beta_0 \rangle \langle \alpha_1, \beta_1 \rangle \cdots \langle \alpha_n, \beta_n \rangle \in (\{0,1\} \times \{0,1\})^\star$$

emits the bit-string

$$\gamma_0 \gamma_1 \cdots \gamma_n$$

that comprises (in reverse order) the low-order $n+1$ bits of the *product* of

$$\alpha_n \alpha_{n-1} \cdots \alpha_0 \text{ and } \beta_n \beta_{n-1} \cdots \beta_0.$$

Hint: Note that we allow leading 0's in the input.

28. Prove that the following language *is* regular:

$$\left\{ a^\ell b^m c^n \mid m \text{ can be written as } m = h+k \text{ for some } h \neq \ell \text{ and } k \neq n \right\}.$$

29. Consider two FAs, M_m and M_n, that share the one-letter input alphabet $\{a\}$ and that accept, respectively, the set of all strings whose length is divisible by m and the set of all strings whose length is divisible by n.

    a. Use the direct-product construction on M_m and M_n to produce a DFA that accepts the set of all strings whose length is divisible by the least common multiple of m and n.

    b. Prove that when m and n are *relatively prime*—i.e., have unit greatest common divisor—then the DFA you just produced from M_m and M_n is the *smallest* DFA (in number of states) that accepts this set.

30. Explain why there is no analogue of the Myhill–Nerode theorem for NFAs. In particular, why can one *not* identify the "states" of an NFA as the classes of an equivalence relation? Your answer should be short and to the point, not a rambling essay.

    (The answer is not abstruse. Look at some sample NFAs, and *think* about what the various terms in this question mean.)

31. Prove that every ε-free NFA M for which $\varepsilon \notin L(M)$ can be converted to an equivalent ε-free NFA M' that has only one accepting state. (It is easy to prove that this is *not* possible in general for DFAs.)

32. Prove that an ε-free NFA that is allowed to have multiple start states can be converted to an equivalent ε-free NFA M' that has only one start state.

33. We showed in Section 11.1.2 that there exist languages whose smallest accepting DFA is exponentially larger than their smallest accepting NFA. What can you say in this regard about languages over a 1-letter alphapet? That is, if M is an NFA with input alphabet $\{a\}$, what can you say about the size of the smallest DFA that accepts $L(M)$?

34. Recall the NFA M that accepts all strings over $\{a, b\}$ whose penultimate letter is a. When we converted M to an equivalent DFA M', we generated the states of M' in a *lazy* manner—since only states accessible from $\{q_0\}$ are of interest. List the states that we did *not* generate.

35. We have shown that the language $L = \{a^k \mid k$ is a perfect square$\}$ is not regular. It follows that the Myhill–Nerode relation $\equiv_L$ has infinitely many classes. Describe all of the classes.

36. Prove whether each of the following sets is regular.

    a. $\{x \in \{0,1\}^* \mid \ell(x)$ is the sum of two perfect squares$\}$
       (**Hint:** Consult any standard number theory book to see which integers are sums of so-and-so many squares.)
    b. $\{x \in \{0,1\}^* \mid x$ begins with a palindrome of length $\geq 3\}$.
    c. $\{x \in \{0,1\}^* \mid x$ contains a subword that is a palindrome$\}$.
    d. $\{a^m b^n \mid m = n^2\}$.
    e. $\{a^m b^n \mid m \equiv n^2 \pmod 7\}$.

37. Consider the following family of finite (hence, regular) languages: $\mathscr{L}^{\neq} = \{L_i^{\neq} \mid i \in \mathbb{N}\}$. For each $n \in \mathbb{N}$,

$$L_n^{\neq} = \{xy \mid x,y \in \{0,1\}^n \text{ and } x \neq y\}.$$

Prove the following assertions.

a. For all n, there is an NFA M, having $O(n^2)$ states, such that $L(M) = L_n^{\neq}$.
   (You may describe M in English, but be sure that you describe in detail the states of M and the transitions among them.)
b. Any *deterministic* FA that accepts $L_n^{\neq}$ must have at least 2^n states.

38. Contrast the family $\mathscr{L}^{\neq}$ of the preceding problem with the family of finite (hence, regular) sets $\mathscr{L}^{=} = \{L_i^{=} \mid i \in \mathbb{N}\}$, where for each $n \in \mathbb{N}$,

$$L_n^{=} = \{xy \mid x,y \in \{0,1\}^n \text{ and } x = y\}.$$

Prove that any NFA that accepts $L_n^{=}$ must have at least 2^n states.

39. Present algorithms that decide, of given FAs M_1 and M_2 that share the input alphabet Σ, whether:

a. $L(M_1) = \emptyset$;
b. $L(M_1) = \Sigma^{\star}$;
c. $L(M_1) = L(M_2)$;
d. $L(M_1) \subseteq L(M_2)$.

40. Prove that the following language L_1 *is not* context free, but that language L_2 *is* context free:

$$L_1 = \{xy \in \{0,1\}^{\star} \mid x = y\};$$
$$L_2 = \{xy \in \{0,1\}^{\star} \mid \ell(x) = \ell(y) \text{ and } x \neq y\}$$

41. Use pumping arguments to prove the following assertions.

a. The following language of *perfect squares under concatenation* is not context-free:

$$L = \{xx \mid x \in \{a,b\}^{\star}\}.$$

   Each string in L consists of a string of a's and b's, followed by an identical copy of the same string.
b. The set of strings over the alphabet $\{a\}$ whose length is a power of 2 is not regular.
c. The set of strings over the alphabet $\{a\}$ whose length is a power of 2 is not context-free.
d. The following language is not regular:

$$L = \{a^i b^j c^k \mid i = j \text{ or } j = k\}.$$

42. Prove Theorem 6.3.
43. You are given two DFAs, M_1 and M_2. Present—and justify—an algorithm that decides whether $L(M_1) = L(M_2)$.
    Do this problem in the following two distinct ways.

a. Use the Myhill–Nerode theorem to argue that there is a unique smallest DFA equivalent to each DFA M_i. Then use the preceding fact to craft an algorithm that decides whether $L(M_1) = L(M_2)$.

   Remark. The "uniqueness" of the smallest DFA is only up to possible renamings of the states. Your algorithm should take this into account.

b. Use the fact that the regular languages are closed under the Boolean operations to reduce the equivalence problem to an easier decision problem that is phrased in terms of such operations.

44. Use the direct-product DFA $M_{1,2}^{\otimes}$ from the proof of the closure of regular languages under Boolean operations to *prove* that the regular languages are closed under the set-theoretic operation of *symmetric difference* $\oplus$ defined as follows. $L_1 \oplus L_2$ is the set of all strings that belong *either* to L_1 *or* to L_2, but *not to both*.

45. Present—and justify—an algorithm that decides of a given FA $M = (Q, \Sigma, \delta, q_0, F)$ whether $L(M) = \Sigma^*$. Your algorithm should run in time $O(|Q| \times |\Sigma|)$.

46. Repeat our lower-bound argument for the database language of Section 5.5, for an OTM that has one *two-dimensional* tape (with move repertoire UP, DOWN, LEFT, RIGHT, NO-MOVE).

47. Let L be the following "unary" version of the database language of Section 5.5. Each $w \in L$ has the form

$$w = 0^{n_1} 10^{n_2} 1 \cdots 10^{n_r} 20^{m_1} 10^{m_2} 1 \cdots 10^{m_s},$$

where

- each $m_i, n_i \geq 1$,
- $m_s \in \{n_1, n_2, \cdots, n_r\}$.

Find (and verify) upper and lower bounds on the time required for an online TM with one linear worktape to accept L. For full credit, your bounds should be only a constant factor apart.

48. Prove Lemma 7.1. The proof builds directly on our definition of "$\leq$" in terms of injections.

49. This problem considers *pairing functions*, i.e., bijections

$$f(x,y) = \mathbb{N} \times \mathbb{N} \to \mathbb{N},$$

that are *additive*, in the following sense. For each $x \in \mathbb{N}$, there exists a positive integer "stride" s_x such that for all $y \in \mathbb{N}$,

$$f(x, y+1) = f(x,y) + s_x.$$

Prove that there *do not* exist additive pairing functions for which all of the "strides" s_x are equal.

50. Prove that the set of square matrices whose entries are positive integers is countable.

51. Prove that the uncountability of the set of 0-1-valued functions, $\{f : \mathbb{N} \to \{0,1\}\}$, implies the uncountability of the set of integer-valued functions, $\{f : \mathbb{N} \to \mathbb{N}\}$.

52. Prove Corollary 7.2. You may want to attempt this by focusing first on the following simpler version. of the problem.
    Prove that there are functions $f : \mathbb{N} \to \mathbb{N}$ that are not computed by any C program, even if one can employ arbitrarily long numerals in the program.

53. Say that a language L is *recursively enumerable* if there is a surjective total recursive function $f : \mathbb{N} \to L$. (Note that $f(\mathbb{N}) \subseteq L$.) Prove that a language L is recursively enumerable if and only if it is semidecidable. (One direction is rather challenging.)

54. Prove Theorem 9.4. You should frame your response in terms of some programming language (of your choice), indicating how the theorem alters a given program and arguing that after this alteration, the old and new programs will compute the same values.

55. Say that the language $L \subseteq \Sigma^*$ is *enumerable in increasing order*, in the following sense. There is a surjective total recursive function $f : \mathbb{N} \to L$ such that for all integers i, $f(i+1) > f(i)$ when the strings $f(k)$ are interpreted as numerals (in base $|\Sigma|$). (Note that $f(\mathbb{N}) \subseteq L$.) Prove that L is decidable.

56. Prove that every *infinite* semidecidable language L contains an infinite decidable subset.

57. The relation "is mapping-reducible to" (symbolically, $\leq_m$) satisfies the following. For any sets A and B, $[A \leq_m B]$ iff $[\bar{A} \leq_m \bar{B}]$.

58. If A is decidable and $B \notin \{\emptyset, \mathbb{N}\}$, then $A \leq_m B$.
    (Intuitively, a set that needs no "help" is "helped" by any set.)

59. Prove the following theorem. *Every nontrivial property of functions that contains a program for the empty function—i.e., a program that never halts on any input— is not semidecidable.*

60. Prove that the following set is *not semidecidable*: $\{\langle x,y \rangle \mid \text{program } x \equiv \text{program } y\}$. The defining condition here is that programs x and y compute the same function.

61. Prove that for every finite alphabet Σ, there is a *computable injection* $F : \Sigma^* \times \Sigma^* \to \Sigma^*$.

62. State—with a justifying proof—whether each of the following sets is semidecidable, decidable, or neither.

    a. The set S_1 of programs that halt when started with the null string as input.
    b. The set S_2 of programs that halt when the input is the *reversal* of their descriptions.
    c. The set S_3 of strings that are (halting) *computations* by a Turing machine (TM).
    d. The set $S_4 = \{\langle x,y \rangle \mid x \text{ and } y \text{ are equivalent finite automata}\}$.
    e. The set $S_5 = \{\langle w,x,y,z \rangle\}$ such that:
        • y is the inital configuration of program w on input x;
        • z is a terminal (i.e., halting) configuration of program w when w is presented with input x.
       Consider carefully the implications of any ordered quadruple $\langle w,x,y,z \rangle$'s membership in S_5.

63. Classify—with proofs—the following languages as being (*i*) decidable or (*ii*) semidecidable but not decidable or (*iii*) not semidecidable.

   a. The language $\{x \mid \text{program } x \text{ never halts on any input}\}$.
   b. The language $\{x \mid \text{program } x \text{ halts on at least one input}\}$.
   c. The language $\{x \mid \text{program } x \text{ halts on infinitely many inputs}\}$.
      Hint. Prove that one can reduce $\overline{\text{DHP}}$ to L.

64. Classify—with proofs—the following languages as being (*i*) decidable or (*ii*) semidecidable but not decidable or (*iii*) not semidecidable.

   a. The language $\{\langle x, y \rangle \mid \text{programs } x \text{ and } y \text{ halt on precisely the same inputs}\}$.
      Hint. Can you think of any specific programs y for which this problem is undecidable?
   b. The language $\{x \mid \text{program } x \text{ halts on the blank input tape}\}$.

65. We noted in Corollary 7.2 that one could infer the existence of noncomputable functions $f : \mathbb{N} \to \{0, 1\}$ from the fact that the class **F** of such functions is uncountable, while the set **P** of programs in any programming language is countable. Present a *standalone (direct)* proof of the existence of noncomputable such functions. *What you need to present is a formal proof that there is no injection from* **F** *into* **P**.

66. Prove that the "complementary halting problem" $\overline{\text{HP}}$ is not semidecidable.

67. Prove the following theorem.
   Say that the language A is mapping-complete (= m-complete), and say that the language B is decidable. Then it is not possible that $A \leq_m B$.
   You should supply a complete proof of this fact, from definitions and first principles. For instance, it is *not* adequate for you to invoke vague intuitive principles such as "Good things travel up; bad things travel down."

68. a. Present an infinite family **S** of sets such that $S \leq \bar{S}$ for each $S \in$ **S**.
   b. Is every semidecidable set m-reducible to TOT?
   c. Prove that TOT is *productive*, in the following sense. There is a total computable function f with the following property. For every program x whose *domain* is a subset of TOT, we have

   $$f(x) \in \text{TOT} \setminus \text{domain}(\text{program } x).$$

   (This problem is not as hard as it looks if you get beyond the formalism.)

69. We remarked in Section 13.3.4 that there exists an efficient algorithm for *the subset-sum problem*. Recall that each instance of this problem consists of n positive integers, $m_1, m_2, \ldots, m_n$, plus a *target integer* t. Given such an instance, one must decide whether there is a subset of the integers m_i that sum to t. The efficient algorithm $\mathscr{A}$ makes this decision concerning the instance $\langle m_1, m_2, \ldots, m_n; t \rangle$ within $O(nt)$ steps.
   Describe and validate an algorithm that solves the subset-sum problem with efficiency $O(nt)$.

70. Prove that the following sets are in **NP**:

     a. The set of binary representations of composite (i.e., nonprime) integers.

     b. The set of state-transition tables of DFAs that accept at least one string.

71. Prove: The relation "is polynomial-reducible to" ($\leq_{poly}$) is reflexive and transitive.

72. As usual, call a set A *nontrivial* if $A \notin \{\emptyset, \mathbb{N}\}$. *Prove:* If there is a nontrivial set A that is not **NP**-Hard, then $\mathbf{P} \neq \mathbf{NP}$.

    Hint: Prove and then use the following fact: If $A \in \mathbf{P}$ and B is nontrivial, then $A \leq_{poly} B$. (This fact is a complexity-theoretic instance of the intuition that a set that needs no "help" is "helped" by any set.)

73. As usual, let *co*-**NP** denote the class of languages whose complements are in **NP**. *Prove:* If *co*-**NP** contains an **NP**-Complete set, then $\mathbf{NP} = co\text{-}\mathbf{NP}$.

74. *Prove:* For any sets A and B, $[A \leq_{poly} B]$ if and only if $[\overline{A} \leq_{poly} \overline{B}]$.

75. Prove that the language HP is **NP**-Hard.

76. a. What is the "size" (in the complexity-theoretic sense of the term) of a 3-CNF formula that has n clauses and m literals?

     b. What is the size of a graph that has n vertices and m edges? (There are at least two correct answers.)

    Hint. How would you design a TM that could process *any* 3-CNF formula or *any* graph? The only "tough" issue is how to represent "names." You should use the most compact (to within constant factors) fixed-length encoding of "names."

77. *Prove:* For any pair of languages A and B such that neither B nor $\overline{B}$ is empty: If $A \in \mathbf{P}$, then $A \leq_{poly} B$.

78. *Prove:* A language L is accepted by an IRTM that operates in space $S(n) = O(1)$ if and only if L is regular.

79. *Prove:* The following space functions are constructible. (You may describe your construction algorithms via English-language descriptions.)

$$S(n) = \lfloor \log_2 n \rfloor;$$
$$S(n) = n^2;$$
$$S(n) = 2^n.$$

80. Let $t(n) \geq n$ be any nondecreasing integer function. Prove that any *offline* TM M that operates in time $t(n)$ can be simulated by an *online* TM M' that has the same worktape repertoire (number and structure) as M and that operates in time $t'(n) \leq t^2(n)$.

81. Let φ be a propositional formula in CNF. Let us be given an arbitrary assignment of truth values to the propositional variables in φ. Prove that one can (deterministically) determine in time *linear* in the number of symbols in formula φ whether this assignment is a satifying one for φ, i.e., causes φ to evaluate to 1.

    Try to solve this problem using a single stack to "evaluate" formula φ.

References

1. F.E. Allen and J. Cocke (1976): A program data flow analysis procedure. *Comm. ACM 19* 137–147.
2. D.N. Arden (1960): *Theory of Computing Machine Design.* Univ. Michigan Press, Ann Arbor, pp. 1–35.
3. E.T. Bell (1986): *Men of Mathematics.* Simon and Schuster, New York.
4. F. Bernstein (1905): Untersuchungen aus der Mengenlehre. *Math. Ann. 61*, 117–155.
5. G. Birkhoff and S. Mac Lane (1953): *A Survey of Modern Algebra*, Macmillan, New York.
6. E. Bishop (1967): *Foundations of Constructive Analysis*, McGraw Hill, New York.
7. W.W. Boone, F.B. Cannonito, R.C. Lyndon (1973): *Word Problems: Decision Problem in Group Theory*, North-Holland, Amsterdam.
8. R. Buyya, D. Abramson, J. Giddy (2001): A case for economy Grid architecture for service oriented Grid computing. *10th Heterogeneous Computing Wkshp.*
9. G. Cantor (1874): Über eine Eigenschaft des Inbegriffes aller reellen algebraischen Zahlen. *J. Reine und Angew. Math. 77*, 258–262.
10. G. Cantor (1878): Ein Beitrag zur Begründung der transfiniter Mengenlehre. *J. Reine Angew. Math. 84*, 242–258.
11. A.L. Cauchy (1821): *Cours d'analyse de l'École Royale Polytechnique, 1ère partie: Analyse algébrique.* l'Imprimerie Royale, Paris. Reprinted: Wissenschaftliche Buchgesellschaft, Darmstadt, 1968.
12. N. Chomsky (1956): Three models for the description of language. *IRE Trans. Information Theory 2*, 113–124.
13. N. Chomsky (1959): On certain formal properties of grammars. *Inform. Contr. 2*, 137–167.
14. A. Church (1941): *The Calculi of Lambda-Conversion. Annals of Math. Studies 6*, Princeton Univ. Press, Princeton, NJ.
15. A. Church (1944): *Introduction to Mathematical Logic, Part I. Annals of Math. Studies 13*, Princeton Univ. Press, Princeton, NJ.
16. W. Cirne and K. Marzullo (1999): The Computational Co-Op: gathering clusters into a metacomputer. *13th Intl. Parallel Processing Symp.*, 160–166.
17. E.F. Codd (1970): *A relational model of data for large shared data banks. Comm. ACM 13*, 377–387.
18. S.A. Cook (1971): The complexity of theorem-proving procedures. *ACM Symp. on Theory of Computing*, 151–158.
19. I.M. Copi, C.C. Elgot, J.B. Wright (1958): Realization of events by logical nets. *J. ACM 5*, 181–196.
20. T.H. Cormen, C.E. Leiserson, R.L. Rivest, C. Stein (2001): *Introduction to Algorithms (2nd ed.).* MIT Press, Cambridge, MA.
21. H.B. Curry (1934): Some properties of equality and implication in combinatory logic. *Annals of Mathematics, 35*, 849–850.

22. H.B. Curry, R. Feys, W. Craig (1958): *Combinatory Logic. Studies in logic and the foundations of mathematics*. North-Holland, Amsterdam.

23. M. Davis (1958): *Computability and Unsolvability*. McGraw-Hill, New York.

24. J.R. Douceur (2002): The Sybil attack. *1st Intl. Wkshp. on Peer-to-Peer Systems*.

25. R.W. Floyd (1962): Algorithm 97: Shortest Path. *Comm. ACM 5*, 345.

26. R.W. Floyd (1967): Assigning meanings to programs. *Proc. Symposia in Applied Mathematics 19*, 19–32.

27. R. Fueter and G. Pólya (1923): Rationale Abzählung der Gitterpunkte. *Vierteljschr. Natur-forsch. Ges. Zürich 58*, 380–386.

28. M.R. Garey and D.S. Johnson (1979): *Computers and Intractability*. W.H. Freeman and Co., San Francisco.

29. A. Gilat (2004): *MATLAB: An Introduction with Applications (2nd ed.)*. J. Wiley & Sons, New York.

30. K. Gödel (1931): Über Formal Unentscheidbare Sätze der Principia Mathematica und Ver-wandter Systeme, I. *Monatshefte für Mathematik u. Physik 38*, 173–198.

31. B. Goldberg (1996): Functional programming languages, *ACM Computing Surveys 28*, 249–251.

32. O. Goldreich (2006): On teaching the basics of complexity theory. In *Theoretical Computer Science: Essays in Memory of Shimon Even. Springer Festschrift series, Lecture Notes in Computer Science 3895*, Springer, Heidelberg.

33. G.H. Golub, C.F. Van Loan (1996): *Matrix Computations* (3rd ed.) Johns Hopkins Press, Baltimore.

34. P.R. Halmos (1960): *Naive Set Theory*. D. Van Nostrand, New York.

35. D. Harel (1987): *Algorithmics: The Spirit of Computing*. Addison-Wesley, Reading, MA.

36. J. Hartmanis and R.E. Stearns (1966): *Algebraic Structure Theory of Sequential Machines*. Prentice Hall, Englewood Cliffs, NJ.

37. L.S. Heath, F.T. Leighton A.L. Rosenberg (1992): "Comparing queues and stacks as mechanisms for laying out graphs." *SIAM J. Discr. Math. 5*, 398–412.

38. F.C. Hennie (1966): On-line Turing machine computations. *IEEE Trans. Electronic Computers, EC-15*, 35–44.

39. F.C. Hennie and R.E. Stearns (1966): Two-tape simulation of multitape Turing machines. *J. ACM 13*, 533–546.

40. S. Homer and A.L. Selman (2001): *Computability and Complexity Theory*. Springer, New York

41. J.E. Hopcroft, R. Motwani, J.D. Ullman (2001): *Introduction to Automata Theory, Languages, and Computation* (2nd ed.). Addison-Wesley, Reading, MA.

42. J.E. Hopcroft and J.D. Ullman (1979): *Introduction to Automata Theory, Languages, and Computation* (1st ed.) Addison-Wesley, Reading, MA.

43. N. Immerman (1988): Nondeterministic space is closed under complementation. *SIAM J. Comput. 17*, 935–938.

44. *The Intel Philanthropic Peer-to-Peer program.* ⟨www.intel.com/cure⟩.

45. K.E. Iverson (1962): *A Programming Language*. J. Wiley & Sons, New York.

46. J. Jaffe (1978): A necessary and sufficient pumping lemma for regular languages. *SIGACT News*, 48–49.

47. R.M. Karp (1967): Some bounds on the storage requirements of sequential machines and Turing machines. *J. ACM 14*, 478–489.

48. R.M. Karp (1972): Reducibility among combinatorial problems. In *Complexity of Computer Computations* (R.E. Miller and J.W. Thatcher, eds.) Plenum Press, NY, pp. 85–103.

49. S.C. Kleene (1936): General recursive functions of natural numbers. *Math. Annalen 112*, 727–742.

50. S.C. Kleene (1952): *Introduction to Metamathematics*. D. Van Nostrand, Princeton, NJ.

51. S.C. Kleene (1956): Realization of events in nerve nets and finite automata. In *Automata Studies* (C.E. Shannon and J. McCarthy, Eds.) *[Ann. Math. Studies 34]*, Princeton Univ. Press, Princeton, NJ, pp. 3–42.

52. D. König (1936): *Theorie der endlichen und unendlichen Graphen.* Lipzig: Akad. Verlag.
53. D.E. Knuth (1973): *The Art of Computer Programming: Fundamental Algorithms* (2nd ed.) Addison-Wesley, Reading, MA.
54. E. Korpela, D. Werthimer, D. Anderson, J. Cobb, M. Lebofsky (2000): SETI@home: massively distributed computing for SETI. In *Computing in Science and Engineering* (P.F. Dubois, Ed.) IEEE Computer Soc. Press, Los Alamitos, CA.
55. P.J Landin (1964): The mechanical evaluation of expressions. *Computer J. 6*, 308–320.
56. L. Levin (1973): Universal search problems. *Problemy Peredachi Informatsii 9*, 265–266. Translated in, B.A. Trakhtenbrot (1984): A survey of Russian approaches to perebor (brute-force search) algorithms. *Annals of the History of Computing 6*, 384–400.
57. J.S. Lew and A.L. Rosenberg (1978): Polynomial indexing of integer lattices, I. *J. Number Th. 10*, 192–214.
58. J.S. Lew and A.L. Rosenberg (1978): Polynomial indexing of integer lattices, II. *J. Number Th. 10*, 215–243.
59. H.R. Lewis and C.H. Papadimitriou (1981): *Elements of the Theory of Computation.* Prentice-Hall, Englewood Cliffs, NJ.
60. P. Linz (2001): *An Introduction to Formal Languages and Automata* (3rd ed.) Jones and Bartlett Publ., Sudbury, MA.
61. A.A. Markov (1949): On the representation of recursive functions (in Russian). *Izvestiya Akademii Nauk S.S.S.R. 13*. English translation: Translation 54, Amer. Math. Soc., 1951.
62. W.S. McCulloch and W.H Pitts (1943): A logical calculus of the ideas immanent in nervous activity. *Bull. Mathematical Biophysics 5*, 115–133.
63. R. McNaughton and H. Yamada (1964): Regular expressions and state graphs for automata. In *Sequential Machines: Selected Papers* (E.F. Moore, ed.) Addison-Wesley, Reading, MA, PP. 157–176.
64. G.H. Mealy (1955): A method for synthesizing sequential circuits. *Bell Syst. Tech. J. 34*, 1045–1079.
65. K. Mehlhorn (1984): *Data Structures and Algorithms 2: Graph Algorithms and NP-Completeness.* Springer-Verlag, Berlin.
66. C.F. Miller (1991): Decision problems for groups – survey and reflections. In *Algorithms and Classification in Combinatorial Group Theory*, Springer, New York, pp. 1–60.
67. M. Minsky (1967): *Computation: Finite and Infinite Machines.* Prentice-Hall, Inc., Englewood Cliffs, NJ.
68. E.F. Moore (1956): Gendanken experiments on sequential machines. In *Automata Studies* (C.E. Shannon and J. McCarthy, eds.) *[Ann. Math. Studies 34]*, Princeton Univ. Press, Princeton, NJ, pp. 129–153.
69. B.M. Moret (1997): *The Theory of Computation.* Addison-Wesley, Reading, MA.
70. J. Myhill (1957): Finite automata and the representation of events. WADD TR-57-624, Wright Patterson AFB, Ohio, pp. 112–137.
71. T. Naur (2006): Letter to the Editor. *Comm. ACM 49*, 13.
72. A. Nerode (1958): Linear automaton transformations. *Proc. AMS 9*, 541–544.
73. I. Niven and H.S. Zuckerman (1980): *An Introduction to the Theory of Numbers.* (4th ed.) J. Wiley & Sons, New York.
74. *The Olson Laboratory Fight AIDS@Home project.* ⟨www.fightaidsathome.org⟩.
75. G.H. Ott, N.H. Feinstein (1961): Design of sequential machines from their regular expressions. *J. ACM 8*, 585–600.
76. C.H. Papadimitriou (1994): *Computational Complexity.* Addison-Wesley, Reading, MA.
77. M.O. Rabin (1963): Probabilistic automata. *Inform. Control 6*, 230–245.
78. M.O. Rabin (1964): The word problem for groups. *J. Symbolic Logic 29*, 205–206.
79. M.O. Rabin and D. Scott (1959): Finite automata and their decision problems. *IBM J. Res. Develop. 3*, 114–125.
80. H. Rogers, Jr. (1967): *Theory of Recursive Functions and Effective Computability.* McGraw-Hill, New York. Reprinted in 1987 by MIT Press, Cambridge, MA.
81. A.L. Rosenberg (1971): Data graphs and addressing schemes. *J. CSS 5*, 193–238.

82. A.L. Rosenberg (1974): Allocating storage for extendible arrays. *J. ACM 21*, 652–670.

83. A.L. Rosenberg (1975): Managing storage for extendible arrays. *SIAM J. Comput. 4*, 287–306.

84. A.L. Rosenberg (2003): Accountable Web-computing. *IEEE Trans. Parallel and Distr. Systs. 14*, to appear.

85. A.L. Rosenberg (2003): Efficient pairing functions—and why you should care. *Intl. J. Foundations of Computer Science 14*, 3–17.

86. A.L. Rosenberg (2006): State. In *Theoretical Computer Science: Essays in Memory of Shimon Even* (O. Goldreich, A.L. Rosenberg, A. Selman, eds.) *Springer Festschrift series, Lecture Notes in Computer Science 3895*, Springer, Heidelberg, pp. 375–398.

87. A.L. Rosenberg and L.S. Heath (2001): *Graph Separators, with Applications*. Kluwer Academic/Plenum Publishers, New York.

88. A.L. Rosenberg and L.J. Stockmeyer (1977): Hashing schemes for extendible arrays. *J. ACM 24*, 199–221.

89. J.B. Rosser (1953): *Logic for Mathematicians*. McGraw-Hill, New York.

90. *The RSA Factoring by Web Project.* ⟨http://www.npac.syr.edu/factoring⟩ (with Foreword by A. Lenstra). Northeast Parallel Architecture Center.

91. B. Russell, Bertrand (1903). Principles of Mathematics. Cambridge: Cambridge University Press.

92. W. Savitch (1969): Deterministic simulation of non-deterministic Turing machines. *1st ACM Symp. on Theory of Computing*, 247–248.

93. M. Schönfinkel (1924): Über die Bausteine der mathematischen Logik. *Math. Annalen 92*, 305–316.

94. A. Schönhage (1980): Storage modification machines. *SIAM J. Computing 9*, 490–508.

95. E. Schröder (1898): Ueber zwei Definitionen der Endlichkeit und G. Cantor'sche Sätze. *Nova Acta Academiae Caesareae Leopoldino-Carolinae (Halle a.d. Saale) 71*, 303–362.

96. E. Schröder (1898): Die selbständige Definition der Mächtigkeiten 0, 1, 2, 3 und die explicite Gleichzahligkeitsbedingung. *Nova Acta Academiae Caesareae Leopoldino-Carolinae (Halle a.d. Saale) 71*, 365–376.

97. M. Sipser (1997): *Introduction to the Theory of Computation*. PWS Publishing, Boston, MA.

98. R.E. Stearns, J. Hartmanis, P.M. Lewis, II (1972): Hierarchies of memory limited computations. *J. Symbolic Logic 37*, 624–625.

99. L.J. Stockmeyer (1973): Extendible array realizations with additive traversal. IBM Research Report RC-4578.

100. R. Szelepcsényi (1987): The method of forcing for nondeterministic automata. *Bull. EATCS 33*, 96–100.

101. R.E. Tarjan (1972): Sorting using networks of queues and stacks. *J. ACM 19*, 341–346.

102. M. Taufer, D. Anderson, P. Cicotti, C.L. Brooks (2005): Homogeneous redundancy: A technique to ensure integrity of molecular simulation results using public computing. *19th Intl. Parallel and Distributed Processing Symp.*

103. R.P. Tewarson (1973): *Sparse Matrices*. In *Mathematics in Science & Engineering*. Academic Press, New York.

104. A.M. Turing (1936): On computable numbers, with an application to the Entscheidungsproblem. *Proc. London Math. Soc.* (ser. 2, vol. 42) 230–265; Correction *ibid.* (vol. 43) 544–546.

105. S. Warshall (1962): A theorem on Boolean matrices. *J. ACM 9*, 11–12.

106. Wikipedia: The Free Encyclopedia (2005):
    http://en.wikipedia.org/wiki/Pumping_lemma

107. R. Zach (1999): Completeness before Post: Bernays, Hilbert, and the development of propositional logic. *Bull. Symbolic Logic 5*, 331–366.

Index

$(u \overset{d}{\Rightarrow} v)$: depth-$d$ ancestry in a rooted tree, 290

$(u \Rightarrow v)$: ancestry in a rooted tree, 26

$(u \rightarrow v)$: arc in a digraph, 25

2-D OTM, 199

$E(q)$: ε-reachability set of state q, 222

$L(M)$: language accepted by M, 36

$L(M, \theta)$, 72

ONE: a program that computes the constant function $f(n) \equiv 1$, 159, 173

$S \cap T$: set intersection, 15

$S \cup T$: set union, 15

$S \setminus T$: set difference, 15

$S \times T$, 15

$S - T$: set difference, 15

Γ: a OTM's working alphabet, 84

Γ: a TM's working alphabet, 45

$\mathbb{N}$: set of nonnegative integers, 13

$\mathbb{N}^+$: set of positive integers, 13

$\Sigma^\star$: all finite-length strings over Σ, 22

Σ^k: all length-k strings, 22

$\mathbb{Z}$: set of all integers, 13

$\mathscr{A}_M(n)$, 79

$\ell(w)$: length of word w, 23

$\emptyset$: the empty set, 14

$\overset{\text{def}}{=}$: "is defined to be", 18

$\equiv$: an equivalence relation, 19

$\equiv_M^{(t)}$, 83

$\equiv_L$, 24, 63–66, 79

$\equiv_M$, 36, 64, 66, 79, 83

$\equiv_\delta$, 66, 67

κ'_L: semicharacteristic function of language L, 24

κ_L: characteristic function of language L, 24

$\mathscr{L}(\mathscr{R})$: language denoted by regular expression $\mathscr{R}$, 224

$\lceil x \rceil$: the ceiling of number x, 27

$\leq_R$: m-reducible within resource R, 264

$\leq_m$: "is m-reducible to", 161

$\leq_{\text{poly}}$: poly-time reducible to, 264

$\lfloor x \rfloor$: the floor of number x, 27

$\log(x)$: base-2 logarithm of x, 28

$\log_b(x)$: base-b logarithm of x, 27

$\ominus$: positive subtraction, 128

$\overline{T}$: set complement, 16

$\overline{X}$: logical not, 17

$\mathscr{P}(S)$: the power set of S, 14, 212, 218

π: the ratio of the circumference of a circle to its diameter, 59, 173

$\mathscr{T}_M(x)$: nondeterministic computation tree, 287

ε: the null string, 22

ε-NFA, 222

ε-nondeterministic finite automaton, 222

ε-reachability set, 222

ε-transition, 222

ε-free language, 106

$\vee$: logical or, 17

$\wedge$: logical and, 17

$\hat{\delta}$: the function δ extended to strings, 35

$\widehat{\mathscr{T}}_M(x)$: truncated nondeterministic computation tree, 289

$\{0, 1\}^\star$: set of all finite-length binary strings, 14

e: the base of the natural logarithm, 59

k-tape OTM, 195

$o(1)$, 29

r-equivalent OTM configurations, 87

r-indistinguishable words, 87

$\text{HP}^{(\text{poly})}$: the poly-time halting problem, 270

NP-completeness of, 271

$\boxed{\text{B}}$: the blank symbol, 45, 84

"–adic" positional number system, 117

"–ary" positional number system, 117

"Milk and Honey", 31

"eventual" bound, 79

"infinitely often" bound, 79
"polynomial-time" as a synonym for "efficient",
    254
"reasonable" notion of digital computer, 148
"standard" universal sets: $\mathbb{N}$, $\mathbb{N}^+$, $\{0,1\}^\star$, 153
"universal" bound, 79
"universal" set U, 14, 153
NP-Complete, 269
NP-Hard, 269
NP-vs.-coNP problem, 297
NP: nondeterministic polynomial-time
    languages, 254
NP: nondeterministic polynomial-time
    languages, 250
P-vs.-NP problem, 7, 216, 250, 254
P: polynomial-time languages, 250, 255
and (logical), 17
nand: not-and, 8
not (logical), 17
or (logical), 17
A-POP: the register machine analogue of the
    POP operation on a stack, 191
A-PUSH: the register machine analogue of the
    PUSH operation on a stack, 191
PUSH: add a symbol to a stack, 179
PUSH: remove a symbol from a stack, 179
xor: exclusive or, 266
NPSPACE, 297

a database problem as a language, 83, 86
acceptance bound, 287
acceptance bounding function, 287
acceptance threshold: θ, 72
accepted language
    FA, 36
    NFA, 212
    NOA, 212
    NTM, 235
    OA, 36
accepting states
    OA, FA, 34
accountability in volunteer computing, 140
acyclic digraph, 26
adjacency matrix (of a graph), 267
Allen, Frances E., 52
alphabet, 22, 34
an information-retrieval problem as a language,
    83, 86
ancestor (node) in a rooted tree, 26
Anderson, David, 126, 139
APF: additive pairing function, 140
arc (of a digraph), 25
array-storage mappings
    column-major, 131

dimension-order, 6, 131
    row-major, 131
associative (operation), 23
asymptotic notation, 28

back-subsitution in a linear system, 108
base-b logarithm, 27
behavior
    FA, 36
    OA, 36
behavior of an OA, an FA, a TM, 35
Bernstein, Felix, 115
big-Ω notation, 28
big-Θ notation, 28
big-O notation, 28
bijection, 22
bijective function, 21, 22
binary relation, 17
binary strings: strings of 0's and 1's, 14
Bishop, Erret, 174
bistable device, 46, 52, 79
bit string, 14
bit: binary digit, 14
bivariate function, 154
blank symbol, $\boxed{B}$, 84
block (of a partition), 19
Boole, George, 16
Boolean minimization problem, 246
Boolean operations
    logical analogues, 16, 246
    on sets/languages, 16, 103, 246

Cantor's diagonalization argument, 121
Cantor, Georg, 5, 112, 113, 119, 148
cardinality of $\mathscr{P}(S)$, 14
cardinality of a finite set, 14
carry-ripple adder, 51, 53
Cauchy, Augustin, 119
ceiling (of a number), 27
ceiling of number x: $\lceil x \rceil$, 27
CFG: context-free grammar, 96
CFL: context-free language, 98
characteristic function, 24
    and (un)decidability, 151
characteristic vector, 15, 117, 122
child (node) in a rooted tree, 26
Chomsky, Noam, 51, 96, 111, 149, 215
Church, Alonzo, 149
Church–Turing thesis, 149, 172, 174, 233
clique (in a graph), 266
CLIQUE problem, 266
closure (under an algebraic operation), 17
closure of regular languages
    under union, 225

closure of space-bounded families of languages under complementation, 297
closure properties of classes of languages, 101, 297
CNF satisfiability problem, 266
CNF: conjunctive normal form, 265
Cocke, John, 52
combinatory logic, 149
compact pairing function, 133
complement of a language, 102
complete problem, 6, 9, 150, 169, 268
  in complexity theory, 268
  in computability theory, 150, 169
Complete problems within a language family, with respect to a resource bound, 268
complex numbers, 14
complexity theory, 215, 245, 248
computability theory, 148
  limitations, 172
computation
  as a labeled forest, 211
  as a labeled graph, 211
  as a string of states, 36, 211
  as an extended state-transition function, 35, 213
  tableau, 272
computation theory
  sources, 10
computational logicians, 247
computational problems as formal languages, 151
computing the classes of $\equiv_L$, 59
concatenation
  of languages, 102, 223, 230
  of words, 23
conceptual axiom, 13, 14, 24
configuration (of a OTM), 87
congruence, 57
conjunction, 17
conjunctive normal form, 265
constant function, 165
constraint satisfaction, 7
context-free grammar, 96
  "syntactic categories", 96
  derivation relation, 97
  derivation relation $\Rightarrow^*_G$ of CFG G, 97
  derivation tree, 98
  derivation under, 98
  generated language, 98
  language generated by CFG G: $L(G)$, 98
  nonterminal symbols, 96
  productions, 96
  rewriting relation $\Rightarrow_G$ of CFG G, 97
  rewriting rules, 97

sentence symbol, 96
terminal symbols, 96
context-free language, 96, 98, 215
context-sensitive language, 215, 297
continuation lemma, 36
control-flow graph, 52
Cook, Stephen A., 215
Cook, Stephen A., 8, 248, 276
Cook–Levin theorem, 10
countability, 74, 113
  infinite set, 115
  of $\mathbb{N}^*$: all finite integer sequences, 118
  of Σ^*, 116
  of all finite subsets of $\mathbb{N}$, 117
  of the set of regular languages, 74
  set, 115
countable set, 74, 115
countably infinite set, 115
Craig, William, 149
Curry, Haskell B., 149
cycle (in a digraph), 26

data graph, 84
data-flow graph, 52
database language L_{DB}, 86
Davis, Martin, 150
De Morgan's laws, 16
De Morgan, Auguste, 16
decidability
  language, 151
  property, 151
  set, 151
decidability vs. semidecidability, 152
degree
  of a tree, 26
  of a tree node, 26
descendant (node) in a rooted tree, 26
determinism, 7
deterministic online automaton, 212
DHP: the diagonal halting problem, 155, 160
  m-completeness of, 171
diagonal halting problem, 155
  partial solvability, 158
  semidecidability, 158
  unsolvability, 156
diagonal shells of $\mathbb{N}^+ \times \mathbb{N}^+$, 128, 133
diagonalization, 121
difference of sets, 15
digraph, 25, 34
Diophantine analysis, 129
Diophantus, 129
direct product of sets, 15
directed graph, 25
Dirichlet's box principle, 64

Dirichlet, Johann P.G. Lejeune, 64
disjunction, 17
disjunctive normal form, 277
DNF: disjunctive normal form, 277
DOA, 212
domain of a function, 153
doubleton (set), 14

Edmonds, Jack, 246
elevator, 31
empty set: ∅, 14
EMPTY: programs that never halt, 165, 166
    nonsemidecidability, 168
encoding, 3, 4, 111
encoding function, 115
envelope (for a function), 28
equivalence class, 19
equivalence of FAs, 63
equivalence relation, 18
    class, 19
    index, 19
    left-invariant, 75
    refinement, 19
    right-invariant, 23
Euclid, 118
Euclid's theorem on primes, 118
extendible arrays/tables, 130
    hashing schemes, 138
    storage mappings, 125, 130
extension of a function, 35

FA, 33, 43, 51
    accepted language, 36
    as a labeled digraph, 34
    behavior, 36
    equivalence, 63
    recognized language, 36
    state-minimization, 66
    state-minimization algorithm, 67
Feinstein, Neil H., 217
Feys, Robert, 149
FIFO queue (data structure), 182
final states
    OA, FA, 34
final, or, accepting states: F
    OA, 34
finite automata, 33, 43, 51
    direct-product construction, 104
    probabilistic, 70
finite state machine, 51
finite-index lemma, 64
floor (of a number), 27
floor of number x: ⌊x⌋, 27
Floyd, Robert W., 52

fooling set, 64, 65
free semigroup, 23
Freud, Sigmund, 111
FSM, 51
function, 20
    bijective, 21
    bivariate, 154
    computable, 153
    domain of, 153
    extension, 35
    injective, 21
    multivariate, 154
    one-to-one, onto, 21
    partial, 20
    partial or total, 20
    source, 20
    surjective, 21
    target, 20
    total, 20
    univariate, 154
functions
    composition, 18
fundamental theorem of arithmetic, 118

Gödel, Kurt, 5, 111, 112, 147, 155, 235, 247
game tree, 238
Garey, Michael R., 247

halting problem, 6, 47, 149, 155
    NP-complete version, 269
    consequences of unsolvability, 158
    partial solvability, 158
    poly-time version, 270
    semidecidability, 158
    Turing machine, 6
    unsolvability, 156
hard problem, 169, 268
    in complexity theory, 268
    in computability theory, 169
Hard problems within a language family, with
        respect to a resource bound, 268
Harel, David, xiii
hashing schemes for extendible arrays/tables,
        138
Heath, Lenwood S., 185
Hennie, Fred C., 84, 199
Herman, Jerry, 31
Hilbert, David, 147
HP: the halting problem, 47, 155, 172
    m-completeness of, 170
hyperbolic shells of $\mathbb{N}^+ \times \mathbb{N}^+$, 133, 138

Immerman, Neil, 297
Immerman-Szelepcsényi theorem, 297

incompleteness theorem, 147, 247, 277
index (of an equivalence relation), 19, 57
information-theoretic bound, 89
injection, 21
injective function, 21
input-recording TM, 257
integer lattice points, 113
integer part (of a number), 27
interpreter (of a program), 155
intersection
    of languages, 102
intersection of sets, 15
IRTM, 257

Jacobellis v. Ohio, 148
Jacquard loom, 148
Jacquard, Joseph Marie, 148
Johnson, David S., 247

König's infinity lemma, 99
König, Dénes, 99
Karp, Richard M., 8, 79, 248, 276
Kleene operations on languages, 103
Kleene, Stephen C., 9, 149
Kleene–Myhill theorem, 7, 9, 101, 103, 105,
        217, 221
    algebraic statement, 101
    proof, 223, 224
Kronecker, Leopold, 20

lambda calculus, 149
Landin, Peter J., 48
language, 23
language accepted
    FA, 36
    NFA, 213
    NOA, 213
    OA, 36
    PFA, 72
    PFA M with acceptance threshold θ:
        $L(M, \theta)$, 72
language of "squares"
    nonregularity, 65
law of excluded middle, 172
leaf (node) in a rooted tree, 26
left-invariant equivalence relation, 75
Leighton, F. Thomson, 185
length of word w ($\ell(w)$), 23
letter, 22
Levin, Leonid, 8, 248, 276
lexicographic order, 22
linear-bounded automaton, 215
literal: logical variable in true or complemented
        form, 265

complemented form, 265
    true form, 265
logic function, 246
logic minimization, 8
logical equivalence, 57, 246
logical implication, 57
logical product, 17
logical relations via Boolean operations, 102
logical sum, 17
logical variables, 265
low-order bit, 53

m-complete problem, 169, 170
    as "hardest" semidecidable problem, 170
    in computability theory, 169, 170
m-reducibility, 161, 162
    helping decide a language, 162
    helping semidecide a language, 162
    transitivity, 163
machine learning: an FA-related model, 70
mapping reducibility, 156, 160–162
    helping decide a language, 162
    helping semidecide a language, 162
    polynomial time, 264
    transitivity, 163
    within resource-bound R, 264
Markov algorithms, 149
Markov, Andrey A., 149
McCulloch, Warren S., 51
Mealy, George, 51
memory requirements (of nonregular
        languages), 79
minimum-state FA, 66
model independence, 150
    complexity theory, 151, 254
    computability theory, 150
Moore, Edward F., 51, 72
multidimensional OTM tape, 84
multivariate function, 154
Myhill, John, 4, 9
Myhill–Nerode theorem, 4, 9, 42, 56, 57, 76,
        83, 220
    applications, 63
    for OAs, 41, 42

Naur, Thorkil, 6
Nerode, Anil, 4
neurally inspired models, 51
NFA, 217
NFAs vs. DFAs, 217
NOA, 212
NOAs and DOAs: equivalence as language
        acceptors, 213
nondeterminism, 3, 7, 209

as alternative universes, 211
as an existential quantifier, 212
as unbounded search, 213
  algorithmic view, 213
  algorithmic view, 213
  logical view, 213, 241, 242
deterministic simulation of, 217, 237
deterministic simulation of NFA by DFA,
  217
deterministic simulation of NTM by OTM,
  236, 237
in complexity theory, 245
in computability theory, 233
nondeterministic computation vs. online
  computation, 236
nondeterministic finite automaton, 217
nondeterministic OA, 211
nondeterministic online automaton, 211, 212
nondeterministic TM, 233, 234
nondeterministic Turing machine, 233, 234
nondeterministic-complexity puzzle, 252
nonrecursive language, 151
nonrecursive set, 151
nontrivial PoF, 166
nontrivial property of functions, 166
NTM, 233, 234
  accepted language, 235
null string (ε), 22
numeral, 53

OA, 33
  accepted language, 36
  as a labeled digraph, 34
  behavior, 36
  computation, 34
  final, or, accepting states: F, 34
  initial state: q_0, 34
  input alphabet, 34
  recognized language, 36
  state-transition function: δ, 34
Occam's razor, 94, 125
one-to-one function, 21
one-to-one, onto function, 21
online automaton (OA), 33
online TM, 43
online Turing machine, 43, 83
  k-tape, 195
  apparent strengthenings of the model, 195
  apparent weakenings of the model, 176
  autonomous state, 46
  blank symbol: $\boxed{\text{B}}$, 45
  computation, 48
  configuration, 47
  instantaneous description, 47

multitape, 195
paper-tape OTM, 186
polling state, 45
queue-based TM, 182
read/write head, 44
stack-based OTM, 180
state-transition function, 47
tape, 44
tape as a data structure, 44
variants, 174
with "paper" tape, 186
with a 2-dimensional worktape, 199
with a "random-access" worktape, 203
with a FIFO queue (and no tape), 182
with a multidimensional worktape, 199
with one-way tapes, 176
with registers (instead of tapes), 189
with two stacks (and no tape), 179
working alphabet: Γ, 45
onto function, 21
operational name, 59
operations on sets
  complementation, 16
  direct-product, 15
  intersection, 15
  set difference, 15
  union, 15
order-n approximation of an OA, 79
origin square of a 2-dimensional worktape, 199
orthogonal list (data structure), 84
OTM, 43
  with a 2-dimensional worktape, 199
  with a "random-access" worktape, 203
OTM tape as a data structure, 84
OTM working alphabet, 84
Ott, Gene H., 217

pairing function, 5, 166
  additive
    balance computation ease, compactness,
      143
    compact, 144
    design, 141
    easily computed, 142
    samples, 142
  additive (APF), 140
  applications, 125
  compactness, 133
  design for array/table storage, 131
  design via shells, 132
  diagonal-shell, 127
  dovetailing for compactness, 135
  hyperbolic-shell, 138
  square-shell, 132

task-allocation function, 140
palindromes, 61, 95
  memory complexity, 81
  nonregularity, 65
  space complexity, 256
  space requirements on an IRTM, 257
  space requirements on an OTM, 257
parent (node) in a rooted tree, 26
partial function
  semicomputable, 153
partial order, 186
partially solvable problem, 6, 152
partition (of a set), 19
path (in a digraph), 25
PFA, 70
  acceptance threshold: θ, 72
  constant of isolation: κ, 77
  isolated threshold, 76
  language accepted, 72
  state-distribution vector, 71
    initial, 71
  state-transition, 71
    as a vector-matrix product, 71
  state-transition matrix, 71
  states, 71
  string acceptance, 72
pigeonhole principle, 64, 76, 92, 99
pillars, 3
Pitts, Andrew, 6, 155
Pitts, Walter H., 51
PoF: property of functions, 165
  nontrivial, 166
pointed language, 237, 256
pointed palindromes, 256
poly-time, 264
positional numbering system, 59
positive subtraction $\ominus$, 128
power set, 14
powers of a language, 102, 223
precision in computing, 139
predicate calculus, 247
prime-factorization theorem, 118
principle of parsimony, 94, 125
problem
  NP-vs.-coNP, 297
  P-vs.-NP, 250
  Boolean minimization, 246
  CLIQUE, 266
  CNF satisfiability, 266
  halting, 47
    Turing machine, 6
  halting problem, 6
  subset-sum, 4, 270, 309
  traveling salesman, 5, 246

procedure APF-Constructor, 141
procedure PF-Constructor, 132
producing a regular expression from a DFA or
    NFA, 229
producing an NFA from a regular expression,
    225
proof
  by contradiction, 39, 62, 74, 107, 122, 159,
    173
  by diagonalization, 121
  nonconstructive, 74
  of countability/encodability, 116
  of nonregularity, 64
  of uncountability/unencodability, 121
proper prefix, 22
property of functions, 165
  nontrivial, 166
propositional calculus, 247, 277
  completeness of, 247, 277
propositional variables, 277
PTM: paper-tape OTM, 186
pumping, 91
  in a finite semigroup, 91
  in context-free languages, 96
  in finite directed graphs, 92
  in finite, closed systems, 91
  in formal languages, 91
  in regular languages, 93
pumping lemma, 91
  for context-free languages, 100
  for regular languages, 94
    insufficiency, 95
    necessary-and-sufficient version, 96
pushdown automaton, 215

QTM: queue-based TM, 182
quaternary relation, 17
queue (data structure), 182

RA-OTM: random-access OTM, 203
Rabin, Michael O., 52, 70, 217
random-access OTM, 203
rational numbers, 13
read/write head of a OTM, 84
real numbers, 13
recognized language
  FA, 36
  OA, 36
recursive functions, 149
recursive language, 151
recursive set, 151
recursively enumerable
  language, 152
  set, 152

reducibility
  mapping reducibility, 156
reducing one problem to another, 117, 150, 160
reduction, 4
  Cook reducibility, 264
  in complexity theory, 5, 264
  in computability theory, 5, 150, 160
  Karp reducibility, 264
  m-reducibility, 161
  mapping reducibility, 161, 264
  one-one reducibility, 161
  polynomial time, 264
  reducing $\overline{\text{DHP}}$ to EMPTY, 168
  reducing DHP to TOT, 159
  reduction function, 161, 264
  Turing reducibility, 161, 264
  within resource-bound R, 264
refinement (of an equivalence relation), 19
refutable logic expression, 249
refutable logical expression, 278
register machine, 189
regular expression, 103, 105, 223, 224
  from a DFA or NFA, 229
  the denoted language, 224
  via system of linear equations, 105
regular languages, 55
  closure properties, 103
    Boolean operations, 101, 103
    complementation, 55, 103
    concatenation, 103, 226
    intersection, 103
    Kleene operations, 103
    language-reversal, 103
    star-closure, 103, 227
    union, 103, 225
regular set, 55
regular-expression theorem, 7
relation
  $\leq$ on set cardinalities, 114
  binary, 17
  equivalence, 18
  quaternary, 17
  reflexive, 18
  symmetric, 18
  ternary, 17
  transitive, 18
relational databases
  storage mappings, 130
relations
  composition, 18
representation of integers
  base-2 representation, 21
  base-b representation, 21
  binary representation, 21

  low-order bit, 21
  number base, 21
  numeral, 21
  tally encoding, 271
  unary representation, 271
resource bound, 264
  compositionally comprehensive, 264
resource-bounding class, 264
resource-bounding function, 263
reversal, 62
  language, 103
  string, 62, 76
Rice, Henry G., 9
Rice–Myhill–Shapiro theorem, 9, 149, 165,
    166, 172
RM: register machine, 189
Rogers, Hartley Jr., 150
Roman numeral, 59
rooted tree, 26
  ancestor (node), 26
  ancestry in, 26
  branch, 26
  child (node), 26
  depth of a node, 26
  depth-d prefix, 26
  descendant (node), 26
  leaf (node), 26
  parent (node), 26
  prefix, 26
  root (node), 26
run-of-7's function, 173
Russel, Bertrand, 14

s-1-1 theorem, 163, 166
s-2-1 theorem, 171
s-m-n theorem, 163, 164, 166
SAT: the satisfiability problem, 266
  NP-Completeness of, 276, 277
satisfiability problem, 266
satisfiable CNF expression, 266
satisfying assignment, 266, 279
Savitch's theorem, 10, 285, 287, 289
Savitch, Walter J., 10, 285
Schönfinkel, Moses, 149
Schönhage, Arnold, 84
scheduling, 7
Schröder, Ernst, 115
Schröder–Bernstein theorem, 115
Scott, Dana, 52, 217
search-via-counting strategy, 286, 297
semantic completeness, 247
semicharacteristic function, 24
  and semidecidability, 152

semicomputable semicharacteristic function, 152
semidecidability
   language, 152
   problem, 152
   property, 152
   set, 152
semidecidability vs. decidability, 152
semidecidable
   problem, 6, 170
semigroup, 91
set complement, 16
SETI at home, 139
SETI athome, 126
Shapiro, Norman, 9
single-valued, 20
singleton (set), 14
small-o notation, 29
solvable problem, 151
space complexity, 252, 285
   deterministic IRTM, 262
   nondeterministic IRTM, 262
   space-bounding function, 262, 263
   constructible, 288
square shells of $N^+ \times N^+$, 132, 133
stack (data structure), 179
star-closure
   of a language, 103, 224
state, 3, 31, 41
   as equivalence class, 57
   state-distribution vector (PFA), 71
      initial, 71
   state-transition function: δ, 34
   state-transition matrix (PFA): Δ, 71
state-minimization algorithm (for FAs), 219
state-minimization algorithm (for FAs), 66, 219
state-transition system, 3
   probabilistic, 70
Stearns, Richard E., 199
Stewart, Potter, 148
STM: stack-based OTM, 180
storage mappings
   extendible arrays/tables, 6, 125, 130
   relational databases, 125, 130
storage modification machine, 84
string, 22
structure mapping, 7
subset construction (for OAs and FAs), 214, 218, 236
   quality, 219, 220
subset-sum problem, 4, 270, 309
surjection, 21
surjective function, 21
synchronous sequential circuits, 51

system of linear equations, 105
   with languages as coefficients, 105
Szelepcsényi, Robert, 297

tableau, 272
tally code, 195
Tarjan, Robert E., 185
task-allocation function, 140
   additive
      base task-index, 140
      stride, 140
tautology, 247, 277
ternary relation, 17
theorem, 147
   Cook–Levin, 10
   Euclid's theorem on primes, 118
   fundamental theorem of arithmetic, 118
   Kleene–Myhill, 7, 9
   Myhill–Nerode, 4, 9, 56, 57
   prime-factorization theorem, 118
   pumping lemma, 91
      for context-free languages, 100
      for regular languages, 94
   regular-expression, 7
   Rice–Myhill–Shapiro, 9, 165
   Savitch, 10, 287, 289
   Schröder–Bernstein, 115
theory of computability, 148
time complexity, 252
   deterministic OA, 253
   NTM, 253
   time-bounding function, 253, 263
time vs. space complexity
   a qualitative comparison, 285, 286
   a quantitative comparison, 285
TM, 43
TM tape as a data structure, 258
TOT: programs that halt on all inputs, 159, 160, 165, 172
   undecidability, 159
total order, 186
total state (of a OTM), 87
tracks in a worktape, 177
traveling salesman problem, 5, 246
truth values, 265, 277
Turing machine, 43, 125, 148
   apparent strengthenings of the model, 195
   apparent weakenings of the model, 176
   blank symbol: $\boxed{\text{B}}$, 45
   circularity, 6
   halting problem, 6
   online, 43
   read/write head, 44
   tape, 44

tape as a data structure, 44
total state, 48
working alphabet: Γ, 45
Turing reducibility, 161
Turing, Alan M., 5, 25, 43, 112, 148, 149, 155,
    235
type-0 grammar, 149
type-2 grammar, language, 96
type-3 grammar, language, 52

uncomputability, 148
  uncomputable function, 148
  via uncountability, 123
uncountability, 74, 113
  of $\{f : \mathbb{N} \longrightarrow \mathbb{N}\}$, 121
  of $\{f : \mathbb{N} \longrightarrow \{0, 1\}\}$, 121
  of the class of PFA languages, 74
  of the set of (countably) infinite binary
      strings, 121

of the set of *all* subsets of $\mathbb{N}$, 121
uncountable set, 74
uncountable sets, 121
undecidability
  language, 151
  property, 151
  set, 151
undirected graph, 26
union
  of languages, 102, 223
union (inclusive) of sets, 15
univariate function, 154
unsolvable problem, 151

volunteer computing, 126, 139
  vulnerability, 126, 139

William of Occam, 94, 125
word, 22